Models for Parallel and Distributed Computation

Applied Optimization

Volume 67

Series Editors:

Panos M. Pardalos
University of Florida, U.S.A.

Donald Hearn
University of Florida, U.S.A.

Models for Parallel and Distributed Computation

Theory, Algorithmic Techniques and Applications

Edited by

Ricardo Corrêa
Universidade Federal do Ceará

Inês Dutra
Universidade Federal do Rio de Janeiro

Mario Fiallos
Universidade Federal do Ceará

and

Fernando Gomes
Universidade Federal do Ceará

KLUWER ACADEMIC PUBLISHERS
DORDRECHT / BOSTON / LONDON

A C.I.P. Catalogue record for this book is available from the Library of Congress.

ISBN 978-1-4419-5219-6

Published by Kluwer Academic Publishers,
P.O. Box 17, 3300 AA Dordrecht, The Netherlands.

Sold and distributed in North, Central and South America
by Kluwer Academic Publishers,
101 Philip Drive, Norwell, MA 02061, U.S.A.

In all other countries, sold and distributed
by Kluwer Academic Publishers,
P.O. Box 322, 3300 AH Dordrecht, The Netherlands.

Printed on acid-free paper

Contents

Contributing Authors

Benjamín Barán is Research Coordinator of the National Computing Center - CNC, at the National University of Asuncion - UNA, since 1994. He is Full Professor at the Engineering School (UNA) where he has been teaching since 1978. D.Sc. in Computer and System Engineering at the Federal University of Rio de Janeiro - Brazil, 1993; M.Sc. in Computer and Electrical Engineering at Northeastern University, Boston - MA, 1987; Electronic Engineering at UNA, 1982. More than 50 internationally published papers. National Science Prize of Paraguay - 1996.

Valmir C. Barbosa is a Professor at the Systems Engineering and Computer Science Program of the Federal University of Rio de Janeiro. He received his Ph.D. from the University of California, Los Angeles, and has been a visitor at the IBM Rio Scientific Center, the International Computer Science Institute in Berkeley, and the Computer Science Division of the University of California, Berkeley. He is the author of *Massively Parallel Models of Computation* (Ellis Horwood, 1993), *An Introduction to Distributed Algorithms* (The MIT Press, 1996), and *An Atlas of Edge-Reversal Dynamics* (Chapman & Hall/CRC, 2000).

Cristina Boeres is an Associate Professor in the Department of Computer Science, Instituto de Computação, Universidade Federal Fluminense (UFF), Brazil. She obtained her Ph.D. from the University of Edinburgh, United Kingdom in 1997.

Michel Cosnard was born in Grenoble (France) on July 9, 1952. He received his Engineering Degree in Computer Science from ENSIMAG (Grenoble, France), the Master's degree in Applied Mathematics from Cornell University (USA) and Doctorat d'Etat degree in Computer Science from the Université de Grenoble (France) in 1975 and 1983 respectively. In 1979 he joined the Centre National de la Recherche Scientifique (France). In 1987, he became Professor of Computer Science at the Ecole Normale Supérieure de Lyon (France) where he has founded the Laboratoire d'Informatique du Parallélisme (LIP). He has been Chairman of the LIP, director of the LHPC until 1997. He served as chairman of the IFIP Working Group WG10.3 on Parallel Processing from 1989 to 1995 and is currently member of the IFIP Working Group WG10.3 on Parallel Processing and of the IFIP Working Group WG10.6 on Neural Networks. From September 1993 to September 1995, he served as Associate director for Information Science and Technology at the French Ministery of Research. He currently serves as Editor in Chief of Parallel Processing Letters, and is member of the Editorial Board of several scientific journals (Parallel Computing, Mathematical Systems Theory,...) and member of the steering committee of the international conferences PACT and SPAA. From September 1997, he is Director of the Research Unit of INRIA in Lorraine (Nancy, France) and of the LORIA Laboratory. His research interests are in the design and analysis of parallel algorithms, in the complexity analysis of automata and neural nets. Michel Cosnard has published more than 100 papers related to parallel processing. In 1994, he was awarded a national prize from the French National Academy of Science. In 1995, he received the IFIP Silver Core.

Inês de Castro Dutra is a lecturer of the Department of Systems Engineering and Computer Science, Federal University of Rio de Janeiro. She obtained her Ph.D. degree from University of Bristol, Department od Computer Science in 1995. She has published more than 20 papers in international conferences and journals, and edited two special edition books on parallelism and implementation technology for constraint logic languages. She has also organised and participated of workshops and conferences on parallelism in logic programming.

Afonso Ferreira received his M.Sc. from the University of São Paulo, Brazil, in 1986 and the Ph.D. from the Institut National Polytechnique de Grenoble in 1990, both in Computer Science. In October 1990, he joined the French CNRS (*Centre National de la Recherche Scientifique*) as a researcher. Since September 1997, Dr. Ferreira is located at the INRIA Sophia Antipolis, where he works at the joint project MASCOTTE of the CNRS/INRIA/UNSA.

His research focuses on the rich interdependence between theoretical algorithms design, models, and experimental analysis. Areas of interests include computational telecommunications, optimisation, and parallel processing. His current application areas concentrate on optical and satellite networks, and mobile communications.

Dr. Ferreira serves on the editorial boards of the Journal of Parallel and Distributed Computing, Journal of Interconnection Networks, Parallel Processing Letters, and Parallel Algorithms And Applications. Dr. Ferreira was Guest

Editor for special issues of Mobile Networks (MoNet), Parallel Computing, Parallel Processing Letters, Theoretical Computer Science, and Journal of Parallel and Distributed Computing, on discrete algorithms and methods for mobile communications, and for parallel systems. He has published over 100 papers in international journals and conferences, has co-authored one book (on satellite networks, in French), has edited 10 books, and has organized and served in Programme Committees for more than 50 international conferences and workshops.

Afonso Ferreira is General co-Chair for the IEEE IPDPS (International Parallel & Distributed Processing Symposium) in 2001 and 2002, and Programme co-Chair for STACS (Symposium on Theoretical Aspects of Computer Science) in 2001 and Programme Chair for STACS in 2002. He served as Publicity Chair for IPDPS'98, and Workshops Chair for the ACM/IEEE Mobicom'98 (International Conference on Mobile Computing and Networking).

Afonso Ferreira is member of the Steering Committees of STACS, and of the annual Workshops: Dial M for Mobility (Discrete Algorithms and Methods for Mobile Computing and Communications), Irregular (Solving Irregularly Structured Problems in Parallel), and WOCCS (Optical Communications for Computing Systems). Among other activities, he is currently member of the Advisory Boards of the IEEE Technical Committee for Parallel Processing, EuroPar, and PDCS.

Cláudio Geyer is a lecturer of the Department of Computer Science, Federal University of Rio Grande do Sul. He obtained his Ph.D. degree from Joseph Fourier University, Grenoble, Department of Computer Science in 1991. He has published more than 20 papers in international conferences, journals and books. He has also organised and participated of workshops and conferences on parallel programming. He has been the coordinator and participated of several international cooperation projects between Brazil and France, USA and Germany.

Isabelle Lassous received her PhD from the University of Paris 7 in 1999 in Computer Science. In September 1999, she had a post-doctoral position at the project HIPERCOM of the INRIA. In October 2000, she joined the French INRIA (*Institut National de Recherche en Informatique et Automatique*) as a researcher. She works at the joint project REMAP of the CNRS/ENS Lyon/INRIA, located at the ENS Lyon.

Her research focuses on the axis *algorithms design - implementation - experimental analysis*. Her main researches concern the design of parallel/distributed graph algorithms, their implementation on various architectures and the associated experimental analysis. Areas of interests include parallel/distributed processing, algorithmic, computational telecommunications.

Carlos Moura has been a Visiting Professor at the Institute for Computing of the Federal Fluminense University – UFF, Niterói, Brazil, for the last three years (http://www.ic.uff.br/~demoura). After getting a Ph.D. in Applied Math at NYU Courant Institute in 1976, he became one of the founding fathers of LNCC – Brazilian National Laboratory for Scientific Computing, where he was a full researcher until retirement, in 1993. Since then he has worked as a consulting and visitor in different institutions, in Brazil and abroad. An active member of SBMAC – Brazilian Society for Computational and Applied Mathematics, and ABEC – Brazilian Association of Scientific Editors, he has acted as president of the former and VP of the latter. In the Editorial Board of J.UCS – Journal for Universal Computer Science (Springer-Verlag and IICM, Graz) since its birth, he was the founding Editor-in-Chief of Computational & Applied Mathematics, formerly published by Birkhäuser-Boston.

Vinod Rebello is currently an Associate Professor in the Department of Computer Science, Instituto de Computação, Universidade Federal Fluminense (UFF), Brazil. He has also held positions as Visiting Professor in the same department and as Lecturer in the Department of Computer Science, University of Edinburgh, United Kingdom. He obtained his Ph.D. from the University of Edinburgh in 1996.

David Skillicorn is a Professor in the Department of Computing and Information Science at Queen's University. He obtained a Ph.D from the University of Manitoba in 1981. His research interests centre around large-scale, distributed, data-intensive applications. He was involved in the development of several parallel programming models, including the Bird-Meertens Formalism (BMF) and Bulk Synchronous Parallelism (BSP). He has also worked extensively on data mining, both for conventional sequential architectures and also for parallel computers.

Siang Wun Song is a Professor of the Department of Computer Science, University of São Paulo, where he is currently the director of the Institute of Mathematics and Statistics. He holds a PhD in Computer Science obtained at Carnegie Mellon University in 1981. He is the author of approximately 53 papers and one book. He is on the editorial boards of Parallel Processing Letters, Parallel and Distributed Computing Practices, and Journal of the Brazilian Computer Society. He has been a member of the ACM since 1972.

Patrícia Kayser Vargas is a lecturer of the Department of Computer Science, Universidade Federal do Rio Grande do Sul, she obtained her M.Sc. degree from Universidade Federal do Rio Grande do Sul in 1998, and is currently pursuing her Ph.D. studies in the same university. She has published 8 papers and has given 2 tutorials.

Foreword

Parallel and distributed computation has been gaining a great lot of attention in the last decades. During this period, the advances attained in computing and communication technologies, and the reduction in the costs of those technologies, played a central role in the rapid growth of the interest in the use of parallel and distributed computation in a number of areas of engineering and sciences. Many actual applications have been successfully implemented in various platforms varying from pure shared-memory to totally distributed models, passing through hybrid approaches such as distributed-shared memory architectures.

Parallel and distributed computation differs from classical sequential computation in some of the following major aspects: the number of processing units, independent local clock for each unit, the number of memory units, and the programming model. For representing this diversity, and depending on what level we are looking at the problem, researchers have proposed some models to abstract the main characteristics or parameters (physical components or logical mechanisms) of parallel computers. The problem of establishing a suitable model is to find a reasonable trade-off among simplicity, power of expression and universality. Then, be able to study and analyze more precisely the behavior of parallel applications. Some relevant questions in such an analysis are: What is intrinsically parallelizable in a given sequential algorithm? What sequential algorithm leads to a parallel counterpart for a given problem? How to exploit the intrinsic parallelism in real parallel or distributed system?

The chapters in this book answer all of these questions. It focuses on advanced techniques used in the design of efficient parallel programs. It presents a wide variety of different models of parallel and distributed computation and applications of these models to the design of efficient algorithms to solve numerical and non-numerical problems. It contains general and specific texts about advanced algorithms for parallel computation and gathers together the state-of-the-art on parallelism with contributions from researchers actively working with parallel computation. Some of the chapters are based on the very good material used during the *International School on Advanced Algorithmic Techniques for Parallel Computation with Applications* held in Natal, Rio Grande do Norte, Brazil, from September 25th to October 2nd, 1999, and organized by Ricardo Corrêa and Mario Fiallos from Federal University of Ceará, Brazil.

The book chapters cover a broad variety of models, ranging from abstract shared-memory models such as PRAM to more realistic models of distributed memory, including CGM, LogP and BSP. It is intended to be a reference text to practitioners, researchers and graduate students on Computer Science, Mathematics, Engineering and Sciences.

We were very fortunate to have so many important contributions to this special edition of the Kluwer Academic Book Series on Applied Optimization. These contributions comprise ten chapters, divided into two parts. The first part is devoted to present different models of parallel and distributed computation used to determine theoretical results and to explore in practice the intrinsic parallelism of applications and algorithms. Starting with the theoretical aspects, Cosnard presents in Chapter 1 an introduction to the general theory of complexity of parallel algorithms. He first recalls the main results of sequential complexity, then introduces models for parallel computation and compares them to sequential models. He shows that some of these models are not reasonable and explains the parallel computation thesis for reasonable models. Finally he presents the PRAM model, which has been extensively used in the study of intrinsic parallelism of algorithms and problems, and gives basic results of parallel complexity.

In Chapter 2, Barbosa discusses general models of resource-sharing computations, with emphasis on the combinatorial structures and concepts that underlie the various deadlock models that have been proposed, the design of algorithms and deadlock-handling policies, and concurrency issues. He states that these structures are mostly graph-theoretic in nature, or partially ordered sets for the establishment of priorities among processes and acquisition orders on resources. He also discusses graph-coloring concepts as they relate to resource sharing.

The remaining chapters of the first part of the book deals with adequate models of parallelism sufficiently close to existing parallel machines. These are simple models but nevertheless intend to give a reasonable prediction of performance when parallel algorithms on them are implemented. Consequently, such algorithms are expected to have theoretical complexity analyses close to actual times observed in real implementations. Boeres and Rebello, in Chapter 3, discuss the task scheduling problem in the context of the $LogP$ model, that deals with realistic communication models. The chapter describes the $LogP$ model and the influence of its communication parameters on task scheduling. The similarities and differences between clustering algorithms under the delay and $LogP$ models are discussed and a design methodology for clustering-based scheduling algorithms for the $LogP$ model is presented. Using this design methodology, a task scheduling algorithm for the allocation of arbitrary task graphs to fully connected networks of processors under $LogP$ model is proposed. The strategy exploits the replication and clustering of tasks to minimize the ill effects of communication on the makespan.

In Chapter 4, Skillicorn describes the Bulk Synchronous Parallel (BSP) model which, at present, represents the best compromise among programming models for simplicity, predictability, and performance. He describes the model from a software developer's perspective and show how its high-level structure is used to build efficient implementations.

In Chapter 5, Ferreira and Guérin-Lassous show that CGM are well adapted to coarse grained systems and clusters. In particular, algorithms designed for such models can be efficient and portable, and can have their practical performance directly inferred from their theoretical complexity. Furthermore, they allow a reduction on the costs associated with software development since the main design paradigm is the use of existing sequential algorithms and communication sub-routines, usually provided with the systems.

The second part of the book contains several chapters on applications of parallel computing. The first one is Chapter 6, due to S. Song. It discusses several CGM scalable parallel algorithms to solve some basic graph problems, including connected components and list ranking. Ekşioğlu, Pardalos and Resende, in Chapter 7, review parallel metaheuristics for approximating the global optimal solution of combinatorial optimization problems. Recent developments on parallel implementation of genetic algorithms, simulated annealing, tabu search, variable neighborhood search, and greedy randomized adaptive search procedures (GRASP) are discussed.

In Chapter 8, Geyer, Vargas and Dutra summarize the main research work being carried out on the implementation of parallel logic programming systems. Their work concentrates on describing techniques for exploiting and-parallelism and or-parallelism while showing important aspects of some systems on shared-memory, distributed-memory and distributed-shared memory architectures. They also present some important pointers to journals, conferences and sites with related information.

The book is closed with two chapters devoted to numerical applications. In Chapter 9, Barán observes that the solution of today large complex problems may need the combination of several different methods, algorithms and techniques in a distributed computing system with heterogeneous processors, usually interconnected through a communication network. In this context, the Team Algorithm is presented as a general technique used to combine different methods and algorithms in a distributed computing system composed of different workstations, personal computers and/or parallel computers. He also shows that experimental results have proved that in many real world problems, Team Algorithms can benefit from the use of asynchronous implementations, speeding up the whole process with an important synergy effect, in a new appealing technique known as Parallel Asynchronous Team Algorithms.

In Chapter 10, Moura presents a survey of some of the methods designed for the numerical solution of differential equations that present an efficient implementation within a parallel environment. Included among these are: the Domain Decomposition and Multigrid hyper-algorithms, the Piecewise Parabolic method, the spectral (frequency) approach, some strategies for finite-element, and finite-difference higher accuracy techniques.

We would like to thank all contributors to this special edition, the Advanced School organizers, and the Kluwer editorial support, in particular Ms. Angela Quilici. Without the help of these people, this edition would not be possible.

This book was processed using the LaTeX text processor package using the Kluwer kapedbk book style.

We hope this book can reach a wide audience and that you can benefit from the very good quality material kindly submmited by our contributors.

Enjoy!

The editors.

I

MODELS OF PARALLEL COMPUTATION

Chapter 1

INTRODUCTION TO THE COMPLEXITY OF PARALLEL ALGORITHMS

Michel Cosnard

LORIA - INRIA, France[*]

Michel.Cosnard@loria.fr

Abstract   We present an introduction to the general theory of complexity of parallel algorithms. We first recall the main results of sequential complexity, then introduce
models for parallel computation and compare them to sequential models. We
show that some of these models are not reasonable and explain the parallel computation thesis for reasonable models. Finally we study the $PRAM$ model and
give basic complexity results for parallel algorithms.

Keywords: Complexity, parallel algorithms, PRAM.

1.   Introduction

Resorting to parallelism in order to speed up computation is not a recent idea.
Even in the earliest years of computing, computer designers thought about
making several processors work simultaneously to solve the same problem.
From the modelling point of view, it has been the same. Models for parallel
computation were introduced shortly after sequential models. However it is
only in the late seventies that the theory of complexity of parallel algorithms
was developed. From this time on, many results have been obtained which
constitutes now the core of the theory. But much work has still to be done since
most of the fundamentals problems are open and since new models continue to
appear.

In this paper we present an introduction to the general theory of complexity
of parallel algorithms. We first recall the main results of sequential complexity,
then introduce models for parallel computation and compare them to sequential

---

[*]LORIA, 615 Rue du Jardin Botanique, 54602 Villers Les Nancy, FRANCE

*R. Corrêa et al. (eds.), Models for Parallel and Distributed Computation. Theory, Algorithmic Techniques
and Applications, 3–25.*
© 2002 *Kluwer Academic Publishers.*

models. We show that some of these models are not reasonable and explain the parallel computation thesis for reasonable models. Finally we study the *PRAM* model and give basic complexity results for parallel algorithms.

2.   Sequential complexity

In this section, we recall the necessary definitions and results of sequential complexity that should be known by the reader in order to understand the main purpose of this chapter: the study of complexity of parallel algorithms.

2.1.   Turing machines

The Turing machine (TM) model [2] is a device which consists of a finite set of states as in the case of finite automata, a semi-infinite tape divided into cells, equiped with a tape head which can move right or left, scanning the cells tape one at a time. When the head is on the leftmost cell, it is not allowed to move left. At each moment the device is in one of the states. Then the device can read the content of the scanned cell, change the content of the cell by writing a new symbol on it, move the head right or left and change its internal state. All these operations are uniquely defined by a transition function. They form a step.

Definition 1.1 *A Turing machine is a five-tuple*

$$M = (Q, \Sigma, \delta, q_0, F)$$

with

- *Q is the finite set of internal states and $q_0 \in Q$ is the initial state,*

- *Σ is a finite alphabet,*

- *$F \subset Q$ is the subset of accepting final states,*

- *$\delta : Q \times \Sigma \to Q \times \Sigma \times \{G, N, D\}$ is the partial transition function, where G, N and D stand for "move left", "do not move" and "move right", respectively.*

Each transition takes unit time. The acceptation time is the number of transitions between the initial state and a final state. The space is the number of used cells between the initial state and a final state.

A machine M starts operating on an input word ω written on the tape and a fixed symbol called blank in the remaining cells. The initial state is q_0. Then M proceeds by applying the transition function as long as possible. Whenever the transition function is undefined, the machine stops. If it stops in a final state, an element of F, then M is said to *accept* ω, otherwise to *reject* ω. Observe that the machine has three possible behaviours: *accept* the word, *reject* it or never stop.

Recall that a *language* L on a alphabet Σ is a subset of Σ^*.

Definition 1.2 *Let $T(n)$ be the maximum time over all words of size n of a language L accepted by a Turing machine M. Let $S(n)$ be the maximum space over all words of size n of a language L accepted by a Turing machine M. $T(n)$ is the time complexity of L and $S(n)$ is the space complexity of L.*

In the previous model, every move is completely determined by the current situation: the state of the machine and the symbol currently scanned by the tape head completely determine the next state and the move of the tape head. This device will be called deterministic. We can relax this condition to obtain nondeterministic devices whose next move can be chosen from several possibilities. We keep the same definition except that we modify the transition function to:

$$\delta : Q \times \Sigma \to P(Q \times \Sigma \times \{G, N, D\})$$

where for any set A, $P(A)$ denotes the power set of A. Hence on a given input there is now not only one computation, but a set of possible computations. Acceptance is defined when at least one of the possible computations stops in an accepting state.

We can now introduce the central classes of complexity. We call:

- $DTIME(t)$ and $NTIME(t)$ the set of languages that are accepted in time t by a deterministic and by a non-deterministic Turing machine, respectively,

- $DSPACE(s)$ the set of languages that can be accepted in space s, with M deterministic, and $NSPACE(s)$ with M non-deterministic,

- P the set of languages such that $T(n)$ is polynomial, with M deterministic,

- $PSPACE$ the set of languages such that $S(n)$ is polynomial, with M deterministic,

- NP the set of languages such that $T(n)$ is polynomial, with M non-deterministic,

- $NPSPACE$ the set of languages such that $S(n)$ is polynomial, with M non-deterministic.

2.2.   Random access machines

The Turing machine model is the basic model of computation. However dealing with Turing machines may be tedious in particular for designing high-level algorithms. Hence one can define more evolveld models.

Definition 1.3 *A Random Access Machine, RAM for short, is a device with a semi-infinite memory, whose cell can contain an integer, a finite number of arithmetic registers and a computation unit. The computation unit has the following set of instructions: read (M_i, R_j), write (M_i, R_j), add (R_i, R_j), sub (R_i, R_j), test (Ri, Rj), conditionals, goto, stop, ... A program is a sequence of instructions. In unit time 1 the RAM can read one or two memory cells or registers, compute and write the result into a memory cell or register.*

As before one can define a non-deterministic RAM by allowing the device to operate non-deterministically: at one moment the machine selects the current computation among several possibilities. The same definitions apply for Time and Space and for the classes P_{RAM}, $PSPACE_{RAM}$, NP_{RAM} and $NPSPACE_{RAM}$.

2.3.   Relations between TM and RAM

We shall now introduce the concept of simulation in order to look at the extend to which the models differ.

Definition 1.4 *Given two models M_1 and M_2, we say that M_1 can be simulated by M_2 in time t and space s if a single computing step of M_1 can be performed on M_2 in time t and space s.*

Theorem 1.5 (Simulation) *RAM and TM can simulate each other in polynomial time and constant space.*

Hence we deduce from the Simulation Theorem that P, NP, $PSPACE$, $NPSPACE$ are identical classes for RAM and TM.

2.4.   Thesis

There are many modifications of the Turing machines or the RAM which are equivalent to the basic models described above in the sense that they both accept the same class of languages: the recursively enumerable sets. Since our main concern is not on calculability but on complexity of parallel models of computation, I shall not go longer along this way. However remark that there are many other models of computation which are equivalent to the Turing machines. Some of them are highly different. Their abundance suggests that Turing machines correspond to a natural and intuitive concept of algorithms.

The intuitive properties of amenability to mechanization and of applicability to data of unbounded size seem to correspond to the class of algorithms defined by the Turing machines. This is the main purpose of the following principle known as Church's Thesis.

Church's Thesis: Arithmetically solvable problems are TM solvable problems.

Writing programs for Turing machines is tedious. Hence we introduced the RAM model and showed the equivalence with the Turing machines within a polynomial increase in time and a constant increase in space. It is a general fact that for every two computation models taken from a wide family of reasonable devices, they simulate each other within the same previous bounds. This is generally admitted as an extension of Church's Thesis, that this fact holds for all reasonable models of sequential computation. This principle has been called the Invariance thesis.

Invariance Thesis: Two reasonable computation models can simulate each other in polynomial time and constant space.

Let us point out that most of the models that we use in daily life are reasonable. From this point of view, we deduce that the complexity class that we already introduced are robust, in the sense that they are the same for a very wide class of computation models. We say that these classes are universal.

However, we shall see later within this chapter that this is not a general fact and that we could define simple models for which the invariance thesis does not hold.

Theorem 1.6 *There exist non reasonable models.*

2.5. Relations between DSPACE and NSPACE

Clearly a deterministic machine can be simulated by a non-deterministic one. Just take the same machine. Obviously the reverse is not true. Hence we have to work out in order to decide if non-determinism is really advantageous. Thus we should study the relations between the classes of complexity. Let us begin with space. In fact the notions of deterministic and non-deterministic space are closely related and this has been proved by Savitch.

Theorem 1.7 (Savitch, 1970 [14])

$$DSPACE(S(n)) \subseteq NSPACE(S(n)) \subseteq DSPACE(S(n)^2)$$

$$PSACE = NPSPACE$$

2.6. Central problem

What happens if we consider time instead of space? This is still an open problem and has led to an incredible amount of work on what is known as the central problem of complexity:

$$P = NP \text{ or } P \subset NP \text{ ?}$$

Hence the class NP is a very interesting class of problems, those that can be solved in polynomial time using a non-deterministic procedure. The main interest lies in the fact that many practical problems belong to this class. They can be identified by the following property: there is no known way to compute a solution in polynomial time, but there is a procedure to check in polynomial time whether a potential solution is an actual solution. Hence one may either want to find polynomial algorithm to solve a problem of NP, or prove that this is not possible. This is in fact a very difficult question. This has been originally studied in the early seventies by Cook [4] and Karp [10]. They introduced the class of NP-Complete problems, problems of NP which are the most difficult to solve in the sense that any problem of NP can be transformed into them in polynomial time.

Definition 1.8 *Prob is said to be NP-Complete if any problem of NP can be transformed in $Prob$ in polynomial time and if $Prob$ is in NP.*

From the definition, it is easy to deduce that if a problem of NP can be solved in polynomial time then $P = NP$.

Lemma 1.9 *Let $Prob$ be NP-Complete. If $Prob$ belongs to P then $P = NP$.*

Hence NP-Complete problems are the best candidates for not belonging to P. The first natural NP-Complete problem has been found by Cook. Recall that the satisfiability problem, SAT for short, is to determine, given a boolean formula, whether it is satisfiable or not.

Theorem 1.10 (Cook, 1971 [4]) *The problem SAT is NP-Complete.*

From this first example, many problems have been shown to be NP-Complete. We refer the reader to the excellent survey of the NP-Completeness results in the book by Garey and Johnson [7].

3. PRAM model

Let us now introduce models for parallel computation. There are many such models but since we are primarily interested in complexity of parallel algorithms we shall begin by the easiest to define and to use: the PRAM model.

Definition 1.11 *A Parallel Random Access Machine (PRAM) is a set of independent sequential processors RAM, each with a private memory and sharing a common memory M. Let P_0, P_1, ... , P_{p-1} be the processors. In time 1, each processor executes the sequence: read(M_i), compute(f), write (M_j) synchronously.*

All these atomic operations are executed synchronously. In other words, all the processors read, compute and write at the same time. Some restrictions may be imposed on computations. The most often used is to assume that only arithmetic operations can be executed indivisibly.

3.1. Concurrent or exclusive accesses

The common memory can be accessed in various ways. Snir [16] proposed to classify memory accesses according to whether a single processor or several processors could read from or write to the same memory cell simultaneously.

If only one processor can read a memory cell at a given time it is called *Exclusive Read (ER)* and if several processors can read a memory cell at the same time it is called *Concurrent Read (CR)*. It is obvious that we are talking about the same memory cell, and that in both cases different memory cells can be read simultaneously by different processors.

A similar classification has been proposed for write accesses. Write access is exclusive (*EW = Exclusive Write*) if several procesors can write the contents of a memory cell. In the latter case (*CW = Concurrent Write*), the processors may have different data, which will cause a conflict. Several policies have been proposed to resolve this conflict. Here are some of the main ones:

- *Common*: all the processors must write the same value in the same cell,

- *Priority*: the processors are classified according to an order of priority and only the processor with the highest priority, amongst all the processors that want to write to the memory cell, will access the memory,

- *Random*: any processor can succeed in writing, but whatever happens the algorithm must execute successfully,

- *Majority*: the majority value is written amongst all the values that the processors want to write into the memory cell,

- *Maximum* or *Minimum*: the maximum or minimum value is written amongst all the values that the processors want to write into the memory cell.

All these writing policies are clearly not equivalent ! By combining the various read and write options, we define different models. Generally, we differentiate between the $EREW\,PRAM$, the $CREW\,PRAM$ and the $CRCW\,PRAM$. For this last case we will consider the common $CRCW$ model.

Consider the following program where C is an array of n integers, M and m are memory arrays.

Algorithm: computing with a common $CRCW$ model

> For all $i \in \{1, \ldots, n\}$ in parallel do $M(i) \leftarrow 0$
> For all $i \in \{1, \ldots, n\}$ and $j \in \{1, \ldots, n\}$ **do in parallel**
>   **If** $(C(i) < C(j))$ **then** $M(j) \leftarrow 1$
> For all $i \in \{1, \ldots, n\}$ and $j \in \{1, \ldots, n\}$ **do in parallel**
>   **If** $(M(i) = 0$ and $i < j)$ **then** $M(j) \leftarrow 1$
> For all $i \in \{1, \ldots, n\}$ in parallel **do**
>   **If** $(M(i) = 0)$ **then** $m \leftarrow C(i)$

It is not so easy to see that the previous algorithm computes $Max(C)$ on a common $CRCW\,PRAM$. The parallel time is $O(1)$, the parallel space is $O(n)$ but the number of processors is $O(n^2)$.

3.2.   Computing with PRAM

Let us take an example in order to study how one can perform computations with PRAM. We call prefix computation the following problem. Given n integers x_k, compute the following values:

$$s_0 = x_0;\, s_1 = x_0 + x_1;\, s_i = x_0 + x_1 + \ldots + x_i;\, s_{n-1} = x_0 + x_1 + \ldots + x_{n-1}.$$

For each computation model, we want to find an algorithm and compute its complexity. Remark first of all that the sequential complexity is $O(n)$ since all the s values can be computed using a single recurrence:

Algorithm: sequential prefix computation

> $s(0) \leftarrow x(0)$;
> For all $i \in \{1, \ldots, n-1\}$ **do** $s(i) \leftarrow s(i-1) + x(i)$

On a $CREW$ and $CRCW\,PRAM$ the complexity is $O(log(n))$ with n processors as can be seen on the following algorithm. We assume that in addition to the memories $M(i)$ which contain the initial data we have a further set $S(i)$ which contains the prefixes when the algorithm terminates. The procedure is based on a recursive construction of the prefixes. At step d, the data and processors that possess them are grouped in blocks of size $n/2^{d+1}$. In each block, the upper half of processors add the maximum prefix of the lower half to their current values with i_d being the dth bit of i.

Algorithm: $CREW\,PRAM$ **prefix computation, code of** P_i

For $d \in \{0, \ldots, log(n) - 1\}$ **do**
    If $i_d = 1$ **then** $M(i) = M(i) + M(2^{d+1}.\lfloor i/2^{d+1} \rfloor + 2^d - 1)$

One can argue that the previous algorithm make an extensive use of the concurrent read capabilities and hence that an $EREW\,PRAM$ algorithm will have a much worse complexity. In fact, this is not true, and as long as the order of magnitude is concerned the $EREW\,PRAM$ complexity remains unchanged: $O(log(n))$ with n processors. The procedure is as follows: neighbouring processors begin by computing their prefixes, then processors at a distance 2, then 4, and so forth. At the end of the computation the i-th prefix is $M(i)$ and S contains the sum of all the elements.

Algorithm: $EREW\,PRAM$ **prefix computation, code of** P_i

write($M(i)$) **in** $S(i)$
For $d \in \{0, \ldots, log(n) - 1\}$ **do**
    If $i_d = 1$ **then** $M(i) = M(i) + S(i \oplus d)$
    $S(i) = S(i) + S(i \oplus d)$

3.3. PRAM separation

Our aim here is to study the power of each of the $PRAM$ models, the possibility of moving from one model to another and the relations between them. Before the various $PRAM$ models can be compared, the criteria on which the comparison is made have to be defined. As far as the three main $PRAM$ models are concerned, comparison will focus on computation time. More specifically, we shall use the following definitions.

Definition 1.12 *Let M_1 and M_2 be two models and $\mathcal{P}$ be the family of problems solvable with $PRAM$. M_1 is said to be strictly less powerful than M_2, if M_2 solves all problems in $\mathcal{P}$ at least as fast as M_1, and strictly faster than M_1 for one of them. The relation is written $M_1 << M_2$.*

To simplify the description, we shall consider only one $CRCW$ model and adopt the common model. Of course, an algorithm designed for an $EREW$

$PRAM$ can be executed on a $CRCW\ PRAM$ and any algorithm designed for a $CREW\ PRAM$ can be executed on a $CRCW\ PRAM$. Snir showed that the contrary is not true.

Theorem 1.13 ($PRAM$ **Separation [16]**) *Assume that the three models have p processors. We have:*

$$EREW\ PRAM << CREW\ PRAM << CRCW\ PRAM$$

The problems that separate the models are computing the maximum of n bits for $CREW$ and $CRCW$, and searching in a sorted list for $CREW$ and $EREW$.

3.4.   PRAM simulation

We shall now introduce the concept of simulation in order to look at the extend to which the models differ.

Definition 1.14 *Let M_1 and M_2 be two models. M_1 can be simulated by M_2 in time t, if a single computing step of M_1 can be performed on M_2 in time t. The relation is represented by $M_2 = t.M_1$.*

Theorem 1.15 ($PRAM$ **Simulation**) *Assume that the three models have p processors. We have:*

$$EREW\ PRAM = O(log(p)).CREW\ PRAM$$

$$CREW\ PRAM = O(log(p)).CRCW\ PRAM$$

$$EREW\ PRAM = O(log(p)).CRCW\ PRAM$$

The proof can be found in [17]. It is based on the fact that, when several processors access the same memory cell, the value stored in the cell has to be duplicated to simulate the access. A broadcast algorithm that can be executed in logarithmic time can be used for this purpose.

4.   Boolean and arithmetic circuits

Let us now introduce another model of computation. As will be seen later the boolean circuits model has been mainly used to define classes of complexity, both in sequential and in parallel. This model is based on the combinatorial circuits used in computer architecture.

Definition 1.16 *A boolean circuit, BC for short, is a directed acyclic graph (DAG) such that each vertex executes a boolean function within {enter, constant, and, or, non, exit}. Nodes are organized in levels according to their depth in* *the DAG: the first level contains only entry and constant nodes, the last level contains only exit nodes. The computation is performed level by level. Hence a BC with n entries and m exits compute a function from $\{0,1\}^n$ to $\{0,1\}^m$.*

The computational model in this case is also synchronous. Each node, also called a gate by analogy with models of electronic circuits atomically reads its inputs, evaluates its boolean function and transmits on its output links in one unit of time. For the sake of simplicity, we shall assume in the remaining text that $m = 1$. An arithmetic circuit is a boolean circuit in which boolean functions are replaced by arithmetic functions.

Definition 1.17 *Consider a boolean circuit BC computing a boolean function f. The* cost *(or* size*) of BC, $c(BC)$, is the number of gates of BC. Given a boolean function f, its* boolean cost $c(f)$ *is the size of the smallest circuit computing it.*

The depth *of BC, $d(BC)$ is the length of the longest path in BC. Given a boolean function f, its* boolean depth $d(f)$ *is the depth of the minimal depth circuit computing it.*

It is easy to see that, since the model is atomic and synchronous, $c(f)$ is the sequential time and $d(f)$ is the parallel time.

Relations between TM and BC. In this section we shall relate circuit complexity and depth complexity to time and space complexity of Turing machines. However as indicated above, boolean circuits have a fixed number of input bits. Hence, when faced with an infinite set, we should use a different circuit for each length.

Definition 1.18 *Let $A \subseteq \{0,1\}^*$. For any given n, we define the circuit (resp. depth) complexity of A at length n as the boolean cost (resp. depth) of the caracteristic function f_n of $A \cap \{0,1\}^n$.*

Observe that as functions of n both $c(A)$ and $d(A)$ only depend on A. Correspondingly, if we consider the function family associated to A and call it f_n we can generalize the notion of cost and depth.

The next theorem relates time and space complexity of a Turing machine computation to size and depth complexity.

Theorem 1.19 *(TM Simulation by BC [2]) If f_n is computed in $T(n)$ and $S(n)$ by a Turing machine, then it can be computed in $C(T^2)$ and $D(S^2)$ by a boolean circuit.*

The converse is more difficult. In fact, as we shall see later, leaving the freedom to design a specific circuit for each value of n gives to boolean circuits a strong power. Hence in order to better understand the relationships between the two models, let us begin with a circuit of given size.

Definition 1.20 *The circuit value problem CVP is the following. Let BC of size c, with n entries, and x a boolean vector of size n. If x is put at the entries of BC, is the exit 1 ?*

The next theorem shows that the circuit value problem belongs to P.

Theorem 1.21 ([2]) *CVP of size c can be solved in $Poly(c)$ on a deterministic Turing machine.*

Let us now come back to a given problem family, depending on parameter n. We can associate to the problem family a family of BC: BC_n solves $Prob_n$. But remember that the circuit for the problem of size n is completely independent of the circuit for the problem of size $n - 1$. Such a family will be called non-uniform. Hence we can prove the fundamental theorem of non-uniform families of boolean circuits.

Theorem 1.22 (Non-Reasonable model) *The model of non-uniform families of boolean circuits IS NOT reasonable.*

In order to go further and to understand which class of languages are computed with non-uniform families of boolean circuits of polynomial size, we first introduced a variant of Turing machines. We call *advice functions* any functions from N to Σ^*. We can define complexity classes obtained by assuming that we have an advice function available that helps us in solving our problems, and measuring the amount of resources needed by our algorithms using the appropriate advice function. The main characteristics of advice functions is that the value of the advice does not fully depend on the particular instance we want to decide, but just on its length.

Definition 1.23 *A Turing machine with a polynomial advice function is a Turing machine which can use for each input word of length n an advice $f(n)$ of length polynomial in n : $\mid f(n) \mid \leq Q(n)$ for some ploynomial Q. Such a family of functions is designed by poly.*

P/poly is the class of languages that can be recognized by a Turing machine with a polynomial advice function.

Consider now non-uniform families of boolean circuits of polynomial size. The next theorem shows that with such models of computation we can recognized $P/poly$.

Theorem 1.24 (Non-Uniform P [2]) *A set A has polynomial size circuits if and only if $A \in P/poly$.*

In order to obtain a reasonable model, one should restrict the freedom of building completely different circuits form one problem size to the next.

Definition 1.25 *We call* small *a transformation that could be computed in logarithmic time. A uniform family of circuits is such that we can compute* CB_n *from* CB_{n-1} *by a small transformation*

Theorem 1.26 *The model of uniform families of boolean circuits IS reasonable.*

5.   Parallel computation thesis

As we just saw in the previous section, Turing machines and uniform families of boolean circuits are reasonable sequential models of computation. From the point of view of parallel processing, one can also say that PRAM and uniform families of boolean circuits are reasonable parallel models of computation. In both cases non-uniform families of boolean circuits are not reasonable. Based on the relationships between these models, the following thesis has been proposed.

Parallel Computation Thesis: A problem can be solved in polynomial space on a reasonable sequential computer if and only if it can be solved in polynomial time on a reasonable parallel computer.

Theorem 1.27 (Parallel Computation Thesis) *The Parallel Computation Thesis is satisfied by the UBC and the TM. The Parallel Computation Thesis is satisfied by the PRAM and the RAM.*

6. NC class

If we consider reasonable models for sequential and parallel computations, one may ask whether it is interesting to use parallel computers. Can we solve in parallel all the (computable) problems IN LESS TIME than in sequential ? Surprisingly this question is very hard to answer and remains mainly open. In order to go deeper in it Nick Pippenger introduced new complexity classes known as NC classes [13].

Definition 1.28 *For a given integer* k, *we call* NC^k *the class of problems solvable in* $O(\log^k(n))$ *time on a* $CREW\,PRAM$ *with a polynomial number of processors. We define the class* NC *as the union of the classes* NC^k.

$$NC = \cup NC^k$$

Remark that the class NC is robust to logarithmic changes in the model of computation. The addition of n integers, the prefix computation and the scalar product are examples of problems in NC^1. The transitive closure of a boolean matrix is an example of a problem in NC^2. We have the following direct properties.

Theorem 1.29 (NC Class) *For a given integer k,*

$$NC^{k-1} \subseteq NC^k$$

WE DO NOT KNOW if the inclusion is strict !
Moreover, the NC class is included in P:

$$NC \subseteq P$$

WE DO NOT KNOW if the inclusion is strict !

7.   Central problem of parallel complexity

$NC = P$ is the central problem of parallel complexity. As a first feeling, one could think that the answer would be positive. It would be nice that every problem could be solved faster in parallel than in sequential. However the answer is certainly negative ! In order to better understand this problem we shall behave as for the central problem of complexity ($P = NP$) and try to study $P - NC$.

Definition 1.30 *A problem $Prob$ of P is P-Complete if any problem of P can be transformed in $Prob$ in polylogarithmic parallel time with a polynomial number of processors (NC reduction).*

Hence $Prob$ (if any) is the most difficult problem to solve in parallel. Consider the following problem:

Closure: A finite set X and a binary operator $\oplus$. $T \subset X$ and $x \in X$ are given.
     Question: does x belong to the closure of T by $\oplus$?

Theorem 1.31 ([8]) *Closure is P-Complete !*

Many other P-Complete problems are currently known, for instance the circuit value problem, CVP, Monotone CVP, Maxflow and Linear programming.

Theorem 1.32 (Serna [15]) *If $NC \neq P$ then there exists a language which is neither in NC nor P-Complete.*

Theorem 1.33 (Vollmer [18]) *If $NC \neq P$ then there are infinitely many degrees in $P - NC$.*

For both theorems we still do not have natural examples. Moreover many interesting practical problems are not known to be in NC nor to be P-Complete. 2-Variable linear programming, integer greatest common divisor or modular integer exponentiation are examples of such problems.

8.   Distributed memory PRAM

In the case of distributed memory, the $PRAM$ model must be modified to take memory accesses into account. Among others, Cosnard and Ferreira suggested generalizing the $PRAM$ model by modelling access to local memories using an interconnection network [5]. Hence each processor can only access its own memory cells and has to use the interconnection network in order to access the remaining memory cells.

Definition 1.34 *A $DRAM$ (Distributed RAM) is a set of p processors P_i, of p memory cells M_i and a family of pairs $X = (i, j)$ corresponding to the interconnection network. P_i can access M_j only if $(i, j) \in X$.*

Hence in order to access a distant data, processors have to cooperate. X is the interconnection network of the $DRAM$. In order to read or write a data item located in a memory cell to which it is not directly connected, a processor must obtain the help of other processors which, by reading and writing, will move the data item from the initial memory cell to a cell that can be accessed by the processor that requires the data item.

In order to simplify the presentation let define for a given i, $X_i = \{j \mid (i, j) \in X\}$. The processor P_i can access only the memory areas M_j, whenever j belongs to X_i. For such j the processors P_j are called the neighbours of P_i. As in the case of $PRAM$, time is discrete and at any time interval, all the processors of a $DRAM$ operate synchronously and execute the following operations: Read(M_j), Compute, Write (M_k), for $j, k \in X_i$. Different models $EREW$, $CREW$, $CRCW DRAM$ can also be defined according to how memory is accessed.

Hypercube RAM.   Assumptions about concurrent reading and writing exert an influence on the interconnection network. However, where the inteconnection network has a constant degree, that is the cardinal of X_i is constant for all i and independent of the number of processors, then the concurrent access can be executed sequentially with a constant overhead. It follows that differences in power betwen the models will be hidden by the notation O. This is not the case if the inteconnection netwok is a graph whose degree is a non-constant function of the number of processors. We shall consider a hypercube for the sake of simplicity.

Definition 1.35 *A HypercubeRAM or $HRAM$ is a $DRAM$ with a hypercube interconnection network:*

$$X_i = H_i = \{j \mid binary\ expansions\ of\ i\ and\ j\ differ\ only\ on\ one\ bit\ \}.$$

Before going to generalize theorems for differentiating and simulating $PRAM$, we shall look in a little more detail at the $HRAM$ model by describing an algorithm for the prefix computation. The following algorithm is directly adapted from Nassimi and Sahni's algorithm [12]. Note that it corresponds exactly to the $EREW\,PRAM$ algorithm. Indeed the communication graph maps directly to a hypercube.

> **Algorithm:** $EREW\,HRAM$ **prefix computation, code of** P_i
> **write**$(M(i))$ **in** $S(i)$;
> **For** $d \in \{0, \ldots, log(n) - 1\}$ **do**
>   **If** $i_d = 1$ **then** $M(i) = M(i) + S(i \oplus d)$;
> $S(i) = S(i) + S(i \oplus d)$

The procedure is as follows: neighbouring processors begin by computing their prefixes, then processors at a distance 2, then 4, and so on, with i_d being the d-th bit of i and $i \oplus d$ the index i in which the d-th bit has been complemented. Processor P_i writes to memory cells of indices $i \oplus d$ where $i \oplus d$ belongs to H_i. At the end of the computation the $i^t h$ prefix is the $M(i)$ and S contains the sum of all the elements.

Hence, the complexity of prefix computation on an $EREW\,HRAM$ is $O(log(n))$, with n processors.

Our study of differentiation of $PRAM$ can be applied to $HRAM$. Here, too, we shall consider only one $CRCW$ and adopt the same writing policy, the common policy. Of course, an algorithm designed for an $EREW\,HRAM$ can be executed on a $CREW\,HRAM$, and an algorithm written for a $CREW\,HRAM$ can be executed on a $CRCW\,HRAM$.

Theorem 1.36 ($HRAM$ Separation) *Assume that the three models have p processors. We then have:*

$$EREW\,HRAM \ll CREW\,HRAM \ll CRCW\,HRAM$$

Now let us look at the extend to which the models differ, by using the simulation in the same way as above.

Theorem 1.37 ($HRAM$ Simulation) *Assume that the three models have p processors. We then have:*

$$EREW\,HRAM = O(log(p)).CREW\,HRAM$$

$$CREW\,HRAM = O(log(p)).CRCW\,HRAM$$

$$EREW\,HRAM = O(log(p)).CRCW\,HRAM$$

If we continue this comparison between models, we may want to look at the power of $HRAM$ compared with $PRAM$.

Theorem 1.38 ($HRAM$ versus $PRAM$ Comparison) *Assume that the three models have p processors. We then have:*

$$EREW\,HRAM << EREW\,PRAM$$

$$CREW\,HRAM << CREW\,PRAM$$

$$CRCW\,HRAM << CRCW\,PRAM$$

$CREW\,HRAM$ and $EREW\,PRAM$ are incomparable .

$CRCW\,HRAM$ and $CREW\,PRAM$ are incomparable .

9. Performances

So far we have presented and compared various models of computation. If we restrict to reasonable models, one may want to measure for a given model, say for example the $CREW\,PRAM$, the gain in performance obtained when using p processors. Let us first begin by defining some tools for this measure and also the properties of these tools.

Definition 1.39 (Speedup) *Let T_1 be the execution time of a sequential algorithm and T_p the execution time of a parallel algorithm on a $PRAM$ with p processors. The speedup is defined as:*

$$S_p = T_1/T_p$$

We can make a comment about the speedup notion. Firstly, we must specify what we mean by the sequential execution time. In the preceding definition we assumed implicitly that the algorithm is independent of the number of processors, the same algorithm is used in sequential and for any value of p, which is generally false. Therefore, it seems more reallistic to consider the execution time of the best sequential algorithm. However, one generally admits that the speedup is limited by the number of processors.

Theorem 1.40 (Speedup Limitations) *For any p, $1 \leq S_p \leq p$.*

There is no specific reference for this theorem, which is generally used without any justification being given. In this sense, it is part of the folklore of parallelism. The formulation of this theorem is very imperfect, because some hypotheses have not been specified. In particular, there is an important point about the resources available to the processors. It is evident that p processors can store p times more data that one can. The speedup limitation theorem therefore assumes sufficient resources for each processor.

Various authors have attempted to specify more practical speedup limits. Amdahl proposed considering the inherent quantity of sequential program, which can be executed by a single processor.

Theorem 1.41 (Amdahl's Law [1]) *Let* $T_1 = T_{seq} + P \times T_{par}$, *then*

$$S_p \leq T_1/T_{par},$$

where T_{seq} corresponds to the time spent executing the sequential part of the program code, T_{par} corresponds to the time spent by the slowest processor executing the parallel part of the code, and P corresponds to the total number of processors.

Therefore, according to Amdahl's law the speedup is, for any p, less than the proportion of sequential code. For example, if 10% of the program is sequential, then for any p the speedup will be less than 10. Hence the speedup is limited by a limit which is independent of the number of processors and the structure of the machine.

The preceding approach is reasonable in the case of traditional parallel computers. However, in the case of massively parallel machines with distributed memory, this approach is insufficient. The speedup thus defined is generally not very good because of the cost of communication between processors and does not take into account the fact that the size of the problems processed can be much larger which is often crucial in many real applications. We can then define a new speedup by referring to the largest possible size of problems that can be solved using the set of processors. Gustafson proposes to calculate the ratio between sequential time and parallel time, by extrapolating the sequential time on an ideal processor with a memory whose size is equivalent to that of all the processors.

Theorem 1.42 (Gustafson's Law [9]) *Let* $T_1 = T_{seq} + T_{par}$, *then*

$$S_p \geq p.T_p/T_1$$

where T_{seq} corresponds to the time spent executing the sequential part of the program code and T_{par} corresponds to the total time spent by all processors executing the parallel part of the code.

A parallel algorithm with p processors is optimal, if its execution time $t_{opt}(p)$ is optimal. $t_{opt}(p)$ is the time complexity of the problem with p processors. Call t_{opt} the minimum of the $t_{opt}(p)$, for a given problem size. Several algorithms executed in t_{opt} may exist and each requires a different number of processors. The minimum number of processors required to solve the problem in optimal time will be called p_{opt}. Using one processor $t_{opt}(1)$ is the sequential complexity of the problem. The theorem on speedup limitations implies that the upper bound of the product $p.t_{opt}(p)$, called the cost of the algorithm, is $t_{opt}(1)$. Where the cost of a parallel algorithm is the same as that of an optimal sequential algorithm, the parallel algorithm is said to be *cost optimal*.

Fine granularity is the term used if the number of processors is of the same order as the number of data items. If not, coarse granularity is the term used. Granularity is a simple, rather than strictly mathematical, method of classifying algorithms. One result that is part of the folklore of parallel algorithms was proved by Brent [3]. It enables a parallel algorithm with p processors to be constructed from a parallel algorithm with a greater number of processors. A coarse-gran algorithm can thus be deduced from a fine-grain algorithm. Known as Brent's Lemma, it can be applied to any algorithm and any $PRAM$ model.

Lemma 1.43 (Brent's Lemma) *Let there be given a $PRAM$ model. If a parallel algorithm requires t time units and q operations for solving a given problem, then there exists an algorithm with p processors solving the same problem in time $t(p) \leq t + (q - t)/p$.*

Note that the number of processors is not specified for the initial algorithm, although it is very important for the algorithm. Operation is understood to mean that a processor is active (a processor is either inactive or it is executing an operation). Thus the number of operations gives the total active time for all the processors. As a very important consequence of Brent's Lemma, remark that if t is small compared to q/p and if $q = t_{opt}(1)$ is the sequential complexity, we get an optimal algorithm:

$$t(p) = t_{opt}(p) = t_{opt}(1)/p$$

As an example, let us apply the preceding result to the prefix computation. Recall that the sequential complexity is $t_{opt}(1) = O(n)$ and that on an $EREW\,PRAM$ the complexity is $t_{opt} = O(log(n))$. Such a complexity result has been obtained with n processors. Applying Brent's Lemma we get the following complexity for p processors:

$$t_{opt}(p) = O(n/p) \text{ if } p > n/log(n)$$

$$t_{opt}(p) = O(log(n)) \text{ if } p = n/log(n)$$

Remark the incredible power of Brent's Lemma since for the same complexity of $O(log(n))$, we reduce the number of processors from n to $n/log(n)$. Hence obtaining a cost optimal algorithm.

10.   Evaluation of arithmetic expressions

Consider a $CREW\,PRAM$ with binary arithmetic operations. As an example of the use that can be done of the previous results, let us study the complexity of the parallel evaluation of arithmetic expressions. Recall that an arithmetic expression of size n is a problem with n data and one result and with only binary arithmetic operations. The expression is simple if a variable is used only once: $X_i + X_j + X_k$ is a simple expression but X^3 is not simple.

A very powerful method in order to study the complexity of the evaluation in parallel of arithmetic expressions is to use labeled binary trees.

Definition 1.44 *Let p be a given integer. We denote by $A(p)$ the set of labeled binary trees constructed as follows:*

1 The tree with 1 node labeled 0 is in $A(p)$ at depth 0.

2 The binary tree with three vertices whose leaves are labelled 0 and whose root is labelled 1 is a member of $A(p)$ and is called the basic tree.

3 For $d \geq 0$, an element of $A(p)$ at depth $d + 1$ can be obtained from an element at depth d by adding 1 to all the labels and replacing at most p leaves by the basic tree.

Lemma 1.45 (Relations between Expressions and Trees) *Any parallel algorithm for evaluating an expression of size n with p processors can be represented by a tree $A(p)$ with at least n leaves.*

Any simple expression of size n can be represented by a tree $A(p)$ with n leaves. We can associate to it a parallel evaluation algorithm for a $CREW\,PRAM$ with p processors.

For any tree $A(p)$ with n leaves, there exists a simple expression of size n represented by this tree.

There is a closed connection between binary trees and boolean or arithmetic circuits. As one may think, the depth of the tree is strongly correlated with the parallel evaluation time of the associated expression. Hence we should try to compute the minimum depth of a tree with n leaves.

Definition 1.46 *We call $F(p, d)$ the maximum number of leaves of the trees of $A(p)$ with depth d and $D(p, n)$ the minimum depth of the trees of $A(p)$ with n leaves.*

Lemma 1.47 *We get:*

1 $F(p, 0) = 1$.

2 $F(p, d + 1) = F(p, d) + min(p, F(p, d))$

3 *If n is such that $F(p, d-1) < n \leq F(p, d)$ then $D(p, n) = d$.*

4 *$F(p, d) = 2^m + p.max(0, d - k)$ where $k = \lceil log(p) \rceil$ and $m = min(k, d)$.*

5 *$D(p, n) = min(k, \lceil log(n) \rceil) + max(0, \lceil (n - 2k)/p \rceil)$*
 $= log(n) + n/p + o(log(n) + n/p)$

Using the above two lemmas, a general lower bound can be obtained for the complexity of the evaluation, in parallel, of any expression. Note that the complexity of an expression is always greater than or equal to that of a simple expression of the same size. One method of associating a simple expression to any other expression is to replace all occurences of variables other than the first one with new variables. The next theorem is due to Munro and Paterson [11].

Theorem 1.48 (Complexity of Parallel Expression Evaluation) *The minimum number of steps for evaluating an expression of size n with p processors is $D(p, n)$, on a CREW PRAM.*

The minimum number of steps for evaluating an expression of size n is $\lceil log(n) \rceil$.

Clearly, if more precise lower bounds or complexity results are required then the general expressions have to be particularized in some way. An initial approach would be to assume that the same operation is repeated in the expression.

Theorem 1.49 *Let $\oplus$ be an associative operation. On a CREW PRAM with p processors $D(p, n)$ is the complexity of the evaluation with p processors of the simple expression $E_n = a_1 \oplus a_2 \oplus \ldots \oplus a_n$.*
 If $p \leq O(n/log(n))$ then the parallel algorithms are cost optimal.

For more detailed results on the complexity of parallel evaluation of arithmetic expressions we refer the reader to [6].

11. Solving linear systems

In this section we whall introduce the basic results on the parallel complexity of solving linear systems. Let us first begin by triangular systems.

Theorem 1.50 (Triangular Systems) *Let L be a lower triangular matrix of size n. On a CREW PRAM:*

1 *For solving $Lx = b$ in parallel, we need at least $2log(n) + o(log(n))$ steps.*

2 *There exists an algorithm for solving $Lx = b$ in $O(log^2(n))$*

Let us now study gereral systems.

Theorem 1.51 (Linear Systems) *Let A be a matrix of size n. On a CREW PRAM:*

1 For solving $Ax = b$, in parallel, we need at least $2log(n) + o(log(n))$ steps.

2 There exists an algorithm for solving $Ax = b$ in $O(log^2(n))$.

12.　　Perspectives and conclusion

In this survey paper, we have presented an introduction to the complexity theory of parallel algorithms: from models of computation to practical parallel algorithms. The purpose was not to go in details concerning some specific algorithms (there are numerous results on this subject and making a survey of all of them is out of scope), but rather to present the fundamentals of the theory and to show how they can be applied in the real world.

Note that the complexity of parallel algorithms is still in its early stages. Many of the basic problems are open: we do not know if parallel processing, in the general sense, is worth using it for any sequential algorithm. We did not give a detailed presentation of communication complexity which deserves a complete presentation. We still do not have a largely accepted model for communication complexity.

There appear now new models of computation that may seem exotic but which are very promising from the point of view of complexity results:

- computing with real numbers

- quantum computing

- bio-computing

References

[1] G.M. Amdahl, Validity of a single processor approach of achieving large scale computing capabilities. *AFIPS Conf. Proc. 30*, 483–485, 1967.

[2] J.L. Balcazar, J. Diaz and J. Gabarro, *Structural Complexity 1*. Springer 1995.

[3] R.P.Brent, The parallel evaluation of general arithmetic expressions. *Journal of ACM*, 21(2):201—206, 1974.

[4] S. Cook, The complexity of theorem proving procedures. *3rd. ACM Symp. on Theory of Computing*, 151–158, 1971.

[5] M. Cosnard and A. Ferreira, On the real power of loosely coupled parallel architectures. *Parallel Processing Letters* 1(2):103–112, 1991.

[6] M. Cosnard and D. Trystram, *Parallel Algorithms and Architectures.* Thomson Computer Press, 1995.

[7] M.R. Garey and D.S. Johnson, *Computers and Intractability - A guide to the theory of NP-completeness.* Freeman and Co. 1979.

[8] A. Gibbons and W. Rytter, *Eficient parallel algorithms.* Cambridge University Press, 1988.

[9] J.L. Gustafson, Reevaluating Amdahl's law. *Communications of ACM* 31,5:532–533, 1988.

[10] R.M. Karp, *Reducibility among combinatorial problems.* in Complexity of Computer Computations (R. Miller and J.Thatcher eds.), 85–104, Plenum Press New York, 1972.

[11] I. Munro and M. Paterson, Optimal algorithms for parallel polynomial evaluation. *Journal Comput. System Sci.* 189–198, 1973.

[12] D. Nassimi and S. Sahni, Data broadcasting in SIMD computers. *IEEE Transactions on Computers*, 30:101–106, 1981.

[13] N. Pippenger, On simultaneous resource bounds. *20th. IEEE Symp. on Foundations of Computer Science*, 307–311, 1979.

[14] W.J. Savitch, Relationships between nondeterministic and deterministic tape complexities. *Journal Comput. System Sci.*, 4:177–192, 1970.

[15] M.J. Serna, *The parallel approximability of P-complete problems.* PhD Thesis, Universitat Politechnica de Catalunya, 1990.

[16] M. Snir, On parallel searching. *SIAM Journal of Computing*, 14(3):688–708, 1985.

[17] U. Vishkin, Implementation of simultaneaous memory address access in models that forbids it. *Journal of Algorithms*, 4:45–50, 1983.

[18] H. Vollmer, The gap-language-technique revisited. *Proc. Computer Science and Logic 91*, LNCS, Springer-Verlag, 1991.

Chapter 2

THE COMBINATORICS OF RESOURCE SHARING

Valmir C. Barbosa

*Universidade Federal do Rio de Janeiro**

valmir@cos.ufrj.br

Abstract   We discuss general models of resource-sharing computations, with emphasis
on the combinatorial structures and concepts that underlie the various deadlock
models that have been proposed, the design of algorithms and deadlock-handling
policies, and concurrency issues. These structures are mostly graph-theoretic in
nature, or partially ordered sets for the establishment of priorities among processes and acquisition orders on resources. We also discuss graph-coloring concepts as they relate to resource sharing.

Keywords: Deadlock models, deadlock detection, deadlock prevention, concurrency measures.

1.   Introduction

The sharing of resources by processes under the requirement of mutual exclusion is one of the most fundamental issues in the design of computer systems,
and stands at the crux of most efficiency considerations for those systems. When
referred to with such generality, processes can stand for any of the computing
entities one finds at the various levels of a computer system, and likewise resources are any of the means necessary for those entities to function. Resources
tend to be scarce (or to get scarce shortly after being made available), so the
designer of a computer system at any level must get involved with the task of
devising allocation policies whereby the granting of resources to processes can
take place with at least a minimal set of guarantees.

---

*Programa de Engenharia de Sistemas e Computação, COPPE, Caixa Postal 68511, 21945-970 Rio de
Janeiro - RJ, Brazil. This author is supported by the Brazilian agencies CNPq and CAPES, the PRONEX
initiative of Brazil's MCT under contract 41.96.0857.00, and by a FAPERJ BBP grant.

*R. Corrêa et al. (eds.), Models for Parallel and Distributed Computation. Theory, Algorithmic Techniques
and Applications, 27–52.*

One such guarantee is of the so-called *safety* type, and in essence forbids the occurrence of *deadlock* situations. A deadlock situation is characterized by the permanent impossibility for a group of processes to progress with their tasks due to the occurrence of a condition that prevents at least one needed resource from being granted to each of the processes in that group. Another guarantee one normally seeks is a *liveness* guarantee, which imposes bounds on the wait that any process must undergo between requesting and being granted access to a resource, and thereby ensures that *lockout* situations never happen.

There are difficulties of various sorts associated with designing and analyzing resource-sharing policies. Some of these difficulties refer to the choice and use of mathematical models that can account properly for the relevant details of the resource-sharing problem at hand. Similarly, there are difficulties that stem from the inherent asynchronism that typically characterizes the behavior of processes in a computer system. This asynchronism, though essential in depicting most computer systems realistically, tends to introduce subtle obstacles to the design of correct algorithms.

In this paper, we are concerned with the several combinatorial models that have proven instrumental in the design and analysis of resource-sharing policies. The models that we consider are essentially of graph-theoretic nature, and relate closely to the aforementioned safety and liveness issues. The essential notation that we use is the following. The set of processes is denoted by $\mathcal{P} = \{P_1, \ldots, P_n\}$, and the set of resources by $\mathcal{R} = \{R_1, \ldots, R_m\}$. For $P_i \in \mathcal{P}, \mathcal{R}_i \subseteq \mathcal{R}$ is the set of resources to which P_i may request access. Similarly, for $R_p \in \mathcal{R}, \mathcal{P}_p \subseteq \mathcal{P}$ is the set of processes that may request access to R_p. Clearly, for $1 \leq i \leq n$ and $1 \leq p \leq m$, $P_i \in \mathcal{P}_p$ if and only if $R_p \in \mathcal{R}_i$. Also, we let $\mathcal{R}_{ij} = \mathcal{R}_i \cap \mathcal{R}_j$ for $1 \leq i, j \leq n$, and $\mathcal{P}_{pq} = \mathcal{P}_p \cap \mathcal{P}_q$ for $1 \leq p, q \leq m$.

Example 2.1 *If* $\mathcal{P} = \{P_1, P_2, P_3, P_4, P_5\}$ *and* $\mathcal{R} = \{R_1, R_2, R_3, R_4, R_5, R_6\}$ *with* $\mathcal{R}_1 = \{R_1, R_2\}$, $\mathcal{R}_2 = \{R_2, R_3, R_6\}$, $\mathcal{R}_3 = \{R_3, R_4, R_6\}$, $\mathcal{R}_4 = \{R_4, R_5, R_6\}$, *and* $\mathcal{R}_5 = \{R_1, R_5\}$, *then* $\mathcal{P}_1 = \{P_1, P_5\}$, $\mathcal{P}_2 = \{P_1, P_2\}$, $\mathcal{P}_3 = \{P_2, P_3\}$, $\mathcal{P}_4 = \{P_3, P_4\}$, $\mathcal{P}_5 = \{P_4, P_5\}$, *and* $\mathcal{P}_6 = \{P_2, P_3, P_4\}$. *In addition, we have the nonempty sets shown in Table 2.1.*

The following is how the remainder of the paper is organized. In Section 2, we provide an outline of the generic computation that is carried out by the members of $\mathcal{P}$ in order to share the resources in $\mathcal{R}$. Such an outline is given as an asynchronous distributed algorithm, and aims at emphasizing the communication that must take place among processes for resource sharing. This communication comprises at least messages for requesting and granting access to resources. Depending on how such messages are composed and handled by the processes, one gets one of the various *deadlock models* that have appeared in the literature. These models are our subject in Section 3. The two sections that follow (Sections 4 and 5) are devoted to the combinatorics underlying the

Table 2.1. Resource and process sets for Example 2.1.

$$
\begin{aligned}
\mathcal{R}_{12} &= \{R_2\} & \mathcal{P}_{12} &= \{P_1\} \\
\mathcal{R}_{15} &= \{R_1\} & \mathcal{P}_{15} &= \{P_5\} \\
\mathcal{R}_{23} &= \{R_3, R_6\} & \mathcal{P}_{23} &= \{P_2\} \\
\mathcal{R}_{24} &= \{R_6\} & \mathcal{P}_{26} &= \{P_2\} \\
\mathcal{R}_{34} &= \{R_4, R_6\} & \mathcal{P}_{34} &= \{P_3\} \\
\mathcal{R}_{45} &= \{R_5\} & \mathcal{P}_{36} &= \{P_2, P_3\} \\
& & \mathcal{P}_{45} &= \{P_4\} \\
& & \mathcal{P}_{46} &= \{P_3, P_4\} \\
& & \mathcal{P}_{56} &= \{P_4\}
\end{aligned}
$$

two broad classes of deadlock-handling policies, namely those of *detection* and *prevention* strategies, respectively. We then move, in Section 6, to a prevention policy that generalizes one of policies discussed in Section 5 and for which an abacus-like graph structure is instrumental. This generalized policy is for the case of high demand for resources by the processes. Section 7 discusses the relationship that exists between concurrency in resource sharing and the various chromatic indicators of a graph. Concluding remarks follow in Section 8.

In this paper, all lemma and theorem proofs are omitted, but references are given to where they can be found.

2. Resource-sharing computations

The model of computation that we assume in this section is the standard *fully asynchronous* (or simply *asynchronous*) model of distributed computing [2]. In this model, every member of $\mathcal{P}$ possesses a local, independent clock, having therefore a time basis that is totally uncorrelated to that of any other process. In addition, all communication among processes take place via point-to-point message passing, requiring a finite (though unpredictable) time for message delivery. Messages are sent over bidirectional communication channels, of which there exists one for every $P_i, P_j \in \mathcal{P}$ such that $\mathcal{R}_{ij} \neq \emptyset$. That is, every two processes with the potential to share at least one resource are directly interconnected by a bidirectional communication channel. If we let $\mathcal{C}$ denote the set of such channels, then the undirected graph $G = (\mathcal{P}, \mathcal{C})$, having one vertex for each process and one edge for each channel, can be used to represent the system over which our resource-sharing computations run. In G, and for $1 \leq p \leq m$, the vertices in $\mathcal{P}_p$ induce a completely connected subgraph (a *clique* [6]). We assume that G is a connected graph, as processes belonging to different connected components never interfere with each other. In the context of Example 2.1, G is the graph shown in Figure 2.1.

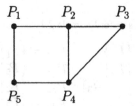

Figure 2.1. The graph G for Example 2.1.

In the computations that we consider, a process executes the following four procedures.

- REQUEST;

- CHECK_PRIORITY;

- COMPUTE;

- CLEAN_UP.

Each of these procedures is executed atomically in response to a specific event, as follows. When the need arises for the process to compute on shared resources, it executes the REQUEST procedure. Typically, this will entail sending to some of its neighbors in G messages requesting exclusive access to resources shared with them. The reception of one such message causes the receiving process to execute CHECK_PRIORITY, whose outcome will guide the process' decision as to whether grant or not the requested exclusive access. If the process does decide to grant the request, then a message carrying this information is sent back to the requesting process, which upon receipt executes COMPUTE. This procedure is a test to see whether the process already holds exclusive access to enough resources to carry out its computation, which it does in the affirmative case; it keeps waiting, otherwise. If and when the resource-sharing computation is completed, the process engages in a message exchange with its neighbors in G by executing the CLEAN_UP procedure. This message exchange may revise priorities and cause previously withheld requests to be granted.

This outline is admittedly far too generic in several aspects, but already it provides the background for the key questions underlying the establishment of a resource-sharing policy. For example: To which resources does a process request exclusive access in REQUEST when in need for shared resources? At which point when executing COMPUTE does it decide it may proceed with its computation? How do the CHECK_PRIORITY and CLEAN_UP procedures cooperate to handle the priority issue properly? Answers to these questions have been given in the context of several application areas, and along with numerous models and algorithms. Addressing them in detail is beyond our intended scope, but in Section 3 we present the abstraction of deadlock models, which summarizes the issues that are critical to our discussion of the combinatorics of resource sharing.

Note that both the execution of REQUEST and the initial test performed by COMPUTE may entail waiting on the part of the calling process. Clearly, then, and depending on how the priority issue is handled, here lies the possibility for unbounded wait, which is directly related to the safety and liveness guarantees we may wish to provide. The approaches here vary greatly, and may be grouped into two broad categories. On the "optimistic" side, one may opt for a somewhat loose priority scheme and risk the loss of those guarantees. In such cases, the loss of safety leads to the need for the capability of detecting deadlocks [11, 17, 23]. The opposing, more "conservative" side is the side of those strategies which "by design" guarantee safety and liveness, thereby preventing their loss beforehand.

As we demonstrate in the remainder of the paper, both categories give rise to interesting combinatorial structures and properties, especially as they relate to the deadlock issue. We then end this section by defining what will be meant henceforth by deadlock, although still somewhat informally. As we go through the various combinatorial structures that relate closely to deadlocks, such informality will dissipate. A subset of processes $S \subseteq P$ is in deadlock if and only if every process in S is waiting for a condition that ultimately can be relieved only by another member of S whose own wait is over. Obviously, then, deadlocks are stable properties: Once they take hold of a group of processes, only the external intervention that eventually follows detection may break them. Prevention strategies, by contrast, seek never to let them happen.

3. Deadlock models

A deadlock model is an abstraction of the rules that govern the wait of processes for one another as they execute the procedures REQUEST, CHECK_PRIORITY, COMPUTE, and CLEAN_UP discussed in Section 2. Deadlock models are defined on top of a dynamic graph, called the *wait-for graph* and henceforth denoted by W.

W is the directed graph $W = (P, W)$, having the same vertex set as G (one vertex per process) and the directed edges in W. This set is such that an edge exists directed from process P_i to process P_j if and only if P_i has sent P_j a request for exclusive access to some resource that they share and is waiting either for P_j to grant the request or for the need for that resource to cease existing as grant messages are received from other processes. For $P_i \in P$, we let $O_i \subseteq P$ be the set of processes towards which edges are directed away from P_i in W.

It follows from the definition of W that the only processes that may be carrying out some computation on shared resources are those that are *sinks* in W (vertices with no adjacent edges directed outward, including isolated vertices). All other processes are waiting for exclusive access to the resources they need. Clearly, then, a necessary condition for a deadlock to exist in W is that W contain a directed cycle.

Fact 2.2 *If a deadlock exists in W, then W contains a directed cycle.*

Example 2.3 *In the context of Example 2.1, suppose a deadlock has happened involving processes P_2, P_3, and P_4. Suppose also that process P_1 is waiting for resource R_2, which is held by P_2, which in turn is waiting for R_3, held by P_3, which is waiting for R_4, held by P_4. If, in addition, R_6 is held by P_2 and awaited by P_4, then the corresponding W is the one shown in Figure 2.2, with the directed cycle on P_2, P_3, and P_4.*

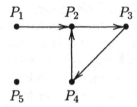

Figure 2.2.   The graph W for Example 2.3.

In Section 4, after we have gone through a variety of deadlock models in the remainder of this section, we will come to the conditions that are sufficient for a deadlock to exist in W.

This is the sense in which graph W is a dynamic structure: Although its vertex set is always the same, as the processes interact with one another by executing the aforementioned procedures, its edge set changes. Normally, given the view we have adopted of the resource-sharing computation as an asynchronous distributed computation, one must bear in mind the fact that it only makes sense to refer to W as associated with some *consistent global state* of the computation [2, 9]. For our purposes, however, such an association does not have to be explicit, so long as one understands the dynamic character of W.

What determines the evolution of W by allowing for changes in the set $\mathcal{W}$ of directed edges is the deadlock model that holds for the computation. The deadlock models that have been investigated to date are the ones we discuss next. In essence, what each of these deadlock models does is to specify rules for vertices that are not sinks in W to become sinks.

The AND model.   In the AND model, a process P_i can only become a sink when its wait is relieved by all processes in $\mathcal{O}_i$. This model characterizes, for example, situations in which a conjunction of resources is needed by P_i [8, 18, 21].

The OR model.   In the OR model, it suffices for process P_i to be relieved by one of the processes in $\mathcal{O}_i$ in order for its wait to finish. The OR model characterizes, for example, some of the situations in which any one of a group of resources (a disjunction of resources) is needed by P_i [8, 18, 20, 21].

The x-out-of-y model.   In this model, there are two integers, x_i and y_i, associated with process P_i. Also, $y_i = |\mathcal{O}_i|$, meaning that process P_i is in principle waiting for communication from every process in $\mathcal{O}_i$. However, in order to be relieved from its wait condition, it suffices that such communication arrive from any x_i of those y_i processes. The x-out-of-y model can then be used, for example, to characterize situations in which P_i starts by requiring access permissions in excess of what it really needs, and then withdraws the requests that may still be pending when the first x_i responses are received [7, 8, 18].

The AND-OR model.   In the AND-OR model, there are $t_i \geq 1$ subsets of $\mathcal{O}_i$ associated with process P_i. These subsets are denoted by $\mathcal{O}_i^1, \ldots, \mathcal{O}_i^{t_i}$ and must be such that $\mathcal{O}_i = \mathcal{O}_i^1 \cup \cdots \cup \mathcal{O}_i^{t_i}$. In order for process P_i to be relieved from its wait condition, it must receive grant messages from all the processes in at least one of $\mathcal{O}_i^1, \ldots, \mathcal{O}_i^{t_i}$. For this reason, these t_i subsets of $\mathcal{O}_i$ are assumed to be such that no one is contained in another. Situations that the AND-OR model characterizes are, for example, those in which P_i perceives several conjunctions of resources as equivalent to one another and issues requests for several of them with provisions to withdraw some of them later [3, 8, 18, 21].

The disjunctive x-out-of-y model.   In this model, associated with process P_i are $u_i \geq 1$ pairs of integers, denoted by $(x_i^1, y_i^1), \ldots, (x_i^{u_i}, y_i^{u_i})$. These integers are such that $y_i^1 = |\mathcal{Q}_i^1|, \ldots, y_i^{u_i} = |\mathcal{Q}_i^{u_i}|$, where $\mathcal{Q}_i^1, \ldots, \mathcal{Q}_i^{u_i}$ are subsets of $\mathcal{O}_i$ such that $\mathcal{O}_i = \mathcal{Q}_i^1 \cup \cdots \cup \mathcal{Q}_i^{u_i}$. In order to be relieved from its wait condition, P_i must be granted access to shared resources by either x_i^1 of the y_i^1 processes in $\mathcal{Q}_i^1$, or x_i^2 of the y_i^2 processes in $\mathcal{Q}_i^2$, and so on. Of course, it makes no sense for $\mathcal{Q}_i', \mathcal{Q}_i'' \in \{\mathcal{Q}_i^1, \ldots, \mathcal{Q}_i^{u_i}\}$ to exist such that $\mathcal{Q}_i' \subseteq \mathcal{Q}_i''$ and $x_i' \geq x_i''$, which is then assumed not to be the case. This model characterizes situations similar to those characterized by the x-out-of-y model, and generalizes that model by allowing for a disjunction on top of it [8, 18].

As one readily realizes, these five models are not totally uncorrelated and a strict hierarchy exists in which a model generalizes the previous one in the sense that it contains as special cases all the possible wait conditions of the other. For example, the x-out-of-y model generalizes the AND model with $x_i = y_i$ and the OR model with $x_i = 1$ for all $P_i \in \mathcal{P}$. Likewise, and also for all $P_i \in \mathcal{P}$, the AND-OR model also generalizes the AND model with $t_i = 1$ and the OR model with $|\mathcal{O}_i^1| = \cdots = |\mathcal{O}_i^{t_i}| = 1$.

Despite this ability of both the x-out-of-y model and the AND-OR model to generalize both the AND and OR models, they are not equivalent to each other. In fact, the AND-OR model is more general than the x-out-of-y model, while the converse is not true. In order for the AND-OR model to express a general x-out-of-y condition, it suffices that, for all $P_i \in \mathcal{P}$, $t_i = \binom{y_i}{x_i}$ and $|\mathcal{O}_i^1| = \cdots = |\mathcal{O}_i^{t_i}| = x_i$.

Example 2.4 *Suppose that we have, for some $P_i \in \mathcal{P}$, $\mathcal{O}_i = \{P_j, P_k, P_\ell\}$. In the x-out-of-y model, $y_i = 3$. If $x_i = 2$, then in the AND-OR model we have, equivalently, $t_i = \binom{3}{2} = 3$, $\mathcal{O}_i^1 = \{P_j, P_k\}$, $\mathcal{O}_i^2 = \{P_j, P_\ell\}$, and $\mathcal{O}_i^3 = \{P_k, P_\ell\}$.*

To finalize our discussion on how the five deadlock models are related to one another, note that the AND-OR model and the disjunctive x-out-of-y model are equivalent to each other. In order to see that the AND-OR model generalizes the disjunctive x-out-of-y model, let $t_i = v_i^1 + \cdots + v_i^{u_i}$, where

$$1 \leq v_i^1 \leq \binom{y_i^1}{x_i^1}, \ldots, 1 \leq v_i^{u_i} \leq \binom{y_i^{u_i}}{x_i^{u_i}},$$

for all $P_i \in \mathcal{P}$. In addition, v_i^1 of the sets $\mathcal{O}_i^1, \ldots, \mathcal{O}_i^{t_i}$ must have cardinality x_i^1 and be subsets of $\mathcal{Q}_i^1$, the same holding for the other superscripts $2, \ldots, u_i$, and explicit care must be exercised to avoid any of the sets $\mathcal{O}_i^1, \ldots, \mathcal{O}_i^{t_i}$ being a subset of another.

That the disjunctive x-out-of-y model generalizes the AND-OR model is simpler to see. For such, it suffices that, for all $P_i \in \mathcal{P}$, we let $u_i = t_i$ and $\mathcal{Q}_i^1 = \mathcal{O}_i^1, \ldots, \mathcal{Q}_i^{u_i} = \mathcal{O}_i^{t_i}$, along with $x_i^1 = y_i^1, \ldots, x_i^{u_i} = y_i^{u_i}$.

Example 2.5 *Let $\mathcal{O}_i = \{P_j, P_k, P_\ell, P_t\}$ for some $P_i \in \mathcal{P}$. In the disjunctive x-out-of-y model, suppose we have $u_i = 2$, $\mathcal{Q}_i^1 = \{P_j, P_k\}$, and $\mathcal{Q}_i^2 = \{P_k, P_\ell, P_t\}$, yielding $y_i^1 = 2$ and $y_i^2 = 3$. If $x_i^1 = x_i^2 = 2$, then in the AND-OR model we have $t_i = \binom{2}{2} + \binom{3}{2} = 4$, $\mathcal{O}_i^1 = \{P_j, P_k\}$, $\mathcal{O}_i^2 = \{P_k, P_\ell\}$, $\mathcal{O}_i^3 = \{P_k, P_t\}$, and $\mathcal{O}_i^4 = \{P_\ell, P_t\}$. Had we started out with this AND-OR setting, then for the disjunctive x-out-of-y model we would have $\mathcal{Q}_i^1 = \{P_j, P_k\}$, $\mathcal{Q}_i^2 = \{P_k, P_\ell\}$, $\mathcal{Q}_i^3 = \{P_k, P_t\}$, and $\mathcal{Q}_i^4 = \{P_\ell, P_t\}$. We would also have $x_i^1 = y_i^1 = x_i^2 = y_i^2 = x_i^3 = y_i^3 = x_i^4 = y_i^4 = 2$. Clearly, this is equivalent to the disjunctive x-out-of-y scenario of the beginning of this example.*

4. Graph structures for deadlock detection

As we remarked in Section 2, computations that make no *a priori* provisions against the occurrence of deadlocks must, if the need arises, resort to techniques for the detection of deadlocks. Detecting the existence of a deadlock in the wait-for graph W can become the detection of a graph-theoretic property on W if we are able to characterize conditions on W that are sufficient for the existence of deadlocks. As we discuss in this section, such a property exists for all the deadlock models of Section 3. However, not always is it the case that detecting this graph-theoretic property directly is the most efficient means of deadlock detection. When this happens not to be the case, alternative approaches must be employed, usually based on some form of simulation of the sending of grant messages.

We start with the AND model, and recognize immediately that the presence of a directed cycle in W is not only a necessary condition for the existence of a deadlock in W but also a sufficient condition. This is so because, in the AND model, every process requires grant messages to be received on all edges directed away from it, which clearly is precluded by the existence of a directed cycle.

Fact 2.6 *In the AND model, a deadlock exists in W if and only if W contains a directed cycle.*

In Figure 2.3, we show two wait-for graphs in the AND model. Circular arcs joining edges directed away from vertices are meant to indicate that the AND model is being used. By Fact 2.6, there is deadlock in the W of Figure 2.3(a), but not in Figure 2.3(b).

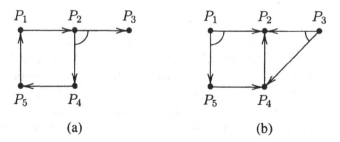

Figure 2.3. Two wait-for graphs in the AND model.

The case of the OR model is more subtle, and it is instructive to start by realizing that the presence of a directed cycle in W is no longer sufficient for the existence of deadlocks. Clearly, so long as a directed path exists in W from every process to at least one sink, then no deadlock exists in W even though a directed cycle may be present. Formalizing this notion requires that we consider the definition of a *knot* in W.

For $P_i \in \mathcal{P}$, let $\mathcal{T}_i \subseteq \mathcal{P}$ be the set of vertices that can be reached from P_i through a directed path in W. This set includes P_i itself, and is known as the *reachability set* of P_i [6]. We say that a subset of vertices $\mathcal{S} \subseteq \mathcal{P}$ is a knot in W if and only if $\mathcal{S}$ has at least two vertices and, for all $P_i \in \mathcal{S}$, $\mathcal{T}_i = \mathcal{S}$. By definition, then, no member of a knot has a sink in its reachability set, which characterizes the presence of a knot in W as the sufficient condition we have sought under the OR model. As it turns out, in fact, this condition is also necessary, being stronger than the necessary condition established by Fact 2.2.

Theorem 2.7 [16] *In the OR model, a deadlock exists in W if and only if W contains a knot.*

The wait-for graphs of Figure 2.4 are for the OR model. A knot is present in Figure 2.4(a) (involving all vertices), but not in Figure 2.4(b). Thence, by Theorem 2.7, there is deadlock in part (a) of the figure but not in part (b), in which P_3 is a sink that can be reached from all other processes.

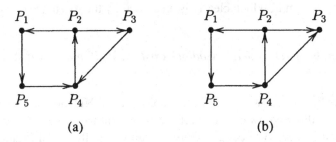

Figure 2.4.   Two wait-for graphs in the OR model.

In order to identify sufficient conditions on W that account for the existence of deadlocks in the remaining deadlock models, we must consider W in a more explicit conjunction with the deadlock model than we have done so far. Let us first consider the x-out-of-y model, and suppose that a subset of vertices $\mathcal{S} \subseteq \mathcal{P}$ can be identified having the property that, for all $P_i \in \mathcal{S}$, $|\mathcal{O}_i \cap \mathcal{S}| > y_i - x_i$. Under these circumstances, it is clear that no member of $\mathcal{S}$ can ever receive the number of grant messages it requires, because at least one of such messages would necessarily have to originate from within $\mathcal{S}$. In this paper, we let a set such as $\mathcal{S}$ be called a $(y - x)$-*knot*, whose existence in W can also be shown to be necessary for deadlocks to exist. As in the case of the OR model, this condition is stronger than the necessary condition of Fact 2.2.

Theorem 2.8 [18] *In the x-out-of-y model, a deadlock exists in W if and only if W contains a $(y - x)$-knot.*

An illustration is given in Figure 2.5, with an integer in parentheses next to the identification of each vertex to indicate its x value. A $(y - x)$-knot appears in Figure 2.5(a), but not in Figure 2.5(b). The $(y - x)$-knot of Figure 2.5(a) involves the vertices of the square. By Theorem 2.8, there is deadlock in the W of part (a) of the figure, but not in that of part (b). Note that P_3 is a sink reachable from all vertices in both graphs, but this is to no avail in the graph of part (a).

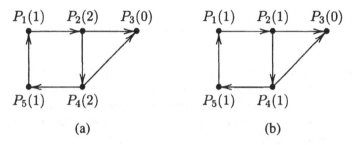

Figure 2.5. Two wait-for graphs in the x-out-of-y model.

We now turn to a discussion of sufficient conditions for deadlocks to exist in W under the AND-OR model. As we discussed in Section 3, the AND-OR model and the disjunctive x-out-of-y model are equivalent to each other, and for this reason the conditions that we come to identify as sufficient under the AND-OR model will also be sufficient under the disjunctive x-out-of-y model if only we perform the transformation described in Section 3.

Our starting point is the following definition. Consider a subgraph W' of W having vertex set $\mathcal{P}$, and for process P_i let $\mathcal{O}'_i \subseteq \mathcal{P}$ be the set of vertices towards which directed edges from P_i exist in W'. In addition, for process P_i let $\mathcal{O}'_i$ be such that $\mathcal{O}'_i \cap \mathcal{O}^1_i, \ldots, \mathcal{O}'_i \cap \mathcal{O}^{t_i}_i$ all have at least one member. We call such a subgraph a *b-subgraph* of W, where the "b" is intended to convey the notion that each directed edge in W' relates to a "bundle" of directed edges stemming from the same vertex in W [3]. An illustration is given in Figure 2.6 of a wait-for graph in part (a) and one of its b-subgraphs in part (b). Circular arcs around vertices in Figure 2.6(a) indicate the "AND" groupings of neighbors that constitute vertices' waits.

Intuitively, a b-subgraph of W represents one of the various "OR" possibilities that are summarized in W under the AND-OR model, provided that we consider such possibilities "globally," i.e., over all processes. As it turns out, the existence of a knot in at least one of the b-subgraphs of W is necessary and sufficient for a deadlock to exist in W. The knot that in this case exists in that b-subgraph is called a *b-knot* in W [3].

Theorem 2.9 [3] *In the AND-OR model, a deadlock exists in W if and only if W contains a b-knot.*

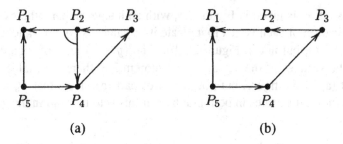

(a)   (b)

Figure 2.6.   *W* and one of its b-subgraphs.

We show in Figure 2.7 another of the b-subgraphs of the wait-for graph *W* of Figure 2.6(a). This b-subgraph has a knot spanning the processes in the triangle, which by Theorem 2.9 characterizes deadlock. In fact, in *W* it is easy to see that P_2 requires a relieve signal not only from P_1 (this one must come eventually) but also from P_4 (which will never come).

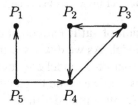

Figure 2.7.   A b-subgraph with a knot.

5.   Partially ordered sets and deadlock prevention

In the remaining sections (Sections 5 through 7), we address the AND model exclusively. For this model, by Facts 2.2 and 2.6 we know that the existence of a directed cycle in the wait-for graph *W* is necessary and sufficient for a deadlock to exist. The fact that this condition is necessary, in particular, allows us to look for design strategies that prevent the occurrence of deadlocks beforehand by precluding the appearance of directed cycles in *W*.

Of course, we also know from Fact 2.2 that directed cycles in *W* are necessary for a deadlock to exist regardless of the deadlock model. However, for deadlock models other than the AND model, we have seen in Section 4 that there may exist a directed cycle in *W* without the corresponding existence of a deadlock. As a matter of fact, we have seen that structures in *W* much more complicated than directed cycles are necessary for deadlocks to exist. Preventing the occurrence of cycles in those other models is then too restrictive, while preventing the occurrence of the more general structures appears to be too complicated. That is why our treatment of deadlock prevention is henceforth restricted to the AND model.

Resource-sharing problems for the AND model are often referred to as the *dining* or *drinking philosophers problem* [10, 13], depending, respectively, on whether every process P_i always requests access to all the resources in $\mathcal{R}_i$ or not. In the remainder of this section, we discuss two prevention strategies for such problems. Both strategies are based on the use of a partially ordered set (a poset), in the first case to establish dynamic priorities among processes, in the second to establish a static global order for resource request.

5.1.   Ordering the processes

Consider the graph G that represents the sharing of resources among processes, and let ω be an acyclic orientation of its edges. That is, ω assigns to each edge in $\mathcal{C}$ (the edge set of G) a direction in such a way that no directed cycle is formed. This orientation establishes a partial order on the set $\mathcal{P}$ of G's vertices, so G oriented by ω can be regarded as a poset.

This poset is dynamic, in the sense that the acyclic orientation changes over time, and can be used to establish priorities for processes that are adjacent in G to use shared resources when there is conflict. More specifically, consider a resource-sharing computation that does the following. A process sends requests for all resources that it needs, and must, upon receiving a request, decide whether to grant access to the resource immediately or to do it later. What the process does is to check whether the edge between itself and the requesting process is oriented outwards by ω. In the affirmative case, it grants access to the resource either immediately or upon finishing to use it (if this is the case). In the negative case, it may either grant access (if it does not need the resource presently) or delay the granting until after it has acquired all the resources it needs and used them. Whenever a process finishes using a group of resources, it causes all edges presently oriented towards itself to be oriented outward, thereby changing the acyclic orientation of G locally. These reversals of orientation constitute priority reversals between the processes involved.

We see, then, that an acyclic orientation establishes a priority for resource usage between every two neighbors in G, and that this priority is reversed back and forth between them as they succeed in using the resources they need. The crux of this mechanism is the simple property that the local changes a process causes to the acyclic orientation always maintain its acyclicity, and therefore its poset nature. If ω' is the acyclic orientation that results from the application of such a local change, then we have the following.

Lemma 2.10 [10] *If ω is acyclic, then ω' is acyclic.*

Note that Lemma 2.10 holds even if ω' results from local changes applied to ω by more than one process concurrently. We show such a pair of orientations in Figure 2.8, where the processes that do the reversal are P_1 and P_4. In the dining-philosopher variant of the resource-sharing computation, such a group of processes does necessarily constitute an *independent set* of G (a set whose members are all nonneighbors) [6].

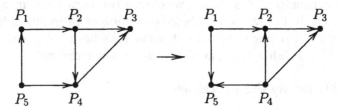

Figure 2.8.   Edge reversal on acyclic orientations.

The acyclicity of the changing orientation of G is crucial in guaranteeing that deadlocks never occur. To see this, consider the subgraph W' of the wait-for graph W that corresponds to processes not being able to grant access to resources immediately or after the (finite) time during which resources are in use. According to the computation we outlined above, this happens when a process, say P_i, receives a request from a neighbor P_j but cannot grant it immediately because it too needs the resource in question and furthermore holds priority over P_j. Clearly, in both W' and the acyclic orientation ω of G that gives priorities, the edge between P_i and P_j is oriented from P_j to P_i. In other words, W' is always a subgraph of G oriented by ω, and by Lemma 2.10 never contains a directed cycle. Because the edges that W has in excess of W' are all directed toward sinks, W is acyclic as well, which by Fact 2.2 implies the absence of deadlocks. If we refer to computations such as the one we described as *edge-reversal computations* [1], then we have the following.

Theorem 2.11 [10] *Every edge-reversal computation is deadlock-free.*

Not only do the orientations of G ensure the absence of deadlocks, but they can be easily seen to ensure liveness guarantees as well. Owing to the relationship of the computations we are considering to the dining-philosopher paradigm, such guarantees are referred to as the absence of *starvation*. That no starvation ever occurs comes also from the absence of directed cycles in W: As the orientations of edges are reversed and W evolves, the farthest sinks for which a process is ultimately waiting come ever closer to it, until it too becomes a sink eventually and its wait ceases.

Theorem 2.12 [10] *Every edge-reversal computation is starvation-free, and the worst-case wait a process must undergo is $O(n)$.*

We note that, in Theorem 2.12, the wait of a process is measured as the length of "causal chains" in the sending of grant messages, as is customary in the field of asynchronous distributed algorithms [2].

5.2. Ordering the resources

The graph G that underlies all our resource-sharing computations has one vertex per process and undirected edges connecting any two processes with the potential to share at least one resource. The undirected graph we introduce now and use throughout the end of the section is, by contrast, built on resources for vertices, and has undirected edges connecting pairs of resources that are potentially used in conjunction with each other by at least one process. This graph is denoted by $H = (\mathcal{R}, \mathcal{E})$, where $\mathcal{E}$ contains an edge between R_p and R_q if and only if $\mathcal{P}_{pq} \neq \emptyset$. H is a connected graph (because G is connected) and contains a clique on $\mathcal{R}_i$ for $1 \leq i \leq n$. The graph H for Example 2.1 is shown in Figure 2.9.

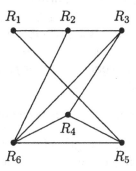

Figure 2.9. The graph H for Example 2.1.

Our interest in graph H comes from the possibility of constructing a poset on its vertices by orienting its edges acyclically, similarly to what we did previously on G. More specifically, let φ be an acyclic orientation of H, and for $R_p, R_q \in \mathcal{R}$ say that R_p *precedes* R_q if and only if a directed path exists from R_p to R_q in H oriented by φ. One acyclic orientation φ for the graph of Figure 2.9 is given in Figure 2.10. Note that the resources in $\mathcal{R}_i$ for any $P_i \in \mathcal{P}$ are necessarily totally ordered by the "precedes" relation.

Now consider the following resource-sharing computation. When a process needs access to a group of shared resources, it sends requests to the neighbors with which it shares those resources according to the partial order implied by φ. The rule to be followed is simple: A process only sends requests for a resource R_p after all grants have been received for the resources that it needs and that precede R_p. The sending of grant responses to requests for a particular resource R_p is regulated by an $O(|\mathcal{P}_p|)$-time distributed procedure on the vertices be-

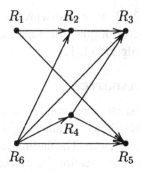

Figure 2.10. The graph H for Example 2.1 oriented acyclically .

longing to the clique in G that corresponds to that resource. This procedure for the acquisition of a single resource must itself be deadlock- and starvation-free [2, 25].

Because φ is acyclic, the evolving wait-for graph W can never contain a directed cycle, so by Fact 2.2 no deadlocks ever occur. The absence of directed cycles in W comes from the fact that such a cycle would imply a "hold-and-wait" cyclic arrangement of the processes, which is precluded by the acyclicity of φ. We call these resource-sharing computations *acquisition-order computations*, for which the following holds.

Theorem 2.13 [19] *Every acquisition-order computation is deadlock-free.*

In addition to the safety guarantee given by Theorem 2.13, and similarly to the case of edge-reversal computations, for acquisition-order computations it is also the case that liveness guarantees can be given. In this case, however, liveness does not come from the shortening distance to sinks in evolving acyclic orientations, but rather from the fact that directed distances as given by an acyclic orientation are always bounded.

In order to be more specific regarding the liveness of acquisition-order computations, let us consider a *coloring* of the vertices of H. Such a coloring is an assignment of colors (natural numbers) to vertices in such a way that neighbors in H get different colors. If H can be colored with c colors for some $c > 0$, then we say that it is *c-colorable* [6].

Lemma 2.14 [12] *If H is c-colorable, then there exists an acyclic orientation of H according to which the longest directed path in H has no more than $c - 1$ edges.*

If process P_i is the only one to be requesting resources in an acquisition-order computation, then it waits for resources no longer than is implied by the size of $\mathcal{R}_i$, the vertex set of a clique in H. So its wait is given at most by the longest directed path in H according to the acyclic orientation φ fixed beforehand. If H is known to be c-colorable, then by Lemma 2.14 P_i's wait is bounded from above by c. When other processes are also requesting resources, then let h be the maximum $|\mathcal{P}_p|$ for $1 \le p \le m$, that is, the maximum number of processes that may use a resource (this is the size of a clique in G). We have the following.

Theorem 2.15 [19] *Every acquisition-order computation is starvation-free, and, if H is c-colorable, then the worst-case wait a process must undergo is $O(ch^c)$.*

In Figure 2.9, the assignment of color 0 to R_3 and R_5, color 1 to R_2 and R_4, and color 2 to R_1 and R_6, makes the graph 3-colorable. One of the acyclic orientations complying with Lemma 2.14 is the one shown in Figure 2.10.

Note, in all this discussion, that it must be known beforehand that H is c-colorable so that φ can be built. Also, by Theorem 2.15, it is to one's advantage to seek as low a value of c as can be efficiently found. Seeking the optimal value of c is equivalent to computing the graph's *chromatic number* (the minimum number of colors with which the graph can be colored) [6], and constitutes an NP-hard problem [15].

6. The graph abacus

In this section, we return to the edge-reversal computations discussed in Section 5.1 and consider a generalization thereof in the special context of high demand for resources by the processes. Such a heavy-load situation occurs when, in the resource-sharing computation, processes continually require access to all the shared resources they may have access to, and endlessly go through an acquire-release cycle. Situations such as this bring to the fore interesting issues (some of which will be discussed in Section 7) that are relevant not only for the remainder of this section but also in the context of our discussion in Section 5.1, in which we addressed edge-reversal computations.

While it is conceivable that, in normal situations, such computations may still be deadlock-free even if the corresponding orientations of G have cycles, the same cannot happen under heavy loads. This is so because what those orientations do is to provide priority. In a light-load regime, a cyclic dependency in the priority scheme may go unnoticed if the pattern of resource demand by the processes happens never to cause a directed cycle in W. Under heavy loads, on the other hand, the acyclicity of G's orientations is strictly necessary.

The generalization we consider is the following. Associated with each process P_i is an integer $r_i > 0$ to be used to control the dynamic evolution of priorities as given by the succession of acyclic orientations of G. These numbers are to be used in such a way that, as the computation progresses, and for any two neighbors P_i and P_j in G, the ratio of the number of times P_i has priority over P_j to the number of times P_j has priority over P_i "converges" to

r_j/r_i in the long run [4, 14]. The special case of Section 5.1 is obtained by setting r_i to the same number for all $P_i \in \mathcal{P}$. In that case, neighbors always have alternating priorities and the ratio is therefore 1.

We refer to computations with this generalized control of priorities as *bead-reversal computations*, in allusion to the following implementation, which views G's edges as the rods along which the beads of a generalized abacus (a graph abacus) are slid back and forth. For $(P_i, P_j) \in \mathcal{C}$, let e_{ij} beads be associated with edge (P_i, P_j). In order for P_i to have priority over P_j, there has to exist at least r_i beads on P_i's side of the edge and strictly less than r_j on P_j's side. When this is the case, the change in priority is performed by moving r_i of those beads towards P_j.

In an bead-reversal computation, the rule for process P_i is the following. Upon terminating its use of the shared resources, send r_i beads to the other end of every edge on which at least r_i beads are on P_i's side. Under the assumption of heavy loads, this must be the case for all edges adjacent to P_i, because under these circumstances processes can only access resources when they have priority over all of their neighbors.

Just as with edge-reversal computations, it is possible to associate an orientation of G's edges to the priority scheme of bead-reversal computations. For such, an edge is oriented towards P_i if and only if there are at least r_i beads on P_i's side of the edge. In order to preserve the syntactic constraints that an edge must be amenable to orientation in any of the two possible directions, and that it has to be oriented in exactly one direction at any time, we must clearly have

$$\max\{r_i, r_j\} \le e_{ij} \le r_i + r_j - 1.$$

But it is possible to obtain a precise value for e_{ij} within this range, and also to come up with a criterion for an initial distribution of the beads along the edges of G in such a way as to provide the desired safety and liveness guarantees. Safety is in this case associated with the acyclicity of the orientations of G as they change, while liveness refers to achieving the desired ratios. As in the case of Section 5.1, we aim at an acyclic wait-for graph W (for deadlock-freedom, by Fact 2.2). As for liveness, since achieving the desired ratios already implies starvation-freedom, what we aim at are computations for which those ratios are achieved, henceforth called *ratio-compliant*.

We begin with the subgraph G_{ij} of G having for vertices the neighbors P_i and P_j in G, along with the single edge between them. In what follows, g_{ij} is the greatest common divisor of r_i and r_j.

Theorem 2.16 [4] *If* $e_{ij} = r_i + r_j - g_{ij}$, *then every bead-reversal computation on* G_{ij} *is deadlock-free and ratio-compliant.*

Theorem 2.16 makes no provisions as to the distribution of the e_{ij} beads on the single edge of G_{ij}, and does as such hold for any of the $(r_i + r_j)/g_{ij}$ possible distributions, as we see in Figure 2.11. In that figure, $r_i = 2$ and $r_j = 3$ (this is indicated in parentheses by the vertices' identifications), and an evolution of bead placements is shown from left to right. For each configuration, the number of beads on each of the edge's ends is indicated by small numbers. The corresponding orientation of the edge is also shown. Note that all five possible distributions of beads appear, and that from the last one we return to the first. When we consider the entirety of G, however, the question of how to place the beads on G's edges becomes crucial.

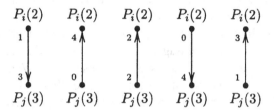

Figure 2.11.   Bead reversal on G_{ij}.

We begin the analysis of the general case by introducing some additional notation and definitions. First, let K denote the set of all the simple cycles in G (those with no repeated vertices). Membership of vertex P_i in $\kappa \in K$ is denoted by $P_i \in \kappa$, and membership of edge (P_i, P_j) in $\kappa \in K$ is denoted by $(P_i, P_j) \in \kappa$. Now, for $\kappa \in K$, let κ^+ and κ^- denote the two possible traversal directions of κ, chosen arbitrarily. We use a_{ij}^+ to denote the number of beads placed on edge (P_i, P_j) on its far end as it is traversed in the κ^+ direction, and a_{ij}^- likewise for the κ^- direction. Obviously, at all times we have $a_{ij}^+ + a_{ij}^- = e_{ij}$.

For $\kappa \in K$, let

$$\rho(\kappa) = \sum_{P_i \in \kappa} r_i$$

and

$$\sigma(\kappa) = \max \left\{ \sum_{(P_i, P_j) \in \kappa} a_{ij}^+, \sum_{(P_i, P_j) \in \kappa} a_{ij}^- \right\}.$$

According to these equations, $\rho(\kappa)$ is the sum of r_i over all vertices P_i of κ, while $\sigma(\kappa)$ denotes the total number of beads found on κ's edges' far ends as κ is traversed in the κ^+ direction or along the κ^- direction, whichever is greatest. In addition, it is easy to see that both $\rho(\kappa)$ and $\sigma(\kappa)$ are time-invariant. We are now ready to state the counterpart of Theorem 2.16 for G as a whole.

Theorem 2.17 [4] *If $e_{ij} = r_i + r_j - g_{ij}$ for all $(P_i, P_j) \in C$, then every bead-reversal computation is deadlock-free and ratio-compliant if and only if $\sigma(\kappa) < \rho(\kappa)$ for all $\kappa \in K$.*

One interesting special case that can be used to further our insight into the dynamics of bead-reversal computations is the case of graphs without (undirected) cycles, that is, cases in which G is a tree. In such cases, $K = \emptyset$ and Theorem 2.17 becomes a simple generalization (by quantification over all of C) of Theorem 2.16.

To finalize this section, we return to the wait-for graph W to analyze its acyclicity. As in the case of edge-reversal computations (cf. Section 5.1), W is related to an oriented version of G, as follows. In bead-reversal computations, the orientation of edge (P_i, P_j) is from P_i towards P_j if and only if there are at least r_j beads placed on the P_j end of the edge (and, necessarily, fewer than r_i on P_i's end). Recalling, as in Section 5.1, that W is the graph the results when processes cannot send grant messages, then the heavy-load assumption implies that W coincides with G oriented as we just discussed. What this means is that Theorem 2.17 is an indirect statement on the acyclicity of W: If a directed cycle exists in W, then obviously for the corresponding underlying κ we have $\sigma(\kappa) \geq \rho(\kappa)$, which characterizes an orientation of G that is not acyclic either.

But the attentive reader will have noticed that violating the inequality of Theorem 2.17 does not necessarily lead to a directed cycle in G's orientation (or in W). The significance of the theorem, however, is that such a cycle is certain to be created at some time if the inequality is violated. What the theorem does is to provide a criterion for the establishment of initial conditions (bead placement) that is necessary even though at first no cycle might be created otherwise.

Example 2.18 *Let G be the complete graph on three vertices, and let $r_1 = 1$, $r_2 = 2$, and $r_3 = 3$. Then $e_{12} = 2$, $e_{13} = 3$, and $e_{23} = 4$. Employing the same convention as in Figure 2.11, and identifying the κ^+ direction of traversal with the clockwise direction for the single simple cycle κ, we show in Figure 2.12 two possibilities for bead placement. The one in part (a) has $\sigma(\kappa) = 5$, while $\sigma(\kappa) = 6$ for part (b), both values determined by the κ^+ direction. We have $\rho(\kappa) = 6$ for this example, so $\sigma(\kappa) < \rho(\kappa)$ in part (a), whereas $\sigma(\kappa) = \rho(\kappa)$ in part (b). Although both orientations are acyclic, the reader can check easily that the evolution of the bead placement in Figure 2.12(b) will soon lead to a directed cycle, while for the other acyclicity will be indefinitely preserved. This is, of course, in accordance with Theorem 2.17.*

7.   Graph coloring and concurrency

From a purely algorithmic perspective, heavy-load situations such as introduced in the beginning of Section 6 provide a simpler means of implementing edge-reversal and bead-reversal computations than the overall scheme discussed

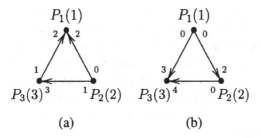

Figure 2.12. Two possible bead placements.

in Section 2. In a heavy-load regime, the need for processes to explicitly request and grant resources becomes moot, because the reversal of priorities (edge orientation or beads) can be taken to signify that permission is granted (or partially granted, in the case of bead reversals) to access shared resources.

Given this simplification, the following is how an edge-reversal computation can be implemented. Start with an acyclic orientation of G. A process computes on shared resources when it is a sink, then reverses all edges adjacent to it and waits to become a sink once again. Similarly, and following our anticipation in Section 6, a bead-reversal computation is also simple, as follows. Start with a placement of the beads that not only leads to an acyclic orientation of G but also complies with the inequality prescribed by Theorem 2.17. A process P_i computes when it is a sink (has enough beads on all adjacent edges), then sends r_i beads to each of its neighbors in G. This is repeated until P_i ceases being a sink, at which time it waits to become a sink again.

Heavy-load situations also raise the question of how much concurrency, or parallelism, there can be in the sharing of resources. While neighbors in G are precluded from sharing resources concurrently, processes that are not neighbors can do it, and how much of it they can do depends on the initial conditions that are imposed on the computation (acyclic orientation of G or bead placement). In the remainder of this section, we discuss this issue of concurrency for edge-reversal computations only, but a similar discussion can be done for bead-reversal computations as well [4].

The simplest means to carry out this concurrency analysis is to abandon the asynchronous model of computation we have been assuming (cf. Section 2) and to assume full synchrony instead. In the *fully synchronous* (or simply *synchronous*) model of distributed computation [2], processes are driven by a common global clock that issues ticks represented by the integer $s \geq 0$. At each tick, processes compute and send messages to their neighbors, which are assumed to get those messages before the next tick comes by.

An edge-reversal computation under the synchronous model is an infinite succession of acyclic orientations of G. If these orientations are $\omega_0, \omega_1, \ldots,$

then, for $s > 0$, ω_s is obtained from ω_{s-1} by turning every sink in ω_{s-1} into a *source* (a vertex with all adjacent edges directed outward). The number of distinct acyclic orientations of G is finite, so the sequence $\omega_0, \omega_1, \ldots$ does eventually become periodic, and from this point on it contains an endless repetition of a number of orientations that we denote by $p(\omega_0)$ (this notation is meant to emphasize that the acyclic orientations that are repeated periodically are fully determined by ω_0). Let these $p(\omega_0)$ orientations be called the *periodic orientations* from ω_0.

Lemma 2.19 [5] *The number of times a process is a sink in the periodic orientations from ω_0 is the same for all processes.*

We let $m(\omega_0)$ denote the number asserted by Lemma 2.19, and let $m_i(s)$ denote the number of times process P_i is a sink in the subsequence $\omega_0, \ldots, \omega_{s-1}$.

Intuitively, it should be obvious that the amount of concurrency achieved from the initial conditions given by ω_0 depends chiefly on the periodic repetition that is eventually reached. In order to make this more formal, let $Conc(\omega_0)$ denote this amount of concurrency, and define it as

$$Conc(\omega_0) = \lim_{s \to \infty} \frac{1}{sn} \sum_{P_i \in \mathcal{P}} m_i(s).$$

That is, we let the concurrency from ω_0 be the average, taken over time and over the number of processes, of the total number of sinks in the sequence $\omega_0, \ldots, \omega_{s-1}$ as $s \to \infty$ (the existence of this limit, which is implicitly assumed by the definition of $Conc(\omega_0)$, is only established in what follows, so the definition is a little abusive for the sake of notational simplicity).

Theorem 2.20 [5] $Conc(\omega_0) = m(\omega_0)/p(\omega_0)$.

Theorem 2.20 characterizes concurrency in a way that emphasizes the dynamics of edge-reversal computations under heavy loads. But the question that still remains is whether a characterization of concurrency exists that does not depend on the dynamics to be computed, but rather follows from the structure of G as oriented by ω_0.

This question can be answered affirmatively, and for that we consider once again the set K of all simple cycles in G. For $\kappa \in$ K, we let $c^+(\kappa, \omega_0)$ be the number of edges in κ that are oriented by ω_0 in one of the two possible traversal directions of κ. Likewise for $c^-(\kappa, \omega_0)$ in the other direction. The number of vertices in κ is denoted by $|\kappa|$.

Theorem 2.21 [5] *If G is a tree, then $Conc(\omega_0) = 1/2$. Otherwise, then*

$$Conc(\omega_0) = \min_{\kappa \in K} \frac{\min\{c^+(\kappa, \omega_0), c^-(\kappa, \omega_0)\}}{|\kappa|}.$$

Except for the case of trees, by Theorems 2.20 and 2.21 we know that the amount of concurrency of an edge-reversal computation is entirely dependent upon ω_0, the initial acyclic orientation. The problem of determining the ω_0 that maximizes concurrency is, however, NP-hard, so an exact efficient procedure to do it is unlikely to exist in general [5].

Example 2.22 *When G is a ring on five vertices, we have a representation of the original dining philosophers problem [13]. For this case, consider the sequence of acyclic orientations depicted in Figure 2.13, of which any one can be taken to be ω_0. We have $m(\omega_0) = 2$, $p(\omega_0) = 5$, and $\text{Conc}(\omega_0) = 2/5$. This concurrency value follows from either Theorem 2.20 or Theorem 2.21.*

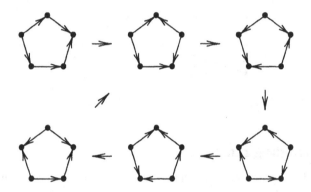

Figure 2.13. A heavy-load case of edge reversal.

Another interesting facet of this concurrency issue is that it relates closely to various forms of coloring the vertices of G. Consider, for example, the *k-tuple coloring* of the vertices of G obtained as follows [24]. Assign k distinct colors to each vertex in such a way that no two neighbors share a color. This type of coloring generalizes the coloring discussed in Section 5.2, for which $k = 1$. The minimum number of colors required to provide G with a k-tuple coloring is its *k-chromatic number.*

In the context of edge-reversal computations, note that the choice of an initial acyclic orientation ω_0 implies, by Lemma 2.19, that G admits an $m(\omega_0)$-tuple coloring requiring a total of $p(\omega_0)$ colors. If these colors are natural numbers, then neighbors in G get colors that are "interleaved," in the following sense. For two neighbors P_i and P_j, let $c_i^1, \ldots, c_i^z$ and $c_j^1, \ldots, c_j^z$ be their colors, respectively, with $z = m(\omega_0)$. Then either $c_i^1 < c_j^1 < \cdots < c_i^z < c_j^z$ or $c_j^1 < c_i^1 < \cdots < c_j^z < c_i^z$.

So the question of maximizing concurrency is, by Theorem 2.20, equivalent to the question of minimizing the ratio of the total number of interleaved colors to the number of colors per vertex (this is the ratio $p(\omega_0)/m(\omega_0)$) by choosing

ω_0 appropriately. The optimal ratio thus obtained, denoted by $\bar{\chi}(G)$, is called the *interleaved multichromatic* (or *interleaved fractional chromatic*) *number of G* [5]. When the interleaving of colors is not an issue, then what we have is the graph's *multichromatic* (or *fractional chromatic*) *number* [22].

Letting $\chi(G)$ denote the chromatic number of G and $\chi^*(G)$ its multichromatic number, we have

$$\chi^*(G) \le \bar{\chi}(G) \le \chi(G).$$

A graph G is shown in Figure 2.14 for which $\chi^*(G) = 5/2$, $\bar{\chi}(G) = 8/3$, and $\chi(G) = 3$, all distinct therefore. One of the orientations that correspond to $\bar{\chi}(G) = 8/3$ is the one shown in the figure.

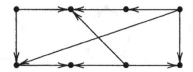

Figure 2.14. A graph G for which $\chi^*(G) < \bar{\chi}(G) < \chi(G)$.

8.   Concluding remarks

Distributed computations over shared resources are complex, asynchronous computations. Performing such computations efficiently while offering a minimal set of guarantees has been a challenge for several decades. At present, though problems still persist, we have a clear understanding of several of the issues involved and have in many ways met that challenge successfully.

Crucial to this understanding has been the use of precise modeling tools, aiming primarily at clarifying the timing issues involved, as well as the combinatorial structures that underlie most of concurrent computations. In this paper, we have concentrated on the latter and outlined some of the most prominent combinatorial concepts on which the design of resource-sharing computations is based. These have included graph structures and posets useful for handling the safety and liveness issues that appear in those computations, and for understanding the questions related to concurrency.

Acknowledgments

The author is thankful to Mario Benevides and Felipe França for many fruitful discussions on the topics of this paper.

References

[1] V. C. Barbosa. *Concurrency in Systems with Neighborhood Constraints.* PhD thesis, Computer Science Department, University of California, Los Angeles, CA, 1986.

[2] V. C. Barbosa. *An Introduction to Distributed Algorithms.* The MIT Press, Cambridge, MA, 1996.

[3] V. C. Barbosa and M. R. F. Benevides. A graph-theoretic characterization of AND-OR deadlocks. Technical Report COPPE-ES-472/98, Federal University of Rio de Janeiro, Rio de Janeiro, Brazil, July 1998.

[4] V. C. Barbosa, M. R. F. Benevides, and F. M. G. França. Sharing resources at nonuniform access rates. *Theory of Computing Systems*, 34:13–26, 2001.

[5] V. C. Barbosa and E. Gafni. Concurrency in heavily loaded neighborhood-constrained systems. *ACM Trans. on Programming Languages and Systems*, 11:562–584, 1989.

[6] J. A. Bondy and U. S. R. Murty. *Graph Theory with Applications.* North-Holland, New York, NY, 1976.

[7] G. Bracha and S. Toueg. Distributed deadlock detection. *Distributed Computing*, 2:127–138, 1987.

[8] J. Brzezinski, J.-M. Hélary, M. Raynal, and M. Singhal. Deadlock models and a general algorithm for distributed deadlock detection. *J. of Parallel and Distributed Computing*, 31:112–125, 1995.

[9] K. M. Chandy and L. Lamport. Distributed snapshots: Determining global states of distributed systems. *ACM Trans. on Computer Systems*, 3:63–75, 1985.

[10] K. M. Chandy and J. Misra. The drinking philosophers problem. *ACM Trans. on Programming Languages and Systems*, 6:632–646, 1984.

[11] K. M. Chandy, J. Misra, and L. M. Haas. Distributed deadlock detection. *ACM Trans. on Computer Systems*, 1:144–156, 1983.

[12] R. W. Deming. Acyclic orientations of a graph and chromatic and independence numbers. *J. of Combinatorial Theory B*, 26:101–110, 1979.

[13] E. W. Dijkstra. Hierarchical ordering of sequential processes. *Acta Informatica*, 1:115–138, 1971.

[14] F. M. G. França. *Neural Networks as Neighbourhood-Constrained Systems.* PhD thesis, Imperial College, London, UK, 1994.

[15] M. R. Garey and D. S. Johnson. *Computers and Intractability: A Guide to the Theory of NP-Completeness.* Freeman, New York, NY, 1979.

[16] R. C. Holt. Some deadlock properties of computer systems. *ACM Computing Surveys*, 4:179–196, 1972.

[17] E. Knapp. Deadlock detection in distributed databases. *ACM Computing Surveys*, 19:303–328, 1987.

[18] A. D. Kshemkalyani and M. Singhal. Efficient detection and resolution of generalized distributed deadlocks. *IEEE Trans. on Software Engineering*, 20:43–54, 1994.

[19] N. A. Lynch. Upper bounds for static resource allocation in a distributed system. *J. of Computer and System Sciences*, 23:254–278, 1981.

[20] J. Misra and K. M. Chandy. A distributed graph algorithm: Knot detection. *ACM Trans. on Programming Languages and Systems*, 4:678–686, 1982.

[21] D.-S. Ryang and K. H. Park. A two-level distributed detection algorithm of AND/OR deadlocks. *J. of Parallel and Distributed Computing*, 28:149–161, 1995.

[22] E. R. Scheinerman and D. H. Ullman. *Fractional Graph Theory: A Rational Approach to the Theory of Graphs*. Wiley, New York, NY, 1997.

[23] M. Singhal. Deadlock detection in distributed systems. *IEEE Computer*, 22:37–48, 1989.

[24] S. Stahl. n-tuple colorings and associated graphs. *J. of Combinatorial Theory B*, 20:185–203, 1976.

[25] J. L. Welch and N. A. Lynch. A modular drinking philosophers algorithm. *Distributed Computing*, 6:233–244, 1993.

Chapter 3

SOLVING THE STATIC TASK SCHEDULING PROBLEM FOR REAL MACHINES

Cristina Boeres
*Universidade Federal Fluminense**
boeres@ic.uff.br

Vinod E.F. Rebello
*Universidade Federal Fluminense**
vinod@ic.uff.br

Abstract   While the task scheduling problem under the delay model has been studied extensively, relatively little research exists for more realistic communication models such as the $LogP$ model. The task scheduling problem is known to be NP-complete even under the delay model (a simplified instance of the $LogP$ model). This chapter describes the $LogP$ model and the influence of its communication parameters on task scheduling. The similarities and differences between clustering algorithms under the delay and $LogP$ models are discussed and a design methodology for clustering-based scheduling algorithms for the $LogP$ model is presented. Using this design methodology, a task scheduling algorithm for the allocation of arbitrary task graphs to fully connected networks of processors under $LogP$ model is proposed. The strategy exploits the replication and clustering of tasks to minimize the ill effects of communication on the makespan.

Keywords: Multicomputer task scheduling, clustering, task replication, $LogP$ model.

1.   Introduction

Recent developments in high performance computing have seen an increased popularity in the use of a network of PCs to tackle computation intensive par-

---

*UFF, Instituto de Computação, Rua Passo da Pátria 156, São Domingos, Niterói 24210-240, RJ, Brazil. The authors are partially supported by research grants from the Conselho Nacional de Desenvolvimento Científico e Tecnológico (CNPq), Brazil.

R. Corrêa et al. (eds.), Models for Parallel and Distributed Computation. Theory, Algorithmic Techniques and Applications, 53–84.

allel applications [8]. Parallel processing using a cluster of processors, also commonly known as *cluster computing*, offers a much larger community of users than before the opportunity to efficiently solve parallel problems on a readily available platform. Cluster computing is potentially seen as a low cost means of harnessing supercomputing performance due to the cost-effectiveness, flexibility and scalability of such systems in comparison to traditional parallel machines.

However, cluster computing still faces the same dilemma as traditional parallel computing, that is, how to design architecturally independent algorithms that execute efficiently on a wide range of existing and future platforms. This involves addressing two principal problems:

1. the design of parallel algorithms and the partitioning of applications into tasks; and

2. the scheduling of these tasks on the processors of the target machine, respecting communication and synchronisation requirements, together with the routing and scheduling of messages on the communication links.

1.1.   Designing parallel algorithms

Typically, two of the most common reasons for devising a parallel program as opposed to a sequential one are to find a solution to the problem more quickly and to solve more complex versions of the given problem. Unfortunately, designing efficient high performance parallel applications is not easy, particularly in view of the lack of a general programming methodology and a standard model of parallel computation. In order to solve certain problems in parallel, it may be necessary to investigate various existing parallel programming strategies and even develop new techniques. Both the design and efficiency of a parallel algorithm, however, depend on various characteristics of the parallel machine upon which the algorithm is to be executed. Given the importance of the portability of parallel algorithms, the relevant characteristics of these machines should therefore be encompassed into a common model or representation for all parallel machines.

A variety of models of parallel computation have been designed, from the theoretical and abstract [2] to the more realistic and practical [56]. While the theoretical models are used mainly to analyse the complexity of parallel algorithms, the realistic models attempt to capture the aspects of machines most relevant to the performance of algorithms. A standard model of parallel computation has not yet been established, due to the variety of parallel systems in existence and the difficulty in identifying the relevant characteristics.

In parallel machines, both processing speed and network capacity influence performance. In general terms, a programmer needs to know if the gains from parallelisation outweigh the communication costs. While early models of parallel computation were simple (e.g. PRAM [27]), choosing to effectively ignore the costs of communication, a number of models have evolved to include, to varying degrees of abstraction, costs of communication. The *LogP* [22, 6] and *BSP* [57, 46, 54] models are two of the most well known parallel computation models.

The *LogP model* considers architectural characteristics such as: the number of processors; the processing cost to send or receive a message; the transmission delay of a message between processors; and the maximum rate that messages can be sent or received. These parameters help represent the dominant communication characteristics of each individual communication as well as the fact that networks have a limited communication capacity. While the *LogP* model does not specify any programming methodology, the *BSP model* can be considered a model for parallel programming, mainly due to the definition of synchronised bands of computation and communication (*supersteps*). The *BSP* model considers characteristics similar to *LogP*, but takes a more collective view of communication by modelling the costs of all of the communications that take place within a superstep as a single entity [54, 15].

Although other models have been proposed, they appear in principle to be extensions or refinements to the *LogP* model rather than completely new models [5, 56]. The *LogGP* model has been proposed as an extension of the *LogP* model, where an additional parameter captures the bandwidth due to long messages [5]. This model considers the fact that parallel machines generally have an alternative way of dealing with long messages. This can be exploited to improve performance by bundling messages together [5, 14]. In an attempt to capture more accurately communication behaviour in clusters of processors, the model proposed in [56] specifies architectural parameters similar to those of *LogGP* but in greater detail.

1.2.   Scheduling tasks

By solving the first problem, the quantity of parallelism available and thus the maximum performance that could be attained by the application can be identified. On the other hand, the hardware components of a computing platform and their characteristics limit the parallelism that can be utilised, and therefore the best performance that can be achieved by the machine. How much performance is actually obtained, however, will be determined by just how well the second problem is addressed.

The effective scheduling of a parallel program on to the processors of a parallel machine is crucial for achieving high performance. Due to the cost of communication, an inappropriate scheduling of the tasks of an application can offset the benefits from parallelisation. The objective of scheduling is to minimise the completion time of a parallel application by judiciously allocating tasks to the available processors and sequencing their execution. The scheduling problem may even be more acute in cluster computing since system upgrades due to technological advances could hinder portability and efficiency. For example, easy changes in hardware components (and thus processor or network speeds) afforded by the flexibility of such systems may change the performance ratio of the system.

The scheduling problem can be classified in a number of ways. General classifications include, for example, scheduling on *homogeneous* and *heterogeneous* systems, and *static* and *dynamic* scheduling. Homogeneous computing uses one or more machines of the same type to provide high performance, whereas heterogeneous computing uses a diverse suite of machines. In static scheduling, which is usually done at compile-time, the characteristics of a parallel program (such as task processing times, communication, data dependencies and synchronisation requirements) are known before program execution [21]. In dynamic scheduling, few assumptions about the parallel program can be made before execution, and thus, scheduling decisions have to be made on the fly [26]. This chapter focuses on the problem of static scheduling in a homogeneous computing environment. Various further sub-classifications or variants of the scheduling problem exist, e.g. scheduling with a bounded or unbounded number of processors, with or without task recomputation, some of which are discussed in the rest of this chapter.

The task scheduling problem is NP-Complete for most of its variants except for a few highly simplified cases [19, 28, 26]. Attempts to find (near-) optimal completion times have focussed on two main approaches. The traditional approach is to design heuristics with polynomial time complexity which attempt to construct a single near optimal solution [17, 26, 43]. The second, less common and more time consuming approach (unless carried out in parallel) is to search for the best schedule in (a restricted part of) the solution space. For example, heuristics based on Genetic Algorithms [41], Tabu Search [50], branch and bound [20].

Numerous heuristics of construction have been proposed, differing not only in the scheduling strategy adopted but also the assumptions made with regard to the structure and characteristics of both the program and the architecture. Given the complexity of the task scheduling problem, it is common for simplifying assumptions (such as uniform task execution times, particular program structures, zero or unit inter-task communication costs, full connectivity between processors) to be made. A more complete classification of scheduling algorithms and their assumptions can be found in [17, 42].

The characteristics of a parallel architecture which influence the design and performance of parallel applications have long been the subject of investigation [55]. The communication characteristics considered relevant to the scheduling problem are often defined in a *communication model*. Currently, the standard communication model in the scheduling community is the *delay model*, where the sole architectural parameter for the communication system is the *latency* or *delay*, i.e. the transit time, for each message transmission [49]. Recent research has shown, however, that the dominant cost of communication in today's architectures is that of crossing the network boundary. This is a cost which cannot be modeled as a latency parameter and thus has motivated the development of new parallel programming models such as $LogP$ [22] and BSP [54].

These architectural characteristics imply that a new class of scheduling heuristics is required to generate efficient schedules for realistic abstractions of today's parallel computers [15]. To date, however, relatively few scheduling algorithms exist for models such as $LogP$.

This chapter describes a design methodology for task replication-based clustering algorithms for the problem of scheduling arbitrary task graphs under the $LogP$ model. The following section describes, in further detail, the communication parameters adopted by existing communication models, while Section 3 outlines the influence of the $LogP$ parameters on the scheduling problem. Some of the most common approaches and techniques adopted by scheduling algorithms are described Section 4. Section 5 then introduces the some additional terminology and definitions used in the remainder of this chapter. A design methodology for clustering-based scheduling algorithms under the $LogP$ model is discussed in Section 6. Section 7 presents a clustering algorithm, based on this design methodology, for scheduling general or arbitrary $DAGs$ on $LogP$ machines (assuming an unbounded number of processors). Section 8 briefly discusses the performance of the new algorithm against other clustering algorithms under both delay and $LogP$ model conditions. Section 9 make some comments with regard to the validation of the methodology while the final section concludes the chapter.

2. Models of communication

Technological developments are leading to a class of parallel computers with similar characteristics. This class, in essence, consists of a collection or cluster of computers (microprocessor(s), cache and memory) connected to each other by a fast communication network. The main differences among parallel computers within this class are the number and type of processors involved and the interface with the network. The relative cost of purchasing such computers added to the freedom of developing (and upgrading) such *clusters of processors* has led many researchers and companies to make these their computing system of choice.

It is thought, however, that such parallel computers will not scale to millions of nodes (as expected with supercomputers) and with the continued increase in memory capacity of each computer or processing node, programs designers will need to implement their parallel algorithms assuming that a large amount of data will be allocated to each processing node. Of course, this assumption has a direct impact on the performance of the system as a whole. Network technology has improved so significantly that the limits on communication performance are due to the interface between the computer and the network, rather than in the network itself. Therefore, an abstract view of the network interface performance, including communication latency and network bandwidth, is crucial for adapting algorithms to the target parallel architecture.

For many years, the communication model used in the scheduling problem only considered the latency to be a significant communication parameter [49]. Now however, with the advances in technology and changes to the parallel machines in use, others communication characteristics such as those specified in the *LogP* model can no longer be neglected [22, 6, 24, 45].

LogP [22] is a model of a distributed-memory MIMD parallel machine where communication is carried out point-to-point. Four architectural parameters are considered by this model:

- L - an upper bound on the *latency* incurred when transmitting a message between communicating processors. This parameter is an upper bound due to the difficulty in predicting the latency accurately;

- o - the processing *overhead* incurred when sending or receiving a message. During this time the processor cannot execute tasks nor send or receive other messages;

- g - the *gap* or interval between two consecutive message sends or receives by a processor (in some machines, this due to the communication co-processor opening a communication channel [5]). The inverse of g is actually the network bandwidth;

- P - the number of processors available.

Furthermore, the *LogP* model assumes that the network has a finite capacity and so there can be most $\lceil L/g \rceil$ messages in transit to or from any processor at any time.

In order to quantify values for the *LogP* communication parameters, the following assumptions are often considered [22, 23, 6]: the latency L can be seen as a function of the diameter of the network; the overhead o is the average of the send and receive overheads; and the gap g can be seen as a function of the bisection bandwidth.

Note that *LogP* only models short messages and does not accurately predict the performance of applications which send long messages. In general, parallel machines have alternative mechanisms for dealing with long messages, e.g. a

DMA device connected to the network interface. Programs may wish to take advantage of this fact to communicate more efficiently by bundling many short messages to a common destination into fewer, longer ones.

The $LogGP$ model [5] is an extension of the $LogP$ model with the objective to represent both short and long messages. This model captures the three communication bottlenecks in architectures: the time during which the processor is busy preparing to send or receive a message, modelled by the overhead o; the network startup, captured by g; and the time per byte for long messages G. The inverse of this additional parameter, G, can be viewed as the available per processor communication bandwidth for long messages. Suppose that w is the underlying message size of the target machine and that K bytes are sent by a processor. Under the $LogP$ model, the overall communication time for K bytes would be $o + (\lceil K/w \rceil - 1) * \max\{g, o\} + L + o$ cycles, while under $LogGP$, the sending of a single K-byte message would take $o + (K - 1)G + L + o$ cycles.

Although the $LogP$ and $LogGP$ models succeed in modeling parallel machines, e.g. CM-5, Dash, Intel Paragon, Berkeley NOW [22, 45], some researchers think that these models are still too simplistic for cluster computing machines. Due to the rapid changes in processor and network technologies, a more realistic model for cluster of processors was recently proposed for both performance analysis and algorithm development [56]. The architectural parameters in this model are based on those of the $LogGP$ model, but are expressed as functions of message length rather than as constants. Addition parameters are also used to model the effects of network contention.

3. Task scheduling under the $LogP$ model

Research has shown that the delay model is not an accurate communication model for a number of available parallel computer systems [5, 6, 45]. The principle reason, from a scheduling point of view, is that this model assumes certain properties (e.g. the ability to *multicast*, i.e. to send a number of different messages to different destinations simultaneously) that may not hold in practice. The model does not encompass the fact that in today's parallel systems, the CPU generally has to manage or at least initiate each communication [22]. This implies that the communication time cannot be completely overlapped with computation time of tasks.

Under the $LogP$ model, the time spent by a processor to carry out each communication event (be it for sending or receiving a message) is the communication overhead o. For the duration of each communication overhead, the processor is effectively *blocked*, unable to execute other tasks in V or even to send or receive other messages. Consequently, any scheduling algorithm must, in some sense, view these send and receive overheads also as "tasks" to be executed by processors. The rate at which these communication events can be executed is limited by the network (or, more likely, the network interface) capacity modeled by the gap g.

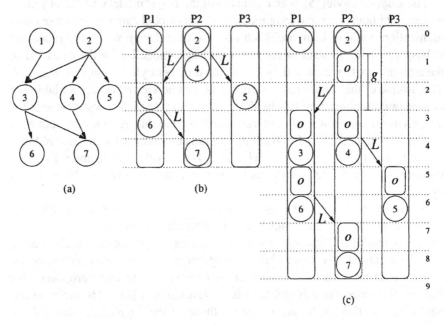

Figure 3.1.   Scheduling the unit cost task graph in (a) on three processors: (b) under the delay model; and (c) under the *LogP* model.

A small example of scheduling under both the delay and *LogP* models is presented in Figure 3.1. One common program representation is a *directed acyclic graph (DAG)* whose vertices or nodes denote *tasks* and whose edges or arcs denote data dependencies (and hence communication, if the tasks are mapped to distinct processors). These program communication costs can be reduced by clustering or grouping tasks together to share processors. In this chapter, we represent a parallel application by a *DAG* $G = (V, E, \varepsilon, \omega)$, where V is the set of tasks, E is the precedence relation among them, $\varepsilon(v_i)$ is the execution cost of task $v_i \in V$, and $\omega(v_i, v_j)$ is the weight associated to the edge $(v_i, v_j) \in E$ representing the amount of data transmitted from task v_i to v_j. In Figure 3.1, the tasks and edges of the *DAG* have unit costs, i.e. $\varepsilon(v_i) = 1, \forall v_i \in V$ and $\omega(v_i, v_j) = 1, \forall(v_i, v_j) \in E$, and the model parameters $L = 1, o = 1$ and $g = 2$.

One important aspect that has to be addressed under a *LogP*-type model is the order in which communication events occur, since two or more messages can no longer be sent simultaneously. It is now possible for one message to delay the sending or receiving of another more *time-critical* message.

Scheduling heuristics which use some form of critical path analysis need to define how to interpret and incorporate these overhead parameters when calculating the costs of paths in the *DAG*. One difficulty is that these overheads are communication costs paid for by computation time.

4.   Scheduling strategies

Due to its intractability, the static scheduling problem has often been subjected to a varying degrees of simplification. These may be in the form of restrictions to the program (in terms of program structure, costs of tasks or the amount of data passed between tasks) or to the communication or architectural model (in terms of the number of processors, and the number or nature of communication parameters). The main objective of a scheduling heuristic is to minimise the completion time or *makespan* of a *DAG* for the given communication model.

The majority of proposed scheduling heuristics are based on the delay model, for example, [58, 39, 25, 51, 34, 30, 49, 44]. These heuristics assume that the communication between two processors can be completely overlapped with the execution of tasks on those processors and that processors are able to multicast messages.

In [39, 7, 29], communication is modelled by the *Linear Model*, where a startup overhead is added to the transmission time of a message. This model assumes the use of a *channel processor* or communication co-processor to carry out communication in parallel with the *task processor* responsible for the execution of the program tasks [7]. Therefore, this startup overhead is not a cost paid for by processing time on the CPU as opposed to the overhead parameter in the *LogP* model.

In comparison with the delay model, identifying scheduling strategies that are based on communication models that specify *LogP*-type characteristics is difficult. Only recently have a few scheduling algorithms appeared in the literature [9, 15, 38, 59, 12]. These algorithms tend to be limited to specific classes or types of *DAG*s and restricted *LogP* models.

The following sections describe some of the most established strategies and techniques used by scheduling algorithms. For a more complete classification and description of various task scheduling algorithms and their relative merits in terms of performance and time-complexity, the reader is referred to [43].

4.1.   List scheduling

The most common technique used in designing scheduling algorithms is called *list scheduling*. The basic mechanism is as follows: initially, priorities are assigned to all tasks; and then, until all tasks have been scheduled, a list of tasks ready for execution is formed from which the task with the highest priority is chosen to be scheduled on the processor upon which it can execute earliest. Numerous methods for assigning priorites to tasks and selecting the most suitable processor have been proposed [1, 19, 17, 7, 25, 52, 31, 32, 53, 26, 36, 3, 40].

The majority of the list scheduling strategies face two difficulties. One is the definition of appropriate task priorities. Dynamic priority list scheduling approaches [40, 43], in which tasks are re-assigned priorities in accordance with the partial schedules produced in each iteration, have improved the quality of schedules at the expense of increased complexity. For example, each iteraction of the algorithm in [3] calculates the priorities based on a critical-path analysis of the partial schedule.

The other difficulty is the fact that list schedulers can produce poor schedules due to their greedy behaviour. Sometimes known as the *horizon effect*, list schedulers generally do not consider the influence of an unready task on the current partial schedule. It has been shown that list scheduling algorithms only produce good results for coarse-grained applications, i.e. when the communication costs are not dominant [10].

4.2.   Clustering algorithms

Most scheduling algorithms perform some sort of *clustering* or grouping of two or more tasks to the same processor. However, the term *clustering algorithms* refers specifically to the class of algorithms which initially consider each task as a cluster (allocated to a unique virtual processor) and then merge clusters (tasks) if the completion time of the parallel program can be reduced. A later stage of the algorithm then allocates the clusters to physical processors to generate the final schedule. The order in which clusters are merged are also normally based on priorities.

Yang and Gerasoulis proposed the clustering algorithm DSC, based on linear clustering [51], in which decisions are based on dynamic priorities [32, 33]. This algorithm produces schedules for an arbitrary *DAG* based on the linear communication model and a fully connected network. The algorithm can find optimal schedules in polynomial time for fork, join, and coarse-grained (inverted) tree *DAGs* and schedules for *DAGs* with granularity γ which are within a factor of $1 + \frac{1}{\gamma}$ of the true optimal schedule[1]. In general, clustering algorithms seem to perform better than list scheduling ones because the clustering decisions are based on a more global view of the state of the partially scheduled *DAG* [33].

---

[1]The term *true optimal schedule* is used here to refer to the schedule which achieves the best attainable makespan for the given communication model but irrespective of the technique(s) used (e.g. replication of tasks, bundling of messages). The term *optimal schedule* is used to refer the schedule which provides the best makespan using the given technique(s) in question.

Zimmerman *et al.* extended the work of linear clustering [37] to propose a clustering algorithm which produces optimal k-linear schedules (given k) for tree-like graphs under the *LogP* model when $o \geq g$ [59]. k-linear schedules allows clusters to contain tasks from upto k direct paths of the *DAG*. The optimal schedule of a tree-like *DAG* is the optimal k-linear schedule with the smallest makespan, generated for each value of k between 1 and the total number of direct paths in the *DAG*. Kort and Trystram have presented an optimal polynomial algorithm for scheduling *fork* graphs on an unbounded number of processors considering $g = o$ and equal sized messages [38].

4.3. Task recomputation

Replicating tasks can be a useful scheduling mechanism, particularly when communication costs are high. As shown in Figure 3.2, the idea is to replicate a common immediate ancestor task, clustering a copy of this task with each of its successors, and thus decreasing the overall communication cost. Jung *et al.* concluded that the best schedule produced by an algorithm that ignores replication may be worse than the true optimal schedule by a multiplicative factor which is a function of the communication delay [35].

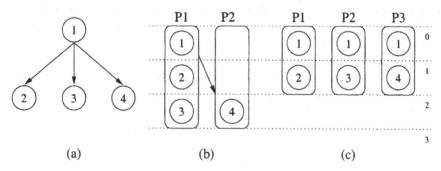

Figure 3.2. Scheduling the unit cost task graph in (a) under the delay model: (b) without replicating tasks; and (c) with replication.

A number of task replication-based algorithms, for an unbounded number of processors under the delay model, have been proposed. [49] present a polynomial-time heuristic which generates a schedule with a makespan within a factor of two of the optimal. For each task in the graph, a cluster is created containing this task and copies of some of its ancestors.

The clustering algorithm (PLW) [48] produces a schedule with a makespan which is at most $(1 + \varepsilon)$ times the makespan of the true optimal (where the granularity of the *DAG* is at least $\frac{1-\varepsilon}{\varepsilon}$ and $0 < \varepsilon \leq 1$). This means that for any given task graph, the schedule produced by PLW will have a makespan of at most twice the optimal. Furthermore, the coarser the granularity of the *DAG*, the closer the makespan will be to the minimum possible.

4.4.   Message bundling

In parallel architectures with high overheads, improved performance can be achieved by the grouping together of messages to a common destination processor into one larger message [5]. In [9, 14], the Multi-stage Scheduling Approach (MSA) for scheduling UET-UDC (unit execution tasks, unit data communication) *DAGs* on a bounded number of processors under the $LogP$ model (also when $o \geq g$) was proposed and later adapted for the BSP model [15]. A novel feature of this approach is the use of task replication and clustering to support message bundling as a scheduling technique to reduce communication costs.

Message bundling can be implemented by clustering together tasks with common ancestors and successors and restricting the communications of the cluster to take place before (for receives) and after (for sends) the execution of all of the tasks belonging to the cluster [14]. This allows messages for the same destination processor, from various tasks within the cluster, to be combined and sent as one. In contrast, other (non-bundling) clustering techniques, for example [30], allow communication to occur as soon as each task within the cluster finishes its execution. This leads to a subtle difference between these clustering approaches: the latter only reduces the total communication cost through the elimination of messages, by allocating source and destination tasks to the same processor; the former, tries to minimise the cost of sending the remaining messages as well.

During the first stage of MSA, a "pseudo"-schedule is produced which specifies well defined *computational intervals* during which clusters are executed. Between them, *communication intervals* exist during which communication between clusters in different computational intervals is carried out (i.e. this is where the communication overheads are scheduled). A second stage is used to map the resulting clusters to the given number of processors. If the number of parallel clusters within the same band is greater than the number of processors, then by Brent's Principle [16] these clusters are merged in such a way that minimises the communication costs.

5.   Terminology

This chapter adopts the overhead definition from the *Cost and Latency Augmented DAG* (CLAUD) model [18] (developed independently of *LogP*) where independent overhead parameters are associated with the *sending* and *receiving* of a message, denoted by λ_s and λ_r, respectively (i.e. $o = \frac{\lambda_s + \lambda_r}{2}$). Throughout this chapter, the term *delay model conditions* is used to refer to situation where the communication overheads (λ_s, λ_r and g) are zero and the term $LogP$ *conditions* used when otherwise.

Papadimitriou and Yannakakis defined $e(v_i)$, the *earliest start time* of a task v_i, to be a lower bound on the optimal or earliest time possible that task

v_i can start executing [49]. It must be emphasized that this is only a bound and that it is not always possible to schedule task v_i at this time. In order to avoid confusion, we introduce the term $e_s(v_i)$, the *earliest schedule time*, to be the earliest time that task v_i can start executing on any of the available processors. Thus, the completion time or *makespan* of the optimal schedule for G is $\max_{v_i \in V}\{e_s(v_i) + \varepsilon(v_i)\}$.

It is common in many scheduling heuristics to prioritise tasks in the graph according to some function of the perceived computation and communication costs along all paths to that task. The costliest path to a task v_i is often referred to as the *critical path* of task v_i. Possible candidate functions include both $e(v_i)$ and $e_s(v_i)$. We refer to the term *graph schedule time*, $gs(v_i)$, as the costliest path to task v_i assuming that each ancestor of v_i is allocated to a distinct processor (thus $gs(v_i) \geq e_s(v_i) \geq e(v_i)$). The choice of cost function is often based on the cost (complexity) of calculating it (of the above three, $gs(v_i)$ is the least costly and $e_s(v_i)$ the most). One should be aware that the critical path of a task v_i may differ depending on the cost function chosen.

6. Design issues for clustering algorithms

This section concentrates on key issues in the design of clustering algorithms for the task-scheduling problem under the *LogP* model. This model is not an extension of the delay model but rather a generalisation, i.e. the delay model is in fact a specific instance of the *LogP* model. Therefore, the underlying principles and the assumptions used in delay model scheduling strategies may no longer be applicable. Thus, it is imperative to understand these principles and identify assumptions which can be adopted by *LogP* clustering algorithms.

Clustering algorithms aim to obtain a near optimal schedule for a *DAG* by attempting to construct clusters so that sink tasks starts executing at the earliest time possible. Where replication is not considered, algorithms initially create a cluster $C(v_i)$ for each $v_i \in V$ containing only the task v_i and then try to merge these clusters if the parallel time does not increase. We focus on algorithms which grow clusters one task at a time, and in particular, those which exploit task duplication to achieve better makespans. In this type of algorithm, a completed cluster $C(v_i)$ will contain the *owner* task v_i and, determined by some cost function, *copies* of some of its ancestor tasks. Replication-based clustering algorithms generally consist of two stages: the first stage searches for the smallest $s(v_i)$, for each $v_i \in V$, by creating $C(v_i)$ and determining the tasks which should be included and their order within the cluster; the second stage identifies which of the clusters are required to implement the input *DAG* G, mapping them to processors.

Algorithm 1 : *mechanics-of-clustering* $(C(v_i))$;

1   $C(v_i) = \{v_i\}$;   /* Algorithm to create a cluster for task v_i */
2   $lcands = iancs(C(v_i))$;
3   **while** *progress condition* **do**
4   let $cand \in lcands$;
5       $C(v_i) = C(v_i) \cup \{cand\}$;
6       $lcands = iancs(C(v_i))$;

The iterative nature of creating a cluster a task at a time can be generalized to the algorithmic description shown in Algorithm 1. In lines 2 and 6, the list ($lcands$) of candidate tasks for inclusion into a cluster usually consists of those tasks which become immediate ancestors of the cluster, i.e. the set of immediate ancestors of tasks already in $C(v_i)$ which themselves are not in $C(v_i)$ ($iancs(C(v_i)) = \{u \mid (u, t) \in E \wedge u \notin C(v_i) \wedge t \in C(v_i)\}$).

How well a clustering algorithm performs (i.e. the quality of the results it produces), whether under the delay or *LogP* model, depends on four crucial design issues: (i) calculating the makespan; (ii) the ordering of tasks within the cluster; (iii) the *progress condition* (line 3); and (iv) the choice of candidate task for inclusion into the cluster (line 4).

The makespan of a cluster $C(v_i)$ can be determined by the costliest path to task v_i (the critical path). Effectively, there are only two types of path to v_i, as shown in Figure 3.3: the *cluster path* which is the path due to *computation* in the cluster (whereas in the delay model, this would just be the costs associated to tasks, under the *LogP* model, the communication overheads must also be considered); and the *ancestor paths*, the paths formed by the immediate ancestor tasks of the cluster. In Figure 3.3, the ancestor paths (a) and (b) are due to the respective tasks $u_2, u_1 \in iancs(C(v_i))$, immediate ancestors of tasks $t_3, t_1 \in C(v_i)$. Note that if an immediate ancestor has two successors in same the cluster, then two separate messages must be sent since the data may be different and message bundling is not considered.

The *cluster path cost*, $m(C(v_i))$, is the earliest time that v_i can start due to the path created by the ancestor tasks of v_i in $C(v_i)$. The order in which these tasks are executed in the cluster has a significant influence on this cost. The problem of finding the best ordering is in itself NP-complete [32]. Ordering the tasks by their $e(v_i)$, for example, is the heuristic-based approach often adopted [48]. The *ancestor path cost*, $c(u, t, v_i)$ is the earliest time that v_i can start due solely to the path from task u, the immediate ancestor of task $t \in C(v_i)$, to v_i. The ancestor path with the largest cost is known as the *critical ancestor path*.

The *progress condition* defines the conditions under which the algorithm should continue to insert tasks into a cluster. When progress condition is not

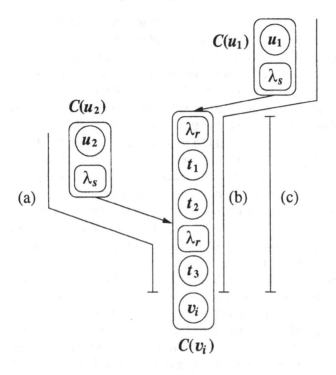

Figure 3.3. The ancestor paths (a) and (b), and cluster path (c) of the cluster $C(v_i)$.

met, the construction of this cluster is complete. Typically, this condition is based on a comparison of the makespan before and after the inclusion of the chosen candidate task. This condition can be simplified to comparing the critical ancestor path cost with the cluster path cost, if an appropriate cost function is being used. If the makespan does not increase, the insertion is successful and the algorithm continues to grow the cluster. Otherwise, the algorithm stops, since it assumes that the makespan can be reduced no further. Note, however, that the mechanics of this algorithm are based on the underlying assumption that the makespan is a monotonically decreasing function (in other words, contains no local minima), with respect to the inserted tasks, until its global minima is reached. This assumption is based on the fourth issue – the choice of candidate has to be the best one in every iteration. But which task is the best choice? The inclusion of a non-immediate ancestor may not decrease the makespan of the cluster in this iteration. It is clear that the inclusion of a completely independent task will never cause the makespan of the cluster to decrease.

Under the delay model, the inclusion of a task has two effects: it will increase the cluster path cost at the expense of removing the critical ancestor path; and introduce a new ancestor path for each immediate ancestor of the inserted task not already in the cluster. Because the cluster path cost increases monotonically, the progression condition can be simplified to a combination of the following two conditions, i.e. the algorithm is allowed to progress while: the cluster path does not become the critical path (the non-critical path condition); and inserting the chosen task does not increase the makespan (the worth-while condition). The makespan of a cluster can only be reduced if the cost of the critical path is reduced. If the critical ancestor path is the critical path of the cluster, the corresponding immediate ancestor task u is the only candidate whose inclusion into the cluster has the potential to diminish the makespan. However, while always selecting this task is on the whole a good choice, it is not always the best [47]. Depending on the cost function be used, the critical path cost could increase the makespan sufficiently to cause the algorithm to stop prematurely, at a local optimal for the makespan, because the inserted task u has a high communication cost with one of its ancestors. While a cost function such as $e(v_i)$ or $e_s(v_i)$ may not reflect this high cost, $gs(v_i)$ would, although it is a poor guide to a good makespan.

The effects of including a task under $LogP$ conditions are similar, except that the cluster path cost may not necessarily increase. For example, in Figure 3.3, if the execution cost of u_2 is less than the receive overhead λ_r, then the inclusion of into $C(v_i)$ will cause the cluster path cost to decrease. Communication overheads need to be treated like the tasks in the cluster, since they are also computation costs incurred by a processor even though they are associated with ancestor paths (i.e. communication). This means that the existence, or not, of an immediate ancestor will influence the cluster path cost and the cost of other earlier ancestor paths of the cluster, in contrast with the delay model. Figure 3.3 shows that the receive overhead associated with ancestor path (a) is also part of both the ancestor path (b) and the cluster path (c). Again, unlike the delay model, not only can the inclusion of even a non-critical ancestor path task reduce the makespan, but improvements to the makespan may be achieved by continuing to include tasks after the non-critical path condition. In the $LogP$ model, in addition to dependencies between program tasks, the problem of scheduling computation within a cluster must now consider the dependencies between these overhead "tasks" and their respective program tasks as well as the throughput restrictions of the network or its interface (i.e. the effects of the $LogP$ parameter g).

All in all, the quality of the makespans produced by a $LogP$ clustering algorithm depends on the quality of the implementation of each of the afore-mentioned design issues. These issues are not completely independent, but interrelated by the role played by the cost function. The cost function can be used both to calculate the path costs and to order tasks within a cluster. The accuracy in calculating the path costs is important since these costs both affect the outcome of progress condition, and are used to calculate the makespan of a cluster. Furthermore, the cost function might also be used as a basis for choosing a candidate task for inclusion. Given this relationship, it may advantageous to use different cost functions. The question then becomes which is the most appropriate function for each design issue.

If a design issue cannot be addressed perfectly, then the implementation of the other issues could try to compensate for this. For example, if the best candidate is not always guaranteed to be chosen, the makespan may not be a monotonically decreasing function. Therefore, progress condition should be designed in such way that will allow the algorithm to escape local minima (i.e. allow the makespan to increase so that further task inclusions can attain an even better makespan), but also not to continue including tasks unnecessarily thus creating a poor cluster.

7.   A new scheduling algorithm

This section describes a task replication-based clustering algorithm for the scheduling of weighted arbitrary task graphs under the $LogP$ model called BNR [11]. The algorithm presented here encompasses just one of a number of possible implementations for each of the design issues discussed in Section 6. These do not necessarily represent the best combination of design issues nor even the best implementation of each of them. Our objective is to simply highlight the merits of basing the design of a clustering algorithm on this methodology.

Unlike traditional delay model approaches, which tend to use the earliest start time $e(v_i)$ as the basis for the cost function, the $LogP$ algorithm presented here uses the earliest schedule time $e_s(v_i)$. Remember that $e(v_i)$ is a lower bound, therefore using $e_s(v_i)$ should produce: better predictions for schedule execution times, since this cost function represents what is achievable; and better schedules, given that cost functions play an important role in the mechanics of clustering algorithms. On the other hand, since finding $e_s(v_i)$ is much more difficult, the algorithm in fact creates the cluster $C(v_i)$ to find a good approximation, $s(v_i)$, to $e_s(v_i)$ which is then used to create the clusters of task v_i's successors. Thus, $s(v_i)$ represents the *scheduled time* or *start time* of task v_i in its own cluster $C(v_i)$.

The implementation of the second design issue also differs from traditional delay model approaches. In this algorithm, tasks within a cluster are ordered neither in non-decreasing order of their earliest start time nor by their earliest schedule time but rather in an order determined by the sequence in which tasks were included in the cluster and the position of their immediate successor task at the time of their inclusion. The merits of this implementation are discussed in [47].

Being replication-based, the BNR heuristic consists of two stages. The first stage tries to creates a cluster $C(v_i)$, $\forall v_i \in V$, identifying the tasks to be included and their order within the cluster, so that $s(v_i)$ is the earliest possible. A number of cluster design restrictions are assumed to help in both finding a good $s(v_i)$ (i.e. close to $e_s(v_i)$ as possible) and reducing the complexity of this stage. The second stage determines which of the clusters, created in stage one, and the number of copies of each cluster that are necessary in order to implement a schedule with a makespan $\max_{v_i \in V}\{s(v_i) + \varepsilon(v_i)\}$ for the input *DAG*.

7.1.   Simplifying *LogP* scheduling

In order to simplify the design of clustering-based *LogP* schedulers, four *cluster design restrictions*, i.e. restrictions with regard to the properties of clusters, are proposed. In *Restriction 1*, in order to minimize the number of send overheads incurred by a cluster, only the owner task may send data to a task in another cluster. This restriction does not adversely affect the makespan since, by definition, an owner task will not start execution later than any copy of itself. Also, if a non-owner task t in cluster $C(v_i)$, $t \neq v_i$, were to communicate with a task in another cluster, the processor to which $C(v_i)$ is allocated would incur a send overhead after the execution of t, which in turn might force the owner v_i to start at a time later than necessary.

Restriction 2 specifies that each cluster can have only one successor. If two or more clusters, or two or more tasks in the same cluster, share the same ancestor cluster, then a unique copy of the ancestor cluster is assigned to each successor irrespective of whether or not the data being sent to each successor is identical[2]. This restriction removes the need to incur the multiple gap costs and send overheads at the end of an ancestor cluster that might unnecessarily delay the execution of successor clusters.

Restriction 3 assumes that receive overheads are executed immediately before their receiving task. This restricts where receive overheads can be scheduled and greatly simplifies the job of ordering computation within the cluster and calculating *path costs*. This simplification comes at the expense of perhaps not being able to achieve the minimal makespan. When $g > \lambda_r$, idle time periods will exist between the successive overheads associated to messages being received by the same task, which cannot be used to scheduled other tasks or receive overheads.

Restriction 4 is related to the behavior of the network interface. This restriction assumes that the receive overheads, scheduled before their receiving task, are ordered by their time of arrival. This is for two reasons: it assumes that the

---

[2]We consider this to be a reasonable assumption since this information is not available in a program graph and would have to be extracted from the user program itself. Also it is not clear that all multicomputer systems support messages with multiple destinations.

network interface implements a queue to receive messages; and it is an efficient ordering given Restriction 3.

These restrictions do not need to be applied by scheduling strategies which are used exclusively under delay model conditions (because of the assumptions made under that model). The purpose of the first two restrictions is to aid in the process of finding the minimal schedule time of a cluster. The price, however, is a potential increase in the number of clusters required, and thus the number of processors needed to implement the corresponding schedule. Since the number of processors used is viewed as a reflection of the cost of implementing a schedule, the second stage of the scheduling algorithm can relax Restrictions 1 and 2 and effectively remove clusters which become redundant (i.e. clusters whose removal will not cause the makespan to increase). Ideally, receive overheads should be treated just like program tasks and be permitted to be scheduled at any valid time or order within the cluster. This involves relaxing Restrictions 3 and 4 which increases the problem of finding the best "tasks" ordering for the cluster.

7.2.   The first stage

In order to construct the cluster $C(v_i)$, the scheduled time of task v_i's immediate ancestors must be known. The clustering strategy, therefore, creates clusters for each task $v_i \in V$ topologically, in the manner described below.

A cluster $C(v_i)$ is created (based on Algorithm 1 and as outlined in Algorithm 2) by including, one at a time, copies of tasks from $iancs(C(v_i))$ together with the owner task v_i. Whether the chosen *candidate* task $u \in iancs(C(v_i))$ is incorporated into cluster $C(v_i)$ depends on the *progress condition*. The progress condition effectively compares the makespans of the following two situations: (a) when $u \notin C(v_i)$; and (b) when $u \in C(v_i)$.

By using the cost function $s(v_i)$, the makespan of a cluster is determined by the *cluster path cost* or *critical ancestor path cost*, whichever is larger. Let $C(v_i) = \langle t_1, \ldots, t_l, \ldots t_k, v_i \rangle$ be tasks of cluster v_i in their chosen execution order. The *cluster path cost*, $m(C(v_i))$, is calculated by summing up all processor computation time along the cluster path including receive overheads, but ignoring any idle periods where a processor would have to wait for the data sent by tasks in $iancs(C(v_i))$ to arrive at $C(v_i)$ (as shown in Procedure 1). Note that since the cluster is only partially complete, the cluster path cost must take into account the fact that start time of a task t_l in the cluster $C(v_i)$ cannot be earlier than its earliest schedule time $e_s(t_l)$ (approximated to $s(t_l)$ as shown in line 5 of Procedure 1). This allows $m(C(v_i))$ to represent a realistic and attainable cost, a cost which will affect whether future iterations are required to complete the cluster.

Procedure 1 : *critical-cluster-path* $(C(v_i))$

1   $m(C(v_i)) := 0$;
2   for $l = 1, \ldots, k$ do   /* $C(v_i) = \langle t_1, \ldots, t_l, \ldots t_k, v_i \rangle$ */
3       $n\_arcs :=$ number of arcs incidents to task t_l;
4       $m(C(v_i)) := m(C(v_i)) + n\_arcs \times \lambda_r$;
5       $m(C(v_i)) := \max\{m(C(v_i)), s(t_l)\} + \varepsilon(t_l)$;

In situation **(a)**, for each task $u \in iancs(C(v_i))$, BNR calculates the earliest time that v_i can start **due solely** to the candidate task u. This *ancestor path cost*, $c(u, t_l, v_i)$, for immediate ancestor task u of task t_l in $C(v_i)$ is (due to Restriction 1) the sum of: the execution cost of $C(u)$ $(s(u) + \varepsilon(u))$; one send overhead (λ_s) (due to Restriction 2); the communication delay $(L \times \omega(u, t_l))$; the receive overhead (λ_r); and the computation time of tasks t_l up to v_i including receive overheads of edges processed after the edge (u, t_l) (due to Restrictions 3 and 4). The critical ancestor task is the task $u \in iancs(C(v_i))$ for which $c(u, t_l, v_i)$ is the largest, i.e. $C(u)$ is the immediate ancestor cluster which most delays start time of task v_i. The ancestor path cost is calculated in Procedure 2, which assigns *ctask* and *maxc* to the critical immediate ancestor and the critical ancestor path cost, respectively (line 10).

Procedure 2 : *critical-ancs-path* $(v_i, C(v_i))$

1    $maxc := 0$;   /* $C(v_i) = \langle t_1, \ldots, t_l, \ldots t_k, v_i \rangle$ */
2   if $(iancs(C(v_i)) \neq \emptyset)$ then
3   for all $u \in iancs(C(v_i))$ do
4   for each $(u, t_l) \in E$ such that $t_l \in C(v_i)$ do
5                $c(u, t_l, v_i) := s(u) + \varepsilon(u) + \lambda_s + L \times \omega(u, t_l)$;
6   for $j = l, \ldots, k$ do
7                    $n\_arcs :=$ no. of arcs incident to t_j after arc (u, t_l);
8                    $c(u, t_l, v_i) := c(u, t_l, v_i) + (n\_arcs + 1) \times \lambda_r + \varepsilon(t_j)$;
                 end do
9   if $(maxc < c(u, t_l, v_i))$ then
10               $maxc := c(u, t_l, v_i)$; $ctask := u$;

If the critical ancestor path cost, $maxc$, is greater than the cluster path cost, $m(C(v_i))$, then task $ctask$ becomes the chosen candidate for inclusion into cluster $C(v_i)$. This is the non-critical path (NCP) condition of the progress condition, as shown in line 5 of Algorithm 2. If this is not the case, the process stops and cluster $C(v_i)$ is now complete.

Before committing the candidate task $ctask$ to $C(v_i)$, the critical ancestor path cost is compared with cluster path cost of $\{C(v_i) \cup \{ctask\}\}$. If $maxc$ is greater than or equal to this new cost (the worth-while (WW) condition of the progress condition, line 8 of Algorithm 2), then task $ctask$ is included, the ancestor path costs are calculated for all tasks in the new set $iancs(C(v_i))$ (line 9), the start time of v_i is updated (line 10) and the loop is repeated. Otherwise, task $ctask$ is not included (line 12), and the formation of cluster $C(v_i)$ is finished.

Algorithm 2 : *construct-cluster* (v_i)

1   if $pred(v_i) = \emptyset$ then $s(v_i) := 0$;

2   else

3      $(maxc, ctask) := critical\text{-}ancs\text{-}path(v_i, C(v_i))$;

4      $m(C(v_i)) := 0$; $s(v_i) := \max\{m(C(v_i)), maxc\}$;

5   while $(m(C(v_i)) < maxc)$ /* NCP condition */

6        $C(v_i) := C(v_i) \cup \{ctask\}$; $m'(C(v_i)) := m(C(v_i))$;

7        $m(C(v_i)) := critical\text{-}cluster\text{-}path(C(v_i))$;

8   if $(m(C(v_i)) \leq maxc)$ then   /* WW condition */

9          $(maxc, ctask) := critical\text{-}ancs\text{-}path(ctask, C(v_i))$;

10          $s(v_i) := \max\{m(C(v_i)), maxc\}$;

       else

11          $s(v_i) := \max\{m'(C(v_i)), maxc\}$;

12          $C(v_i) := C(v_i) - \{ctask\}$;

     end while

   end if

Note that this progress condition can, in some cases, allow the algorithm to proceed even if the makespan increases. This is to allow BNR to climb out of some local makespan minimas as long as the NCP condition is true. The drawback is that the algorithm may end up generating a worse final makespan. This happens to be a consequence of the cost function $s(v_i)$. One solution is to keep a record of the best makespan found so far and rollback [12, 47].

Until now, we have not discussed the influence of the gap parameter g on a receiving cluster. The gap parameter, which represents the rate at which messages can be removed from the network, does not require computation time on a processor like λ_r. However, it does influence when receive overheads can be scheduled within a cluster. Due to the Restriction 3, all receive overheads for messages to a given task must be executed immediately before the recipient task. Given this restriction, the BNR algorithm actually considers the *receiving cost* of a message to be $rc = \max\{\lambda_r, g\}^3$, i.e. BNR also treats g as a blocking parameter, since Restriction 3 does not allow program tasks to be scheduled on the processor during this period anyway. Only when $\lambda_r < g$, does this restriction cause the processor to become unnecessarily blocked.

7.3.　　The second stage

The second stage of this clustering algorithm determines which of the generated clusters are needed to implement the schedule. Recall that these clusters were constructed based on the fact that all of their ancestors send their data at the earliest possible time (due to the application of the design restrictions described in Section 7.1).

The necessary clusters can be identified by simply tracing the dependencies backwards starting with the sink clusters. However, due to Restriction 2, multiple copies of common ancestor clusters may be required. Depending on the type of input DAG, this can cause an explosion in the number of clusters. Note that this is not the case under delay model conditions, since the cluster design restrictions need not be applied.

Relaxing Restriction 2 and reducing the number replicated cluster needed without increasing the makespan can be achieved by realising that clusters do not always require their data to arrive as early as possible. The *latest* time that each cluster can be scheduled so the given makespan is still achieved needs to be calculated. Scheduling a cluster at its latest schedule time and its ancestor at its earliest time creates a period of time or band in which additional communication (or even computation) can take place. The size of the band may be sufficient for an ancestor cluster to send more than one message, each sucessive message be delayed an additional cost of $\max\{\lambda_s, g\}$.

The necessary clusters are then allocated to processors, at worst, one cluster per processor. In some cases, there may be scope to further reduce the number of processors required by mapping clusters whose execution times do not overlap to the same processor. The combination of the way in which the clusters are constructed in Stage 1 and their optimisation and allocation in this stage, permits the $LogP$ network capacity constraint to be met.

8.　　Algorithm performance

Ignoring the differences for $LogP$ scheduling, the clustering algorithms BNR and PLW [48] appear to be similar. Both of these algorithms consist of two stages, utilize task replication to form clusters (creating one for each

---

[3]Note that the term "$n\_arcs \times \lambda_r$" in line 4 of Procedure 1 should be substituted by "$(n\_arcs - 1) \times rc + \lambda_r$" and the term "$(n\_arcs + 1) \times \lambda_r$" in line 8 of Procedure 2 by "$(n\_arcs \times rc + \lambda_r$". This was deliberately not shown earlier for the sake of clarity.

task in a $DAGG$), even growing them by adding a single task at a time. However, the key difference is the use of cost function; BNR uses an approximation to the *earliest schedule time* to determine the start time of owner tasks (i.e. the original tasks in G and not their copies) rather than the *earliest start time* for all tasks including their copies. This difference influences, to some degree, all four of the clustering design issues (Section 6). In the first design issue, the path costs and the makespan of a cluster are calculated differently. For the second design issue, PLW orders tasks in nondecreasing order of their earliest start time. While the third and fourth issues in the two algorithms look similar, due to the use of the two different cost functions, their behaviours are not.

This section briefly compares the quality of the schedules produced by BNR with the clustering heuristics PLW [48] and DSC [33] under delay model conditions, and MSA [14] under $LogP$ conditions. Although the cluster design restrictions cause $LogP$ strategies, such as BNR, to appear expensive with respect to the number of processors required, it is important to show that the makespans produced and processors used are comparable to those of delay model approaches under their model. Scheduling approaches such as DSC and PLW have already been shown to generate good results (optimal or near-optimal) for various types of $DAGs$ [33, 48, 42].

The comparisons were made using a suite of benchmark (UET-UDC) $DAGs$, which includes out-trees, in-trees, diamond $DAGs$ and a set of randomly generated $DAGs$, as well as weighted irregular $DAGs$ taken from the literature.

In the more than 300 experiments [11, 47], the schedules produced by BNR were never worse than those of DSC or PLW. In fact, BNR was better than DSC and PLW in 70% and 85% of the cases, respectively. The performances of both DSC and PLW progressively worsen in comparison with BNR as the $DAGs$ effectively become more fine-grained due to an increasing latency value (as seen from Figures 3.4 and 3.5). DSC generally utilizes the fewest processors since it does not replicate tasks. BNR, on the whole, uses fewer processors than PLW (especially for small latencies), but where this is not true, the compensation is the better makespan. Table 3.1 contains a comparison of makespans produced under the delay model by the three strategies for a group of documented irregular, non-unit cost graphs used by various researchers (denoted in the table as Ir and with the subscript identifying the number of tasks in the graph). M and P represent the makespan obtained and the number of processors needed, respectively.

A comparison of the makespans produced by BNR and MSA, another replication-based clustering algorithm, under various $LogP$ conditions, is shown Figure 3.6. As well as UET-UDC tasks graphs, MSA requires that $g \leq \min\{\lambda_s, \lambda_r\}$. The specific irregular DAG Ir_{41} in this figure represents a molecular physics problem [37].

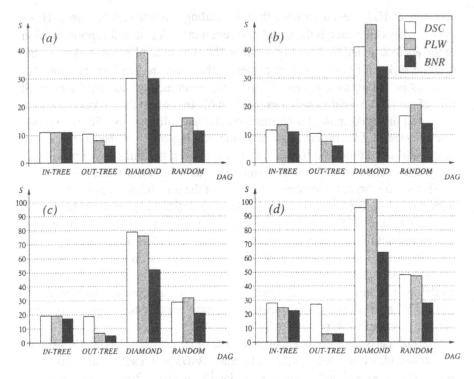

Figure 3.4. Average Makespans for the different classes of fine-grained *DAG*s under the delay model with the latency (a) $L = 1$, (b) $L = 2$, (c) $L = 5$, and (d) $L = 10$.

Table 3.1. Results for non-unit (task and edge) cost irregular *DAG*s ($L = 1$).

DAG	DSC		PLW		BNR	
	M	P	M	P	M	P
Ir_{7a} [31]	9	2	8	3	8	3
Ir_{7b} [34]	8	3	12	5	8	3
Ir_{10} [48]	30	3	27	4	26	3
Ir_{13} [4]	301	7	275	8	246	8
Ir_{18} [40]	530†	5	480	8	370	9

† [40] reports a makespan of 460 on 6 processors.

In 83% of 107 cases, the schedules produced by *BNR* are better than those of *MSA*. Less than 5% of the *MSA* schedules produced makespans smaller than those of *BNR*. The schedules produced by *MSA* are, on average, 17.4% larger. Both algorithm produce optimal solutions for out-tree graphs. However *BNR* requires a significantly greater number of processors than *MSA*, which benefits from the technique of bundling messages.

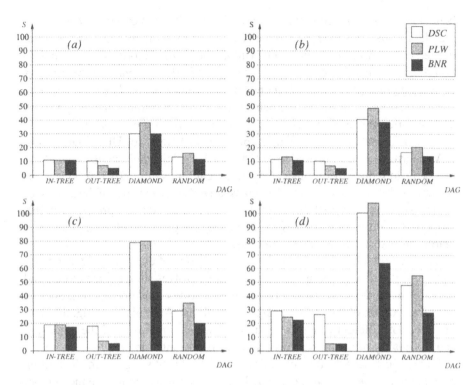

Figure 3.5. Average Makespans for the different classes of coarse-grained DAGs under the delay model with $L = 1$ and task execution costs equal to (a) 1, (b) 2, (c) 5, and (d) 10.

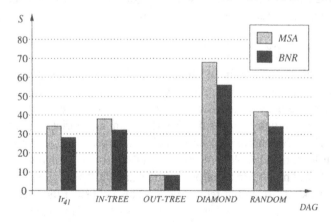

Figure 3.6. Average Makespans for the different classes of UET-UDC DAGs under the $LogP$ model.

9.   Conclusions

Based on these results, BNR compares favorably against traditional cluste-ring-based scheduling heuristics such as DSC and PLW, which are dedicated exclusively to the delay model. It is also necessary is to analyze the heuristic's behavior under $LogP$ conditions, comparing results with other $LogP$ scheduling algorithms. However, due to the scarcity of such algorithms, only a rather limited comparison has been made. Unlike the $LogP$ algorithms cited in Section 4, algorithms based on this clustering methodology place no restrictions on the structure of the input DAG nor on the values of the $LogP$ parameters. Clearly, results of further experiments using other algorithms, and graphs with a more varied range of granularities and connectivities are needed to complete the practical evaluation of the algorithm. Nevertheless, for UET-UDC $DAGs$, BNR perform well compared to MSA, which exploits the benefits of bundling messages to reduce the number of communication events. BNR produces superior schedules with respect to the makespan but at the expense of the processor count. This indicates that the clustering design methodology presented in this chapter is a promising framework to tackle the problem of scheduling task graphs on realistic communication models.

The methodology also identified various aspects of clustering algorithm design for further study and improvement. This involves studying the effects and interactions of the basic design issues, their various implementations, and the use or relaxation of the cluster design restrictions. An initial investigation into the effect of different progress conditions on the complexity of the algorithm and the quality of the makespans produced is presented in [12].

Given the nature of the schedules produced by this algorithm, it is often possible to use the second stage of a clustering algorithm to reduce the number of processors at the expense of a relatively small increase in the makespan. In this way, like MSA, the BNR clustering algorithm can also be applied to the problem of task scheduling on a bounded number of processors.

10.   Summary

Since tackling the scheduling problem in an efficient manner is imperative for achieving high performance from message-passing parallel computer systems, this problem continues to be a focus of great attention in the research community. One fundamental requirement is the the use of an appropriate communication model [13, 15]. This chapter has presented an overview of one approach to the relatively little studied problem of task scheduling for realistic machine models such as the $LogP$ model. In addition to modelling the transit time of a message, $LogP$-type models (including recent models) also consider the processing cost to send or receive a message. A description of how these parameters affect the scheduling problem shows why existing delay model scheduling heuristics are not suitable for high performance parallel systems. A methodology for designing clustering-based $LogP$ scheduling algorithms is presented. Four basic design issues were identified, the implementations of which play a crucial role in the quality of schedules produced. Based on one particular combination of design issue implementations, an algorithm for scheduling arbitrary, non-unit weighted task graphs onto an unbounded number of processors has been described in detail. The design complexity of this algorithm was reduced by the use of four cluster design restrictions. The algorithm performs well compared to other good, and well-known clustering-based scheduling heuristics.

These initial results indicate that this design methodology is a solid foundation upon which task scheduling algorithms can be built to efficiently exploit parallelism on modern computing systems.

Acknowledgments

The authors would like thank Tao Yang and Jing-Chiou Liou for providing the DSC and PLW algorithms, respectively, and for their help and advice. Special thanks also to Aline Nascimento for her work on the design and implementation of the BNR algorithm. This work is funded by research grants from the Fundação de Amparo à Pesquisa do Estado do Rio de Janeiro (FAPERJ).

References

[1] T. Adam, K. Chandy, and J. Dickson. A comparison of list schedulers for parallel processing systems. *Communications of the ACM*, 17(12):685–690, 1974.

[2] A. Aggarwal, A.K. Chandra, and M. Snir. On communication latency in PRAM computations. Technical Report RE RC 14973, IBM T.J. Watson Research Center, Yorktown Heights, NY, USA, September 1989.

[3] I. Ahmad, Y. Kwok, and M. Wu. Analysis, evaluation and comparison of algorithms for scheduling task graphs to parallel processors. In *Proceedings of the 1996 International Symposium on Parallel Architecture, Algorithms and Networks*, Beijing, China, June 1996.

[4] I. Ahmad and Y-K Kwok. A new approach to scheduling parallel programs using task duplication. In K.C. Tai, editor, *Proceedings of the International Conference on Parallel Processing (ICPP'94)*, volume 2, pages 47–51, St. Charles, Illinois, USA., August 1994.

[5] A. Alexandrov, M. Ionescu, K.E. Schauser, and C. Scheiman. LogGP: Incorporating long messages into the LogP model - one step closer towards a realistic model for parallel computation. *The Proceedings of the 7th Annual Symposium on Parallel Algorithms and Architectures (SPAA'95)*, July 1995. Also as TRCS95-09 - Department of Computer Science - University of Santa Barbara, CA, USA.

[6] T.E. Anderson, D.E. Culler, and D.A. Patterson. A case for NOW (Networks of Workstations). *IEEE Micro*, 15(1):23–39, February 1995.

[7] F.D. Anger, J-J Hwang, and Y-C Chow. Scheduling with sufficient loosely coupled processors. *Journal of Parallel and Distributed Computing*, 9:87–92, 1990.

[8] M. Baker and R. Buyya. Cluster Computing at a Glance. In R. Buyya, editor, *High Performance Cluster Computing Architectures and Systems*, chapter 1, pages 3–47. Prentice Hall, 1999.

[9] C. Boeres. *Versatile Communication Cost Modelling for Multicomputer Task Scheduling Heuristics*. PhD thesis, Department of Computer Science, University of Edinburgh, May 1997.

[10] C. Boeres, G. Chochia, and P. Thanisch. On the scope of applicability of the ETF algorithm. In A. Ferreira and J. Rolim, editors, *Proceeding of the 2nd International Workshop on Parallel Algorithms for Irregularly Structured Problems (IRREGULAR'95)*, LNCS 980, pages 159–164, Lyon, France, September 1995. Springer.

[11] C. Boeres, A. Nascimento, and V. E. F. Rebello. Cluster-based task scheduling for LogP model. *International Journal of Foundations of Computer Science*, 10(4):405–424, 1999.

[12] C. Boeres, A. Nascimento, and V. E. F. Rebello. Scheduling arbitrary task graphs on LogP machines. In P. Amestoy, P. Berger, M. Daydé, I. Duff, V. Frayssé, L. Giraud, and D. Ruiz, editors, *Proceedings of the Fifth International Euro-Par Conference on Parallel Processing (Euro-Par'99)*, LNCS 1685, pages 340–349, Toulouse, France, April 1999. Springer.

[13] C. Boeres and V. E. F. Rebello. Versatile task scheduling of binary trees for realistic machines. In C. Lengauer, M. Griebl, and S. Gorlatch, editors, *Proceedings of the Third International Euro-Par Conference on Parallel Processing (Euro-Par'97)*, LNCS 1300, pages 913–921, Passau, Germany, August 1997. Springer-Verlag.

[14] C. Boeres and V. E. F. Rebello. A versatile cost modelling approach for multicomputer task scheduling. *Parallel Computing*, 25(1):63–86, 1999.

[15] C. Boeres, V. E. F. Rebello, and D. Skillicorn. Static scheduling using task replication for LogP and BSP models. In D. Pritchard and J. Reeve, editors, *The Proceedings of the 4th International Euro-Par Conference on Parallel Processing (Euro-Par'98)*, LNCS 1470, pages 337–346, Southampton, UK, September 1998. Springer.

[16] R.P. Brent. The parallel evaluation of general arithmetic expressions. *Journal of the ACM*, 21:201–206, 1974.

[17] T.L. Casavant and J.G. Kuhl. A taxonomy of scheduling in general-purpose distributed computing systems. *IEEE Transactions on Software Engeneering*, 14(2):141–154, February 1988.

[18] G. Chochia, C. Boeres, M. Norman, and P. Thanisch. Analysis of multicomputer schedules in a cost and latency model of communication. In *Proceedings of the 3rd Workshop on Abstract Machine Models for Parallel and Distributed Computing*, Leeds, UK., April 1996. IOS press.

[19] E.G. Coffman Jr. *Computer and Job Shop Scheduling Theory*. John Wiley, 1976.

[20] R. Corrêa and A. Ferreira. A polynomial-time branching procedure for the multiprocessor scheduling problem. In P. Amestoy, P. Berger, M. Daydé, I. Duff, V. Frayssé, L. Giraud, and D. Ruiz, editors, *Proceedings of the Fifth International Euro-Par Conference on Parallel Processing (Euro-Par'99)*, LNCS 1685, pages 272–279, Toulouse, France, August 1999. Springer.

[21] M. Cosnard and M. Loi. Automatic task graph generation techniques. *Parallel Processing Letters*, 5(4):527–538, December 1995.

[22] D. Culler, R. Karp, D. Patterson, A. Sahay, K.E. Schauser, E. Santos, R. Subramonian, and T. von Eicken. LogP: Towards a realistic model of parallel computation. In *The Proceedings of the 4th ACM SIGPLAN Symposium on Principles and Practice of Parallel Programming*, San Diego, CA, USA, May 1993.

[23] B. Di Martino and G. Ianello. Parallelization on non-simultaneous iterative methods for systems of linear equations. In *Parallel Processing (CONPAR'94-VAPP VI)*, LNCS 854, pages 254–264. Springer-Verlag, 1994.

[24] J. Eisenbiergler, W. Lowe, and A. Wehrenpfennig. On the optimization by redundancy using an extended LogP model. In *Proceedings of the International Conference on Advances in Parallel and Distributed Computing (APDC'97)*, pages 149–155. IEEE Comp. Soc. Press, 1997.

[25] H. El-Rewini and T.G. Lewis. Scheduling parallel program tasks onto arbitrary target machines. *Journal of Parallel and Distributed Computing*, 9(2):138–153, June 1990.

[26] H. El-Rewini, T.G. Lewis, and H.A. Ali. *Task Scheduling in Parallel and Distributed Systems*. Prentice Hall Series in Innovative Technology. Prentice Hall, 1994.

[27] S. Fortune and J. Wyllie. Parallelism in random access machines. In *Proceedings of the 10th Annual ACM Symposium on Theory of Computing*, pages 114–118. ACM Press, 1978.

[28] M.R. Garey and D.S. Johnson. *Computers and Intractability*. W.H. Freeman and Co., 1979.

[29] A. Gerasoulis, Jiao J., and T. Yang. A multistage approach to scheduling task graphs. *DIMACS Series in Discrete Mathematics and Theoretical Computer Science*, 22:81–103, 1995.

[30] A. Gerasoulis, S. Venugopol, and T. Yang. Clustering task graphs for message passing architectures. In *Proceedings of the International Conference on Supercomputing*, pages 447–456, Amsterdam, The Netherlands, June 1990.

[31] A. Gerasoulis and T. Yang. A comparison of clustering heuristics for scheduling directed acyclic graphs on multiprocessors. *Journal of Parallel and Distributed Computing*, 16:276–291, 1992.

[32] A. Gerasoulis and T. Yang. List scheduling with and without communication. *Parallel Computing*, 19:1321–1344, 1993.

[33] A. Gerasoulis and T. Yang. DSC: scheduling parallel tasks on an unbounded number of processors. *IEEE Transactions on Parallel and Distributed Systems*, 5(9):951–967, 1994.

[34] J-J. Hwang, Y-C. Chow, F.D. Anger, and C-Y. Lee. Scheduling precedence graphs in systems with interprocessor communication times. *SIAM Journal on Computing*, 18(2):244–257, 1989.

[35] H. Jung, L. Kirousis, and P. Spirakis. Lower bounds and efficient algorithms for multiprocessor scheduling of DAGs with communication delays. In *Proceedings of the ACM Symposium on Parallel Algorithms and Architectures*, pages 254–264, 1989.

[36] A.A. Khan, C.L. McCreary, and M.S. Jones. Comparison of multiprocessor scheduling heuristics. In K.C. Tai, editor, *Proceedings of the Eighth International Conference on Parallel Processing*, volume 2, pages 243–250, Cancun, Mexico, April 1994. IEEE Computer Society Press - ACM SIGARCH.

[37] S.J. Kim and J.C. Browne. A general approach to mapping of parallel computations upon multiprocessor architectures. In *Proceedings of the 3rd International Conference on Parallel Processing*, pages 1–8, 1988.

[38] I. Kort and D. Trystram. Scheduling fork graphs under LogP with an unbounded number of processors. In D. Pritchard and J. Reeve, editors, *Proceedings of the 4th International Euro-Par Conference on Parallel Processing (Euro-Par'98)*, LNCS 1470, pages 940–943, Southampton, UK, September 1998. Springer.

[39] B Kruatrachue and T. Lewis. Grain size determination for parallel programming. *IEEE Software*, pages 23–32, Jan. January 1988.

[40] Y. K. Kwok and I. Ahmad. Dynamic critical-path scheduling: An effective technique for allocating tasks graphs to multiprocessors. *IEEE Transactions on Parallel and Distributed Systems*, 7(5):506–521, May 1996.

[41] Y-K Kwok and I. Ahmad. Efficient scheduling of arbitrary task graphs to multiprocessors using a parallel genetic algorithm. *Journal of Parallel and Distributed Computing*, 47(1):58–77, November 1997.

[42] Y-K Kwok and I. Ahmad. Benchmarking and comparison of the task graph scheduling algorithms. *Journal of Parallel and Distributed Computing*, 59(3):381–422, December 1999.

[43] Y-K Kwok and I. Ahmad. Static scheduling algorithms for allocating directed task graphs to multiprocessors. *ACM Computing Surveys*, 31(4), December 1999.

[44] W. Lowe and W. Zimmermann. On finding optimal clusterings of task graphs. In *Aizu International Symposium on Parallel Algorithm and Architecture Synthesis*, pages 241–247. IEEE Computer Society Press, 1995.

[45] R.P. Martin, A.M. Vahdat, D.E. Culler, and T.E. Anderson. Effects of communication latency, overhead, and bandwidth in a cluster architecture. In *Proceedings of the 24th Annual International Symposium on Computer Architecture*, pages 85–97, June 1997.

[46] W.F. McColl. BSP programming. In G. Blelloch, M. Chandy, and S. Jagannathan, editors, *Proc. DIMACS Workshop on Specification of Parallel Algorithms, Princeton, May 9-11, 1994*. American Mathematical Society, 1994.

[47] A. Nascimento. Aglomeração de tarefas em arquiteturas paralelas com memória distribuída. Master's thesis, Instituto de Computação, Universidade Federal Fluminense, Brazil, Niterói, RJ, Brazil, 1999. (In Portuguese).

[48] M.A. Palis, J.-C Liou, and D.S.L. Wei. Task clustering and scheduling for distributed memory parallel architectures. *IEEE Transactions on Parallel and Distributed Systems*, 7(1):46–55, January 1996.

[49] C.H. Papadimitriou and M. Yannakakis. Towards an architecture-independent analysis of parallel algorithms. *SIAM Journal on Computing*, 19:322–328, 1990.

[50] S.C.S. Porto, J.P. Kitajima, and C.C. Ribeiro. Performance evaluation of a parallel tabu search task scheduling algorithm. *Parallel Computing*, 26(1):73–90, 2000.

[51] V. Sarkar. *Partitioning and Scheduling Parallel Programs for Multiprocessors*. Pitman, London, 1989.

[52] B. Shirazi, M. Wang, and G. Pathak. Analysis and evaluation of heuristic methods for static task scheduling. *Journal of Parallel and Distributed Computing*, 10:222–232, 1990.

[53] G.C. Sih and E.A. Lee. A compile-time scheduling heuristic for interconnection-constrained heterogeneous processor architectures. *IEEE Transactions on Parallel and Distributed Systems*, 4(2):175–187, 1993.

[54] D. B. Skillicorn, J. M. D. Hill, and W. F. McColl. Question and answers about BSP. *Scientific Computing*, May 1997.

[55] H. S. Stone. *High-Performance Computer Architecture*. Electrical and Computer Engineering. Addison-Wesley, 1993.

[56] A. Tam and C. Wang. Realistic communication model for parallel computing on cluster. In *The Proceedings of the First IEEE International Workshop on Cluster Computing*, pages 92–101, Melbourne, Australia, December 1999. IEEE Computer Society Press.

[57] L.G. Valiant. A bridging model for parallel computation. *Communication of the ACM*, 33:103–111, 1990.

[58] W. H. Yu. *LU Decomposition on a Multiprocessor System with Communication Delay*. PhD thesis, Department of Electrical Engineering and Computer Science, University of California, Berkeley, CA, USA., 1984.

[59] W. Zimmermann, M. Middendorf, and W. Lowe. On optimal k-linear scheduling of tree-like task graph on LogP-machines. In D. Pritchard and J. Reeve, editors, *Proceedings of the 4th International Euro-Par Conference on Parallel Processing (Euro-Par'98)*, LNCS 1470, pages 328–336, Southampton, UK, September 1998. Springer.

Chapter 4

PREDICTABLE PARALLEL PERFORMANCE: THE BSP MODEL

D.B. Skillicorn

*Queen's University**

skill@cs.queensu.ca

Abstract   There are three big challenges for mainstream parallel computing: building useful hardware platforms, designing programming models that are effective, and designing a software construction process that builds correctness into software. The first has largely been solved, at least for current technology. The second has been an active area of research for perhaps fifteen years, while work on the third has barely begun. In this chapter, we describe the Bulk Synchronous Parallel (BSP) model which, at present, represents the best compromise among programming models for simplicity, predictability, and performance. We describe the model from the a software developer's perspective and show how its high-level structure is used to build efficient implementations. Almost alone among programming models, BSP has an associated cost model so that the performance of programs can be predicted on any target without laborious benchmarking. Some progress towards software construction has also been made in the context of BSP.

Keywords: BSP, cluster computing, predictable performance, portability, synchronism.

1.   Applying parallel computing

Parallel computing requires at least three distinct components to be in place before it becomes effective. The first is the parallel hardware that will execute parallel applications; the second is the abstract machine, or programming model, in which parallel applications will be written; and the third is the software design process that allows applications to be built from specifications [39]. All three are essential if parallelism is to be a mainstream technology for computation;

---

*Department of Computing and Information Science, Queen's University, Kingston Canada K7L 3N6

R. Corrêa et al. (eds.), Models for Parallel and Distributed Computation. Theory, Algorithmic Techniques and Applications, 85–115.
© 2002 *Kluwer Academic Publishers.*

as it will have to be to meet the demands of high-end commercial and scientific applications such as data mining and simulation.

The past fifty years have seen the development of a wide variety of parallel architectures. In the past decade, this variety has gradually reduced to two main alternatives: shared-memory MIMD systems of modest size (perhaps 30 processors) using cache coherence, sometimes gathered into larger ensembles that have come to be called *clumps* [15]; and distributed-memory MIMD systems based on commodity processor boards and off-the-shelf interconnects, called *clusters* [1]. This picture may change with the widespread deployment of optical computing devices or other new technology, but at present almost all deployed systems are of these two kinds [24].

The second requirement for effective parallel computing is an abstract machine or parallel programming model, forming the bridge between the hardware and the level at which software is written. All computers consume instructions much faster than they can be written. An abstract machine is only useful if some of its instructions can generate much larger sets of instructions for the hardware in a more or less automatic way. In sequential computing, loops play this role. In parallel computing, we need to include abstract machine instructions that can generate activity in many processors at the same time. A parallel computer can consume trillions of instructions in a short period of time, and also move data between processors; we clearly need abstractions that can insulate software developers from expressing this amount of detail directly.

Finding such abstractions has proved to be enormously difficult, because the following properties are essential [35]:

- Portability – parallel computers, even of the same general kind, differ in their internal arrangements; it is important to be able to move programs from one to another without having to redesign them.

- Efficiency – parallel programs exist for performance, so they must exploit a significant amount of the performance available from the platforms on which they execute.

- Simplicity – the speed with which software can be constructed seems relatively invariant in terms of semantic units; the larger these can be, the faster it is to build effective code.

- Predictable performance – there are many ways to implement any given specification and it is not possible to try all of them; therefore software developers must be able to judge the (true) complexity of their code as it is being developed.

The third requirement for effective parallel computing is the existence of a process for moving from specification to program. We have managed without such a process in sequential computing, at considerable cost. In parallel computing, the wide range of possible platforms means that testing cannot more than skim the surface of a program's behaviour. Hence, it is much more important that programs be correct by construction.

So far, three major advances towards a complete parallel computing environment have taken place. The first was the development, in the 1980s, of higher-order functional programming implemented using graph reduction. This was the first parallel programming model that provided a significant abstraction from the underlying hardware, and which also provided a way to build software (equational transformation) [27, 22, 21]. This model met the requirements for portability and simplicity. Unfortunately, it could not achieve either efficiency or predictability, although progress towards the former continues to be made.

The second advance was the development of a model that was both portable and efficient across a range of architectures, the Bird-Meertens theory of lists [33]. This model met the requirements of portability, simplicity, efficiency, and predictability, and also possessed a software construction methodology, again based on equational transformation and initiality [35]. However, it was limited because it assumed that interconnection topology was important, and also because it applied only to problems whose data could be represented using some particular, though fairly general, structures.

The third advance was the development of Bulk Synchronous Parallelism (BSP), [40, 38, 14] which also meets all four requirements, but does not have the same restrictions on the patterns of data or communication. It can reasonably claim to be the first model that is expressive, while still achieving portability, efficiency, and predictable performance. BSP is by no means the final answer to the problem of parallel programming, but it is a significant step forward. Its major weaknesses are: it is relatively low-level in the sense that programs have to explicitly partition the data and the computations; and it is cannot naturally expressed dynamic and irregular computations. The open problem in parallel programming is how to take the next step to a model that retains the good properties of BSP while simplifying further the programming framework.

2.   Introduction to BSP

BSP is a programming model whose abstract machine consists of a set of processors and their local memories, connected by a network. In other words, it is a distributed-memory abstraction. This means that almost any real parallel computer can readily simulate the BSP abstract machine with little loss of performance. It means that some features of shared-memory computers cannot be directly exploited [38].

The essence of the BSP model is the notion of a *superstep* which can be considered the basic parallel instruction of the abstract machine. A superstep

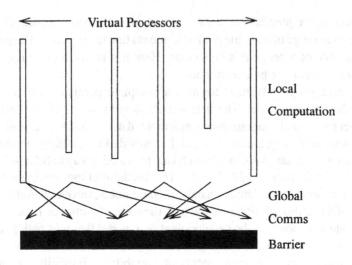

Figure 4.1.   A superstep.

is a global operation of the entire parallel computer, and is decomposed into three phases:

Local computation within each processor on local data.

Non-blocking communication between processors.

A barrier synchronisation of all processors, after which all communication is guaranteed to have completed, and the moved data becomes visible to the destination processors.

Communication and synchronisation are completely decoupled. This avoids one of the big problems with e.g. message passing, where each communication action overloads both data movement and synchronisation. A single process is easy to understand; but the global state that results from the parallel composition of more than one process is extremely hard to understand. When a *send* is written in a process, the <u>entire</u> global state must be understood in order to know whether the operation is safe or will cause deadlock. Thus the send-receive abstraction of message passing is easy to understand in the abstract but hard to use in practice.

Barrier synchronisation collapses the global state at the end of every superstep. Every process that exits the barrier knows that every other process has reached it. This control of the state makes software easier to write and understand. It can also be exploited by the runtime system to make optimisations that are not possible in other systems.

The first question that is always asked about supersteps is this: isn't it ridiculously inefficient to wait until the end of the communication phase to start moving data? Overlapping computation and communication seems like an obvious optimisation. In fact, however, the most that can be saved by a perfect overlap is a factor of two, and it is often much less in practice. Postponing communication allows many optimisations that produce much larger improvements in performance.

The second question that is always asked about supersteps is: aren't barrier synchronisations ridiculously expensive? The answer is that they are if they are done poorly, as they are in many implementations, even by manufacturer's who ought to know better. BSP works hard to make barriers extremely fast. (In fact, barriers are one of the easiest things to arrange physically – the Papers project [9] at Purdue implements barriers in 10s of nanoseconds. There has not been enough demand from users for manufacturers to include barrier hardware in their systems.)

3. Implementation of BSP

Although BSP is an MIMD programming model, its most popular implementation is an SPMD one – the BSPLib library, sometimes called the Oxford BSP Toolset [19]. This library was designed by an international group and implemented at Oxford. It is freely available and runs on a wide variety of platforms. See www.bsp-worldwide.org for more details.

Two styles of communication are available in the library:

- **Direct Remote Memory Access (DRMA)** in which processes can copy data from or to a remote process's memory, without the active participation of the remote process.

- **Bulk Synchronous Message Passing (BSMP)** in which processes send a packet to a queue in a remote process, which can be accessed by the remote process in the following superstep.

The BSPLib library is small compared to message passing libraries. It contains 20 operations, compared to 129 in MPI-1, 202 in MPI-2, and 90 in PVM. The BSPLib operations are shown in Figure 4.1

The standard way of communicating between processors is to use the DRMA operations of the library. These allow data copies to be initiated by one processor to (put) or from (get) the memory of another processor without interrupting the remote processor.

To simplify the writing of computational code, puts and gets may appear textually anywhere within a superstep. In normal use, data are copied from program data structures at the occurrence of a put so that data structures may be reused during the superstep (transfer of the data between processors does not take place until the end of the superstep). High-performance variants exist which do not buffer data, and which may begin communication before the end

Table 4.1.   Basic BSPLib primitives

Class	Operation	Meaning
Initialisation	bsp_begin	Start of SPMD code
	bsp_end	End of SPMD code
	bsp_init	Simulate dynamic processes
Halt	bsp_abort	One process halts all
Enquiry	bsp_pid	Find my process id
	bsp_nprocs	Number of processes
	bsp_time	Local time
Superstep	bsp_sync	Barrier synchronisation
DRMA	bsp_push_reg	Make region globally visible
	bsp_pop_reg	Remove global visibility
	bsp_put	Copy to remote memory
	bsp_get	Copy from remote memory
BSMP	bsp_set_tag_size	Choose tag size
	bsp_send	Send to remote queue
	bsp_qsize	Number of messages in queue
	bsp_get_tag	Match tag with message
	bsp_move	Move from queue
High Performance	bsp_hpput	Unbuffered versions
	bsp_hpget	of communication
	bsp_hpmove	primitives

of the superstep. These are harder to use, since program data structures are off-limits to the program for the remainder of the superstep, once they have been mentioned in a data transfer operation. A simple BSP program is shown in Figure 4.2.

An alternative set of data transfer operations, bulk synchronous message passing, exist for the situation where data distribution is irregular and only the data recipient knows where incoming data may be safely stored.

The message-passing style typical of MPI and PVM has the following drawbacks:

- It is possible to write programs that deadlock, and it is difficult to be sure that a given program will not deadlock;

- The ability to use a send or receive anywhere leads to spaghetti communication logic – message passing is the communication analogue of goto control flow;

- There is no global consistent state which makes debugging and reasoning about programs extremely difficult;

- There is no simple cost model for performance prediction, and those that do exist are too complex to use on programs of even moderate size;

Example: "Hello World"

```
void main(void) {
  int i;
  bsp_begin(bsp_nprocs());
    for(i=0; i<bsp_nprocs(); i++) {
      if (bsp_pid()==i)
        printf("Hello from process %d of %d\n",
               i,bsp_nprocs());
      fflush(stdout);
      bsp_sync();
    }
  bsp_end();
}
```

Figure 4.2.　The simplest BSP program.

- On modern architectures with powerful global communication, they cannot use the full power of the architecture.

BSP avoids all of these drawbacks. Provided BSP's apparent inefficiencies can be avoided, it is clearly a more attractive programming model, at least for algorithms that are fairly regular.

It is possible to program in a BSP style using libraries such as MPI which provide an explicit barrier operation. However, many global optimisations cannot be applied in this setting, and the library operations themselves are not optimised appropriately. For example, the MPI barrier synchronisation on the SGI Power Challenge is 32 times slower than BSPlib's barrier. Replacing MPI calls in a superstep structured program with BSPLib calls typically produces a 5-20% speedup.

BSPLib is available for the following machines: Silicon Graphics Power Challenge (Irix 6.x), Origin 2000; IBM SP; Cray T3E, T3D, C90; Parsytec Explorer; Convex SPP; Digital 8400 and Digital Alpha farm; Hitachi SR2001; plus generic versions on top of either UDP/IP, TCP/IP, system V shared memory, MPI, and certain clusters.

4.　BSP algorithm design

BSP has a simple but accurate cost model that can be used for performance prediction at a high level, before any code is written. Hence, it provides the basis for parallel algorithm design, since the costs of different potential implementations can be compared without building them all.

There are three components to the cost of a BSP superstep: the computation that occurs locally at each processor, the communication that takes place between them, and the cost of the barrier synchronisation.

Costing the local computation is uses the conventional approach of counting instructions executed. This approach, however, is becoming increasingly fragile, not only for BSP but for other parallel and, for that matter, sequential models. The reason is that modern processors are not simple linear consumers of instructions with predictable execution times. Instead, the time a given instruction takes depends strongly on its context: the instructions that execute around it and their use of functional units, and whether or not its arguments are present in the cache. We have been exploring better models for instruction execution, but this is an issue bigger than parallel programming.

Costing communication is an area in which BSP has made a major contribution to our understanding of parallel computation. Most other cost models for communication are bottom up in the sense that they try to compute the cost of a single message, and then sum these costs to get an overall communication cost. Typically, the cost of a single message is modelled as a startup term followed by a per-byte transmission time term. The problem with this approach is that the dominant term in the per-message cost often comes from congestion, which is highly nonlinear and prevents the costs of individual messages being summed in a meaningful way.

Such communication cost models implicitly assume that the bottleneck in a network is somewhere in the centre, where messages collide. This may be a fair assumption for a lightly loaded network, but it does not hold when a network is busy. In typical parallel computer networks, the bottlenecks are at the edges of the network, getting traffic into and out of the communication substrate itself [13, 16]. This continues to be true even for clusters where many of the protocol and copy overheads of older networks (e.g. TCP/IP) have been avoided [11].

If it is the network boundary that is the bottleneck, then the cost of communication will not be dominated by the distance between sender and receiver (i.e. the transit time), nor by congestion in transit, nor by the communication topology of the network. Instead, the communication that will take the longest is the one whose processor has the most other traffic coming from or to it. This insight is at the heart of the BSP cost model, and has led to its adoption as a *de facto* cost model for other parallel programming models too.

BSP has a further advantage because of its superstep structure – it is possible to tell which communication actions will affect each other: those from the same superstep. The communications actions of a superstep consist of a set of transfers between processors. If the maximum volume of data entering and leaving any processor is h then this communication pattern is called an h-relation. The cost of the communication part of the superstep is given by hg, where g captures the *permeability* of the network to high-load traffic. It

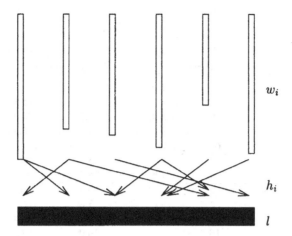

Figure 4.3. Costing a superstep.

is expressed in units of instruction times per byte, so that hg is in units of instruction times. Hence a small value for g is good.

The cost of the barrier synchronisation is similarly expressed in units of instruction times.

The overall cost of a superstep is given by:

$$superstep\ cost = \underbrace{\text{MAX } w_i}_{\text{Computation}} + \underbrace{\text{MAX } h_i \times g + l}_{\text{Communication}}$$

In the BSP cost model, a communication pattern in which one processor sends a value to each of the others, and one in which all processors send a value to all other processors are treated as if they cost the same – they are both p-relations. However, the total amount of data being moved is p in the first case and p^2 in the second. Can these two patterns really have the same cost? In practice, they don't quite; but the time taken versus total amount of traffic is linear and quite flat, not the quadratic that might have been the more intuitive prediction. The BSP cost model doesn't quite hold, but it's much more accurate than other cost models.

Because each superstep ends with a global synchronisation event, the cost of a series of supersteps is just the sum of the costs of each separate superstep.

Some representative performance figures for some common parallel computers are shown in Figures 4.4 and 4.5.

4.1. A BSP design strategy

The BSP cost model makes it clear that the following strategies should be used to write efficient BSP programs:

Machine	s	p	l	l/s	g	g/s
			flops	μs	flops/wd	μs/wd
O2000	101	2	804	7.9	8.3	0.08
		4	1789	17.9	10.2	0.10
		8	3914	39.8	15.1	0.15
		16	15961	158.5	44.9	0.45
		32	39057	386.6	66.7	0.66
		62	138836	1374.2	121.2	1.20
PowerChallenge	74	2	1132	15.3	10.2	0.14
		4	1902	25.7	9.3	0.13
Cray T3E	47	2	269	5.7	2.14	0.05
		4	357	7.6	1.77	0.04
		8	506	10.8	1.64	0.03
		16	751	16.0	1.66	0.04
SP2 switch	26	2	1903	73.2	7.8	0.30
		4	3583	137.8	8.0	0.31
		8	5412	208.2	11.4	0.43
Cray T3D	12	2		5.6	0.3	0.02
		8	175	14.4	0.8	0.07
		64	148	12.3	1.7	0.14
		256	387	32.1	2.4	0.20

Figure 4.4.   Performance for manufactured systems.

Machine	s	p	l	l/s	g	g/s
			flops	μs	flops/wd	μs/wd
SP2	26	2	1903	73.2	7.8	0.30
		4	3583	137.8	8.0	0.31
		8	5412	208.2	11.4	0.43
Alpha farm	10	2	17202	1703.1	81.1	8.0
		3	34356	3401.6	83.0	8.2
		4	47109	4664.3	81.3	8.1
Pentium NOW	61	2	52745	870.4	484.5	8.00
266MHz, 64Mb		4	139981	2309.9	1128.5	18.62
10Mbit shrd enet		8	826054	13631.3	2436.3	32.91
Pentium II NOW	88	2	13683	155.5	37.8	0.43
400MHz, 128Mb		4	27583	313.4	39.6	0.45
100Mbit switch enet		8	38788	440.8	38.7	0.44
Pentium II Krnl-NOW	88	2	5654	64.2	33.5	0.38
400MHz, 128Mb		4	11759	133.6	31.5	0.36
100Mbit switch enet		8	18347	208.5	30.9	0.35

Figure 4.5.   Performance for clusters.

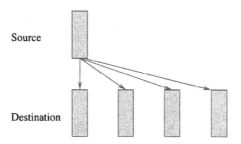

Figure 4.6. Naive broadcast.

- balance the computation in each superstep between processes, since the computation time for a superstep is a *maximum* over the computation times in each process, and the barrier synchronisation must wait for the slowest process;

- balance the communication between processes, since h is a *maximum* over sent and received data; and

- minimise the number of supersteps, since this determines the number of times l appears in the final cost.

As an example of the effectiveness of balance, we consider a basic building block: broadcasting the same value from one processor to all of the others. Suppose that the message to be sent is of size N and that there are p processors.

The simplest implementation (Figure 4.6) is just to send the message out p times (so that the communication is a tree of depth 1 and width p).

The cost of this implementation is:

$$Naive\ broadcast\ cost = pNg + l \qquad (4.1)$$

A second possible implementation (Figure 4.7) is to have the source of the broadcast send it to two other processors. These in turn send it on to two others, and so on (so the communication forms a tree of depth $\log p$).

The cost of this implementation is:

$$Tree\ broadcast\ cost = (\log p)Ng + (\log p)l \qquad (4.2)$$

The communication cost has been reduced, but at the expense of more barrier synchronisations.

There is a third way to implement broadcast (Figure 4.8). In the first step, the source processor sends $1/p$th of the data to each of the other processors. In the second step, all of the processors send their piece of the data to all of the other processors [2].

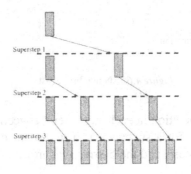

Figure 4.7. Tree broadcast.

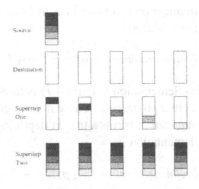

Figure 4.8. Two stage broadcast.

Table 4.2. Costs for broadcasting

Algorithm	Communication	Sync
One-stage	pNg	l
Tree	$(\log p)Ng$	$(\log p)l$
Two-stage	$2Ng$	$2l$

The cost of this implementation is:

$$Two\ stage\ broadcast\ cost\ =\ \left(\frac{N}{p}pg + l\right) + \left(\frac{N}{p}pg + l\right) \quad (4.3)$$

$$=\ 2Ng + 2l \quad (4.4)$$

It is easy to see that this implementation is much better than either of the two previous ones. The communication term is smaller than direct broadcast, while the number of supersteps is a small constant. This technique is an extremely good way to handle broadcasts, but it seems not to have been known before BSP, and it is still not used as often as it should be because conventional cost metrics make it look too expensive. It requires the insight to see that the total exchange in the second step is not ridiculously expensive. Table 4.2 illustrates the different costs of these three implementations.

The only time when it is worth using the one-stage broadcast is when N is small, i.e. $N < l/(pg - 2g)$, and it is never worth using the tree broadcast.

Does the cost model reflect reality for this technique? Figure 4.9 shows the measured cost of the three different implementations on a powerful parallel computer (the Cray T3E).

5. BSP optimisation

BSPLib works quite hard to optimise performance, but it's an unusual kind of optimisation. Instead of optimising programs, it optimises the BSP abstract machine (Figure 4.10). To put it another way, all of the optimisations result in improved values for g and l, rather than improvements to particular parts of particular programs [10].

Using *global* knowledge of the communication patterns in BSPLib programs, improves the *global* communication bandwidth g. Some of the standard techniques used by BSPLib are shown in Table 4.3. They are discussed in the next few subsections.

5.1. Message packing

On most computers, there is a significant startup cost to initiating a transfer of data. Even if the network itself introduces only minor overheads, it's not

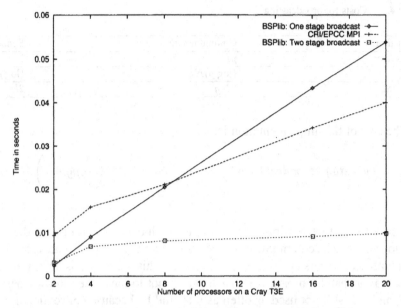

Figure 4.9.   Broadcasting 100,000 doubles on a $p = 20$ Cray T3E.

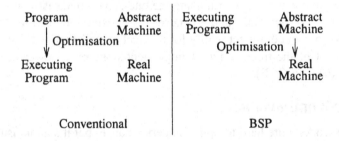

Figure 4.10.   Optimising the model rather than programs.

Table 4.3.   BSPLib optimisations

Optimisation	Kind	Improvement
Message packing	local	small packet bandwidth
Destination scheduling	global	congestion
Pacing	pairwise	congestion
Barrier implementation	global	l

unusual to require a system call with its overheads of saving and restoring state. The BSP cost model claims that sending 1000 1-byte messages from a processor should cost the same as sending a single 1000-byte message, and that seems unlikely.

However, BSP has one advantage over conventional message passing systems – it is not required to begin transmitting data at the point where the program indicates that it should. Instead, a BSP implementation can wait until the end of each superstep before initiating data transfers.

This makes it possible to pack all of the traffic between each pair of processors into a single message so that the overhead is paid only once per processor pair rather than once per message. In other words, no matter how the programmer expresses the messaging requirements, BSP will treat them the same: 1000 1-byte messages become a single 1000-byte message before leaving the sending processor. The communication interface sees a single transfer regardless of the actual sequence of puts in the program text.

5.2. Destination scheduling

Each processor knows that, when it is about to start transmitting data, other processors probably are too.

If all processors choose their destinations in the same way, say Processor 0 first, then Processor 1, and so on, then Processor 0 is going to be bombarded with messages. It will only be able to upload one of them, and the rest will block back through the network. This situation is particularly likely in SPMD programs, where puts to the same destination will tend to be executed simultaneously because each processor is executing the same code.

Because BSPLib does not transmit messages as they are requested, it is able to rearrange the order in which they are actually despatched; in particular each processor may use a different despatch order even though the same order was implied by the puts executed by each processor. Choosing a schedule that makes collisions at the receivers unlikely has a large effect on performance. It can be done by choosing destinations randomly, or by using a latin square as a delivery schedule. Table 4.4 shows this effect on a variety of computers. Notice that in almost every case, the factor of two lost by not overlapping computation and communication is earned back; in the more complex case of the NIC cluster, where the huge increase in time happens because the central switch saturates, using a careful schedule prevents a performance disaster.

5.3. Pacing

Each processor knows that, when it is about to start transmitting data, other processors probably are too.

Table 4.4. 8-processor total exchange in μs. small=16k words per processor pair, big=32k words.

		latin square	dest order	factor
SP2	big	102547	107073	1.0
T3D	big	14658	29017	2.0
PowerChallenge	big	37820	47910	1.3
Cluster(TCP)	small	61336	119055	1.9
	big	134839	248724	1.8
Cluster(NIC)	small	39468	76597	1.9
	big	78683	14043686	178

Networks all contain bottlenecks – as the applied load increases, the throughput eventually drops. It is always useful to control (throttle) input rates to maintain the applied load at (just below) the region of maximum throughput. If the applied load is maintained slightly below the point of maximum throughput, then fluctuations in load create slightly larger throughputs which tend to restore equilibrium. On the other hand, if the applied load goes above the point of maximum throughput, then positive feedback tends to drive the network into congestion and collapse. The ability to control applied load makes it possible to keep the network running at its maximum effective capacity.

This can be done in two ways:

High-level. The target's g value gives its permeability to data under conditions of continuous traffic – and this can easily be converted into units of secs/Mbit, say. There is no point in trying to insert data into the network faster than this value, because g is the <u>best</u> estimate of what the network's ideal throughput value is.

Low-level. When the communication medium is shared (e.g. Ethernet), using a slotting system with a carefully-chosen slot size reduces collisions (a distributed approximation to the latin square scheme). For TCP/IP over Ethernet, this improves performance by a factor of 1.5 over mpich.

5.4.   Barrier implementation

Most standard barrier implementations are quite poor. Because barriers are so important in BSP, BSPLib implements them *very* carefully [17].

On shared-memory architectures, a technique that exploits cache coherency hardware is used to make barriers very fast. Table 4.5 shows some performance data for different barrier implementations on shared-memory systems.

On distributed-memory manufactured architectures, access to the communication substrate is usually possible only through a manufacturer's API that is already far removed from the hardware. This makes it difficult to build fast

Table 4.5. Shared-memory barriers (times in μs, 4 processors.)

	SGI	SUN	DEC
central semaphore	320.8	129.3	97.9
manufacturer	33.5	—	—
tree of semaphores	335.6	167.6	89.7
cache coherence	6.6	2.2	3.8

Table 4.6. Distributed-memory barriers (times in μs, 4 processors).

| | Powerchallenge | | SP2 | |
	SGI MPI	Arg MPI	IBM mpl	BSPLib
provided	73.5	217.7	197.9	—
message tree	74.1	165.9	95.6	—
central message	92.0	234.0	156.2	—
total exchange	105.7	224.9	124.3	137.8

interfaces. Some performance data are shown in Table 4.6. These times are close to the sum of the times required to send individual messages to $p - 1$ destinations.

Total exchange is an attractive way to implement barriers in the context of global exchange of data, although once again this is counterintuitive. Rather than exchanging single bits at the barrier, BSPLib exchanges the sizes of messages about to be communicated between each pair of processors. This costs almost nothing, but makes new optimisations possible because the receiver knows how much data to expect. Distributed-memory implementations of BSPLib execute the barrier immediately after the local computation phase. Since each processor knows how much data to expect, it is simple to preserve the semantics of the program barrier.

The opportunities for optimisation are constrained because BSPLib is a library and does not have access to e.g. the kernel, where further optimisation opportunities exist.

6.   Optimisation in clusters

It is now possible to build a distributed-memory parallel computer from off-the-shelf components (a cluster). Such systems are characterised by: high performance processor boards, a simple interconnection network such as switched Ethernet or Myrinet, and Network Interface Cards (NICs) to handle the interface between processors and network. Protocols such as TCP/IP that were originally designed for long-haul networks are not appropriate for such an environment. Instead, new protocols have been designed to provide low latencies (end-to-end latencies in the 10s of microseconds are typical) and as few data copies as

Table 4.7.   Optimisations for clusters

Optimisation	Kind	Improvement
Hacking protocol stack	local	low-latency messaging
Hole filling error recovery	pairwise	# retransmissions
Resource level acking	pairwise	explicit acks
Barrier acking	global	explicit acks
Shared buffers	local	buffer avail

possible [4, 41, 28, 6]. Changes to the kernel to make it possible to access the network without a system call are also common.

When the target is a cluster, with full access to the kernel, communication hardware, and protocols, further optimisations to BSPLib are possible [12, 11]. Some of them are listed in Table 4.7.

6.1. Hole filling error recovery

TCP/IP is a reliable protocol, but it was designed for an error model that is not relevant to clusters. Clusters do lose packets, but almost invariably because of buffer problems at the receiving end. High-performance protocols need to provide reliability in a different way.

TCP error recovery keeps track of the last packet acknowledged. Once a packet is lost, everything after it is also resent. This is a poor strategy if errors are not in fact bursty.

In the BSPLib cluster protocol, packets contain an extra field with the sequence number after the last one in sequence (i.e. acknowledging both a hole, and the first packet past the end of a hole). Only the actual missing packets need to be resent.

6.2. Reducing acknowledgement traffic

Acknowledgements are piggy-backed onto normal traffic, and therefore use no extra bandwidth, *provided that traffic is symmetric*. However, often it isn't, and so the general strategy is to avoid acknowledgements for as long as possible, preferably for ever. Also, because each processor knows what it is expecting, recovery is driven by the receiver, not by the sender. Because of this, a sender need not know for sure whether a packet has been received for a very long time, unless it needs to free some send buffers.

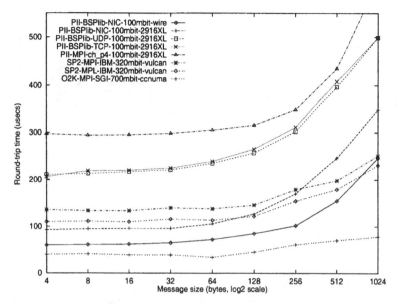

Figure 4.11. Link performance – round-trip times in μs.

6.3. Just in time acknowledgements

Acknowledgements are forced by a sender prediction that it will run out of buffers (because it must keep copies of unacked data). It uses knowledge of the roundtrip message time to send ack requests just in time.

6.4. Managing buffers more effectively

Because the architecture is dedicated, and communication patterns will often be skewed, it makes sense to use a single buffer pool at each processor (rather than separate buffer pools for each *pair* of processors).

The combination of these optimisation makes the BSPLib cluster protocol one of the fastest reliable protocols. Its performance is illustrated in Figures 4.11, 4.12, and 4.13.

These optimisations have a direct effect on application level performance. This is shown in Tables 4.8 and 4.9, using some of the NAS benchmarks.

The combined effect of these communication optimisations is that the communication performance of BSPLib is about a factor of four better than comparable libraries.

7. Software development

The third requirement for successful parallel programming is the existence of a process for developing programs from specifications. This is, of course,

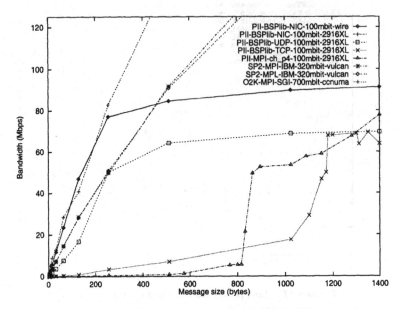

Figure 4.12.   Link performance – bandwidth in *Mbps*.

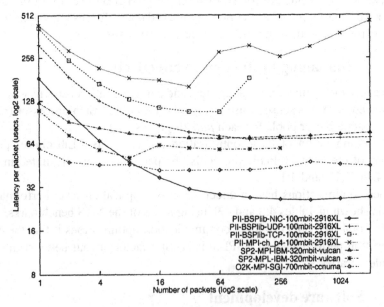

Figure 4.13.   Link performance – latency per packet in *μs*.

Table 4.8. NAS benchmark performance (Mflops/s per process).

	p	SP2 (MPI)	Cluster (MPI)	Cluster (BSPlib)	Origin 2000 (MPI)
BT	4	31.20	51.57	56.18	51.10
SP	4	24.89	33.15	42.36	56.22
MG	8	36.33	21.02	39.62	36.16
LU	8	38.18	46.34	63.09	87.34

Table 4.9. NAS benchmarks communication times

	p	computation time	communication time BSPlib	mpich	
BT	4	724.6s	24.2s	91.1s	3.8
SP	4	458.9s	42.8s	182.2s	4.3
MG	8	11.1s	1.1s	4.6s	4.2
LU	8	205.9s	30.5s	115.9s	3.8

also important for sequential computing; but parallel computing adds an extra spatial dimension to all of the complexities of sequential programming and a software construction process is therefore even more critical.

Formal methods for software development in sequential settings often have large overheads to generate a single program statement. Partly for this reason, they have made little inroad into practical software construction. A big difference in a parallel setting is that the application of a single formal construction can often build the entire parallel structure of a program. Thus the same overhead of using a formal technique results in a much larger gain in program construction.

BSP is particularly attractive from the perspective of a methodology for program construction because:

- Each BSP superstep has a well-defined interface to those that precede and follow it. BSP supersteps resemble a general form of *skeleton* whose details can be instantiated from a few parameters. The extensive skeleton literature has documented the advantages this brings to software development [8, 34, 33].

- BSP programs can be given a simple semantics in which each processor can act arbitrarily on any program variables, with the final effect of a superstep being defined by a *merge* operator which defines what happens if multiple assignments to the same variable are made by different processors. In fact, it is clear that BSP is just one of a family of models defined by different merge functions. This idea has been pursued by He and Hoare [20], and in slightly different way by Lecomber [23].

Too much should not be made of the ability to reason formally about parallel programs. Nevertheless, BSP is noticeable as the only known programming model that is both efficient and semantically clean enough to be reasoned about [36].

8.   Applications of BSP

BSP has been used extensively across the whole range of standard scientific and numeric applications. Some examples are:

- computational fluid dynamics [18, 7],

- computational electromagnetics [25],

- molecular dynamics [26],

- computational geometry [31],

- standard algorithmic building blocks (sorting, queues),

- discrete event simulation [5],

However, BSP's limits with respect to irregular programs handicap it slightly compared to MPI, and so it is far from being the most common parallel programming environment in these applications.

However, the application areas for parallel and high-performance computing are changing – away from these numerical areas to applications that involve more symbolic processing in support of Internet applications. In these newer application areas, there is less pressure on performance as the only variable of interest, and more emphasis on predictability, portability, and ease of maintenance (which is correlated with properties such as lack of deadlock and clean semantics). In these areas, BSP is attracting greater interest. To illustrate the flavour of such applications, we consider *data mining* in the next section.

8.1.   Parallel data mining

Data mining is concerned with extracting useful enterprise knowledge from data that many organisations collect more or less incidentally because of the use of computers as point-of-sale units. Retail businesses are an obvious example; but increasingly business-to-business interactions are computer-mediated and the same opportunities to gather data exist. Interactions at web sites are another opportunity to gather huge amounts of data.

Once this data exists, most organisations feel pressure to use it to improve their interactions with customers (broadly understood). This requires building models to explain the variation that is observed in interactions – the fundamental question is almost always some variant of "why did this potential customer buy, while this one did not". The algorithms required to build such models are often conceptually rather simple, but they face two challenges:

- The datasets involved are huge, typically terabytes, so that the time required simply to move the data around is substantial; and

- Even with these enormous datasets, many algorithms are compute bound, so that many processor cycles are required.

In a sense, almost all of the work goes into controlling the size of the constants, rather than attacking the asymptotic complexity of algorithms. Hence, the ability to get a performance improvement of a factor of p is interesting even if p is in the range 4–30. Data mining is a 'killer application' for parallel computing, especially as those who want the results of parallel programs derive a direct commercial advantage from them, and are thus willing to pay the costs involved [43]. Many data mining algorithms can partition the data evenly, and are thus a good fit with BSP-structured programs [32].

Data mining techniques can be roughly divided into two kinds. The first kind is given data about 'customers' for whom the outcome is known and computes *predictors* that can be used to predict the outcome for future, new customers. The second kind produce *clusters* showing how 'customers' are related to each other. Learning these similarities can be used, for example, to predict something about a new 'customer' based on those customers who are 'similar'.

The most obvious way to parallelize a data mining technique is to parallelize the algorithm itself. It turns out that this is usually not the best approach. Most data mining algorithms have the property that, given some of the available data, they can compute an approximation of their usual output. In other words, if the algorithm is given a *sample* of the data, it will compute an approximation to the predictor, or a partial set of clusters. As the size of the sample grows, the approximation gets more and more similar to the result on the entire dataset.

If there is a way to *combine* these approximations to produce a good global solution, then a much better generic parallelization strategy is: partition the data across processors, execute the approximating versions of the algorithm sequentially, and then use the combining to produce a global answer. This strategy is remarkably effective, partly because the amount of communication it requires is small [37].

8.2. Bagging

There is a generic way in which the results of different algorithms can be combined: if the output is categorical, combine the results by voting, if the output is numeric, combine the results by regression (i.e. averaging). Thus there is always a way to apply the idea described above, an approach that is known as *bagging* [3]. (The deeper question is whether the results of the different algorithms can be combined in a more integrated way, since the cost of voting grows linearly with the number of voters. Bagging is therefore always useful as a *training* technique, but its results may not be attractive as a deployed system.)

Standard bagging algorithms select uniformly random subsets of the training dataset ("bags") and then execute a copy of the base data mining algorithm on each one. Clearly, since these copies are executed independently on their own data, bagging can be trivially parallelized.

Bagging illustrates an important feature of data mining algorithms. It is often possible to achieve much better results by combining the outputs trained on several small bags than to use the output trained on the entire dataset. It's as if the most cost-effective training occurs from the first few examples seen. Hence a bagging algorithm often produces better results from much less input data, and does so in parallel as well.

It is easy to see how to use BSP to implement bagging. It requires only a single BSP superstep. If the bags are chosen to have the same size, then the computations at each processor usually takes similar amounts of time. If the predictors are left at the processors that computed them, then evaluation can also be done in parallel by broadcasting each new data point to the processors and then combining their votes using a gather.

8.3.   Exchanging information earlier

Bagging provides a generic parallelization technique. However, the observation that much of the gain occurs after having seen only a few data points suggest that performance might be improved still further by exchanging information among processors before the end of the algorithm. That way, each processor can learn from its own data, and simultaneously from the data that other processors are using. This turns out to be very productive.

The basic structure of the parallel algorithm is:

- Either partition the data among the processors, or give each processor a bag;

- Execute the approximating algorithm on each processor for a while;

- Exchange the results obtained by each processor with all of the others (a total exchange);

- Update the model being built by each processor to incorporate the information received from other processors;

- Repeat as appropriate.

Of course, this requires the possibility of being able to combine information learned by other processors into the model under construction, so this approach will not work for all data mining algorithms. However, it is effective for several, including neural networks [30, 29], and inductive logic programming [42].

The information that is exchanged is quite small compared to the size of typical datasets, so the communication overhead of this approach is low. Interestingly, superlinear speedup occurs because there are two improvements over executing the plain sequential algorithm. The first is that p processors are being used so we expect roughly a factor of p speedup, less the costs of communication. The second is that each processor finishes its own model faster by integrating the information learned from the other processors, so the overall execution time is even better than expected.

Since this is a true superlinear speedup, not simply some artifact of cache behaviour or whatever, it implies that the sequential algorithm can be improved, and that is indeed the case. Whenever such a superlinear speedup parallelization exists, the sequential algorithm should not attempt to build a model monolithically, but should instead use the same approach of building partial models on small subsets of the data and combining them.

Both of these parallelization strategies are easily implemented in BSP; and BSP's attention to communication patterns such as total exchange means that they execute efficiently.

9.   Summary

BSP has made a number of contributions to both the design of parallel programming models, and implementation strategies. We summarise by listing some of these contributions here:

- BSP has shown that many intuitions about networks and their performance are not valid. In particular, BSP has shown that, even for NIC-based networks with low-overhead protocols, crossing the boundary into and out of a network is the performance bottleneck. Congestion inside the network is negligible by comparison.

  This implies that it takes as long to move data between adjacent processors in the network topology as it does to move data across the network diameter. As a result locality is of limited importance (although co-locality is still critical). This explains the 'death of topology' – almost all modern distributed-memory computers use a hierarchy of crossbar switches (in contrast to the thousands of papers on network topologies that have appeared over the past three decades).

  This property of networks also invalidates the intuition that a dense communication pattern, such as a total exchange, ought to be much more expensive than a broadcast by a single processor; and that a total exchange ought to be quadratically more expensive than a shift. While there is an decrease in network performance for denser communication patterns, it is small.

- BSP has shown that it isn't always best to do things as early as possible. In particular, the ability to wait to carry out data transfers long past the point in the program where they are requested buys considerable opportunities for arranging the pattern and timing of those data transfers to use the network to best effect. We have seen that this can produce a performance improvement of a factor of four over systems such as MPI. Note that this depends on BSP semantics; other messaging systems may create deadlock by postponing communication, and are certain to do so if they reorder it.

- More generally, BSP shows that structured parallel programming is not only a performance win, but it's also a program construction win. Much of the discussion about structured versus unstructured parallel programming resembles the discussion of structured versus unstructured programming in the Sixties. At first glance, it seems as if structure carries a performance cost; the truth is actually the opposite – structure improves performance overall.

- BSP's structure also allows a powerful set of optimisations that rely on global knowledge to be applied. In essence, each processor infers a lot about the global state from its own local state. We have seen how this can be exploited to: arrange the order of communication, control the rate at which data is inserted into the network, and avoid the need for some acknowledgements.

  Furthermore, the BSP approach means that optimisation is a property of the implementation of the model rather than of individual programs. Software developers do not get performance surprises because the compiler has done something they weren't expecting. Instead, they can tell what to expect by examining the values of g and l for their implementation system.

- BSP takes a global approach to performance modelling, measuring macroscopic properties of architectures rather than trying to build them up from measurements of microscopic properties. As a result, BSP's architectural parameters are both stable and simple.

  BSP also uses a set of program parameters that are easily accessible to software developers. As a result, the BSP cost model is tractable in the sense that it can be used to guide software construction well before code is actually written, but also reasonably accurate. In fact, the biggest weakness in the BSP cost model is not related to parallelism at all, but to the difficulty of modelling instruction execution within a single processor.

BSP's cost model makes parallel software design possible, and provides predictable performance without benchmarking.

- The benefits of the BSP model are not theoretical, nor do they apply only at low levels. The benefits of the model play through to the application level, allowing high-performance real applications to be built and maintained.

- The BSP model shows that there are significant benefits, both in simplicity and performance, in using DRMA for data transfer rather than message passing. The biggest weakness of message passing is that it takes two to play – forcing programmers to do a complex matching when they write programs, and processors to synchronise when they wish to transfer data.

Finding effective ways to program parallel computers is difficult because the requirements are mutually exclusive: simplicity and abstraction, but also performance (and preferably predictable performance). Very few models are known that score reasonably well in both dimensions. BSP is arguably the best positioned of known models. Yet it is clear that BSP is still too low level and restricted to become *the* model for parallel programming. However, it is an important step on the way to such a model.

Acknowledgments

A large number of people have been involved in the design and implementation of BSP. In particular, the implementation of BSPLib and much of the performance analysis was done by Jonathan M.D. Hill and Stephen Donaldson.

References

[1] D.J. Becker, T. Sterling, D. Savarese, J. E.Dorbandi, U.A. Ranawak, and C.V. Packer. Beowulf: A parallel workstation for scientific computation. In *Proceedings of the International Conference on Parallel Processing (ICPP)*, pages 11–14, 1995.

[2] Rob H. Bisseling. Basic techniques for numerical linear algebra on bulk synchronous parallel computers. In Lubin Vulkov, Jerzy Waśniewski, and Plamen Yalamov, editors, *Workshop Numerical Analysis and its Applications 1996*, volume 1196 of *Lecture Notes in Computer Science*, pages 46–57. Springer-Verlag, Berlin, 1997.

[3] L. Breiman. Bagging predictors. *Machine Learning*, 24:123–140, 1996.

[4] Philip Buonadonna, Andrew Geweke, and David E. Culler. Implementation and analysis of the Virtual Interface Architecture. In *SuperComputing'98*, 1998.

[5] Radu Calinescu. Conservative discrete-event simulations on bulk synchronous parallel architectures. Technical Report TR-16-95, Oxford University Computing Laboratory, 1995.

[6] Giuseppe Ciacco. Optimal communication performance on Fast Ethernet with GAMMA. In *Parallel and Distributed Processing*, volume 1388 of *Lecture Notes in Computer Science*, pages 534–548. Springer, 1998.

[7] P.I. Crumpton and M.B. Giles. Multigrid aircraft computations using the OPlus parallel library. In *Parallel Computational Fluid Dynamics: Implementation and Results using Parallel Computers. Proceedings Parallel CFD '95*, pages 339–346, Pasadena, CA, USA, June 1995. Elsevier/North-Holland.

[8] M. Danelutto, F. Pasqualetti, and S. Pelagatti. Skeletons for data parallelism in p3l. In C. Lengauer, M. Griebl, and S. Gorlatch, editors, *Proc. of EURO-PAR '97, Passau, Germany*, volume 1300 of *LNCS*, pages 619–628. Springer-Verlag, August 1997.

[9] H.G. Dietz, T. Muhammad, J.B. Sponaugle, and T. Mattox. PAPERS: Purdue's adapter for parallel execution and rapid synchronization. Technical Report TR-EE-94-11, Purdue School of Electrical Engineering, March 1994.

[10] S.R. Donaldson, J.M.D. Hill, and D.B. Skillicorn. Exploiting global structure for performance in clusters. In *Proceedings of IPPS/SPDP'99*, pages 176–182. IEEE Computer Society Press, 1999.

[11] S.R. Donaldson, J.M.D. Hill, and D.B. Skillicorn. Performance results for a reliable low-latency cluster communication protocol. In *PCNOW'99, Workshop at IPPS/SPDP'99*, number 1586 in Lecture Notes in Computer Science, pages 1097–1114. Springer-Verlag, April 1999.

[12] S.R. Donaldson, J.M.D. Hill, and D.B. Skillicorn. BSP clusters: High-performance, reliable, and very low cost. *Parallel Computing*, 26(2–3):199–242, February 2000.

[13] Stephen R. Donaldson, Jonathan M.D. Hill, and David B. Skillicorn. Predictable communication on unpredictable networks: Implementing BSP over TCP/IP. In *Europar'98*, number 1470 in Springer Lecture Notes in Computer Science, pages 970–980, September 1998.

[14] M. Goudreau, K. Lang, S. Rao, T. Suel, and T. Tsantilas. Towards efficiency and portability: Programming the BSP model. In *Proceedings of the 8th Annual Symposium on Parallel Algorithms and Architectures*, pages 1–12, June 1996.

[15] W. W. Gropp and E. L. Lusk. A taxonomy of programming models for symmetric multiprocessors and SMP clusters. In *Programming Models for Massively Parallel Computers*, pages 2–7, October 1995.

[16] J.M.D. Hill and D.B. Skillicorn. Lessons learned from implementing BSP. *Future Generation Computer Systems*, 13(4–5):327–335, April 1998.

[17] J.M.D. Hill and D.B. Skillicorn. Practical barrier synchronisation. In *6th Euromicro Workshop on Parallel and Distributed Processing (PDP'98)*, pages 438–444, Barcelona, Spain, January 1998. IEEE Computer Society Press.

[18] Jonathan M D Hill, Paul I Crumpton, and David A Burgess. The theory, practice, and a tool for BSP performance prediction applied to a CFD application. Technical Report TR-4-96, Programming Research Group, Oxford University Computing Laboratory, Wolfson Building, Parks Road, Oxford, England. OX1 3QD, February 1996.

[19] Jonathan M. D. Hill, Bill McColl, Dan C. Stefanescu, Mark W. Goudreau, Kevin Lang, Satish B. Rao, Torsten Suel, Thanasis Tsantilas, and Rob H. Bisseling. BSPlib: The BSP programming library. *Parallel Computing*, 24(14):1947–1980, December 1998.

[20] C.A.R. Hoare and J. He. *Unified Theories of Programming*. Prentice-Hall International, 1998.

[21] P. Hudak. The conception, evaluation, and application of functional programming. *ACM Computing Surveys*, 21(3):359–411, 1989.

[22] P. Kelly. *Functional Programming for Loosely-Coupled Multiprocessors*. Pitman, 1989.

[23] D. Lecomber. *Methods of BSP Programming*. PhD thesis, Oxford University Computing Laboratory, 1998.

[24] W. F. McColl. General purpose parallel computing. In A. M. Gibbons and P. Spirakis, editors, *Lectures on Parallel Computation*, Cambridge International Series on Parallel Computation, pages 337–391. Cambridge University Press, 1993.

[25] P.B. Monk, A.K. Parrott, and P.J. Wesson. A parallel finite element method for electromagnetic scattering. *COMPEL*, 13, Supp.A:237–242, 1994.

[26] M. Nibhanupudi, C. Norton, and B. Szymanski. Plasma simulation on networks of workstations using the bulk synchronous parallel model. In *Proceedings of the International Conference on Parallel and Distributed Processing Techniques and Applications*, Athens, GA, November 1995.

[27] S.L. Peyton-Jones and David Lester. *Implementing Functional Programming Languages*. Prentice-Hall International Series in Computer Science, 1992.

[28] Loic Prylli and Bernard Tourancheau. A new protocol designed for high performance networking on Myrinet. In *Parallel and Distributed Processing*, volume 1388 of *Lecture Notes in Computer Science*, pages 472–485. Springer, 1998.

[29] R.O. Rogers and D.B. Skillicorn. Using the BSP cost model for optimal parallel neural network training. *Future Generation Computer Systems*, 14:409–424, 1998.

[30] R.O. Rogers and D.B. Skillicorn. Using the BSP cost model to optimize parallel neural network training. *Future Generation Computer Systems*, 14:409–424, 1998.

[31] Constantinos Siniolakis. Bulk-synchronous parallel algorithms in computational geometry. Technical Report PRG-TR-10-96, Oxford University, Computing Laboratory, May 1996.

[32] D. Skillicorn. Strategies for parallel data mining. *IEEE Concurrency*, 7(4):26–35, October–December 1999.

[33] D.B. Skillicorn. Architecture-independent parallel computation. *IEEE Computer*, 23(12):38–51, December 1990.

[34] D.B. Skillicorn. Structuring data parallelism using categorical data types. In *Programming Models for Massively Parallel Computers*, pages 110–115, Berlin, September 1993. Computer Society Press.

[35] D.B. Skillicorn. *Foundations of Parallel Programming*. Number 6 in Cambridge Series in Parallel Computation. Cambridge University Press, 1994.

[36] D.B. Skillicorn. Building BSP programs using the Refinement Calculus. In *Third International Workshop on Formal Methods for Parallel Programming: Theory and Applications (FMPPTA'98)*, Springer Lecture Notes in Computer Science 1388, pages 790–795, March/April 1998.

[37] D.B. Skillicorn. Parallel predictor generation. In *Proceedings of a Workshop on Large-Scale Parallel KDD Systems, KDD'99*, number 1759 in Lecture Notes in Artificial Intelligence, pages 190–196. Springer-Verlag, 2000.

[38] D.B. Skillicorn, J.M.D. Hill, and W.F. McColl. Questions and answers about BSP. *Scientific Programming*, 6(3):249–274, 1997.

[39] D.B. Skillicorn and D. Talia. Models and programming languages for parallel computation. *Computing Surveys*, 30(2):123–169, June 1998.

[40] L.G. Valiant. A bridging model for parallel computation. *Communications of the ACM*, 33(8):103–111, August 1990.

[41] Thorsten von Eicken, David E. Culler, Seth Copen Goldstein, and Klaus Erik Schauser. Active Messages: A mechanism for integrated com-

munication and computation. In *The 19th Annual International Symposium on Computer Architecture*, volume 20(2) of *ACM SIGARCH Computer Architecture News*. ACM Press, May 1992.

[42] Y. Wang and D.B. Skillicorn. Parallel inductive logic for data mining. In *Workshop on Distributed and Parallel Knowledge Discovery, KDD2000*, Boston, to appear. ACM Press.

[43] M. Zaki. Parallel and distributed data mining: A survey. *IEEE Concurrency*, 7(4):14–25, October–December 1999.

Chapter 5

DISCRETE COMPUTING WITH COARSE GRAINED PARALLEL SYSTEMS: AN ALGORITHMIC APPROACH

Afonso Ferreira

*CNRS – I3S – INRIA**

Afonso.Ferreira@sophia.inria.fr

Isabelle Guérin-Lassous

CNRS – ENS de Lyon – INRIA†

Isabelle.Guerin-Lassous@ens-lyon.fr

Abstract   In this chapter we shall show that *coarse-grained* models are well adapted to coarse grained systems and clusters. In particular, algorithms designed for such models can be efficient and portable, and can have their practical performance directly inferred from their theoretical complexity. Furthermore, they allow a reduction on the costs associated with software development since the main design paradigm is the use of existing sequential algorithms and communication subroutines, usually provided with the systems.

Keywords: Cluster computing, coarse grained parallel computers, parallel algorithms.

1.   Introduction

At the time of this writing, the list of the 500 most powerful machines in the world (http://www.top500.org) shows that most of its entries are distributed memory machines with less than 200 processors. Although 24% of these machines are larger and the maximum number of processors in any of such systems is 9,632, it is very interesting to note that 11 clusters belong to the list and that a "home-made" cluster appears at rank 62. These figures lead to the following conclusions.

---

*MASCOTTE, CNRS – I3S – INRIA, BP 93, Sophia Antipolis, F-06902 France.
†REMAP, CNRS – ENS de Lyon – INRIA, 46, al. d'Italie, F-69346 Lyon, France.

R. Corrêa et al. (eds.), Models for Parallel and Distributed Computation. Theory, Algorithmic Techniques and Applications, 117–143.
© 2002 *Kluwer Academic Publishers.*

1 Nowadays, most of high-performance systems are coarse grained, i.e. the number of processors in the machine is much smaller than the input size.

2 Clusters based on off-the-shelf hardware can yield effective parallel systems for a fraction of the price of machines which use special purpose hardware.

We recall that a cluster is a set of machines (workstations, PC) interconnected by high performance local networks, with raw throughput close to 1Gb/s and latency smaller than $10\mu s$. The local networks are either realized with off-the-shelf hardware (e.g. Myrinet and Fast Ethernet), or application-driven devices, in which case additional functionalities are built-in, mainly at the memory access level. Such cluster-based machines typically utilize some flavour of Unix and any number of widely available software packages that support multi-threading, collective communication, automatic load-balance, and others.

Although a great deal of effort has been undertaken on system-level and programming environment issues for these clusters, less attention has been paid to methodologies for the design of algorithms for this kind of parallel systems. The main theoretical model used in the design of parallel discrete algorithms (the PRAM, to be defined in the next section) assumes that the size of the problem (n) is of the same order as the number of processors (p), and that the size of the processors' local memory is constant and very small. Such kind of model, known as *fine-grained* because of the assumptions above, is unfortunately not well adapted to existent systems, including clusters. One important aim in parallel computing is then to propose discrete algorithms that are:

- Efficient on coarse grained systems and especially on low cost clusters.

- Portable over a wide range of systems. By portable we mean that the code should be the same for supercomputers and for home-made clusters, without loosing too much at the efficiency level.

In this chapter we shall show that *coarse-grained* models are well adapted to coarse grained systems and clusters. In particular, algorithms designed for such models can be efficient and portable, and can have their practical performance directly inferred from their theoretical complexity. Furthermore, they allow a reduction on the costs associated with software development since the main design paradigm is the use of existing sequential algorithms and communication sub-routines, usually provided with the systems.

Our approach will be to study problems from the start of the algorithm design task until the implementation of the algorithms on a coarse grained system. We will give a brief introduction to the different models and especially to the main coarse grained models. Then we shall show some recent work on algorithms for discrete problems arising in the areas of graphs and computational geometry. In particular, we will address prefix sums computation, sorting, connected components computations in the case of dense graphs and interval graphs, and two-dimensional convex hull construction.

2. Parallel models

In fine grained models it is assumed that $p = O(n)$, where p is the number of processors and n the input size. There are essentially two main such models: the PRAM model (Parallel Random Access Machine [22, 29]), and the distributed memory machines models ([15, 30]). Although the PRAM model is often a first step to extract the parallelism of a problem, it is very far from existing parallel machines. Hence, algorithms have to be rewritten according to the system's structure, often leading to poor performances. On the other hand, distributed memory machines models are more realistic because they take into account the network topology. Unfortunately, if they lead to more efficient code, they still do not allow the design of portable code because the algorithms strongly rely on the interconnection network.

As a consequence, coarse-grained parallel models were proposed, where $p << n$. Such models take realistic characteristics of the machines into account while covering at the same time as many parallel platforms as possible. The first to have proposed such a model for discrete algorithm design was Valiant, with the BSP (*Bulk Synchronous Parallel*) model [35]. BSP uses slackness in the number of processors and memory mapping via hash functions to hide communication latency and provide for the efficient execution of PRAM algorithms on coarse-grained hardware. More focused on architecture design, Culler *et al.* introduced the LogP model (acronym to be explained later) which, using Valiant's BSP model as a starting point, focuses on the technological trend from fine grained parallel machines towards coarse-grained systems [6]. Other coarse grained models focus more on utilizing local computation and minimizing global operations. These include the *Coarse-Grained Multicomputer* (CGM) model, proposed by Dehne, Fabri and Rau-Chaplin, which will be developed in this chapter [8].

These three models have in common a set of *processors* that are interconnected by some network. A processor can be a monoprocessor machine, a processor of a multiprocessors machine or a multiprocessors machine itself. The network can be any communication medium (bus, Ethernet, shared memory, etc). Below, we give a simple description of each of these three models.

BSP. A BSP algorithm is composed of a sequence of *supersteps*. In a superstep, a processor can perform local computations and communications. Two supersteps are synchronized by a synchronization barrier. The supersteps are characterized by L the synchronization period, g the bandwidth, h the maximal number of words that a processor can send or receive during a superstep, and s the fixed overhead of any communication. A communication step requires time $gh + s$. The complexity of a superstep is the sum of the cost for local computations, the cost for communications and L. The complexity of a BSP algorithm is the sum of the costs of its supersteps.

LogP.   The acronym LogP stands for L the latency, o the overhead of a communication, g the gap (the minimum time interval between two consecutive sends or receipts of messages on one processor) and P the number of processors. It assumes that the interconnection network has a finite capacity because at each superstep a processor can send or receive at most $\lceil \frac{L}{p} \rceil$ messages. The communications between the processors are point-to-point and processors work asynchronously.

CGM.   This model can be seen as a generalization of BSP, where the parameters L, g, h and s and the synchronization step are hidden or made transparent to the algorithm designer and/or user. It assumes that each processor can hold $O(\frac{n}{p})$ data in its local memory. To qualify this model as coarse grained, it is further assumed that $1 << \frac{n}{p}$. The algorithms are an alternation of supersteps. In a superstep, a processor can send data to and receive data from each other processor, once, and the amount of data exchanged by one processor in total is at most $O(\frac{n}{p})$. In CGM, we have to ensure that the number of supersteps is small compared to the size of the input, and does not have to depend on n the input size. In this way, the algorithms are efficient for a large rage of values for the ratio $\frac{n}{p}$.

We shall develop CGM algorithm design in this chapter, because it offers the simplest realization of our goal: to design simple parallel algorithms for discrete problems which lead to efficient, portable and predictive code.

3.   Basic algorithms in CGM

We start by discussing two wide used procedures in parallel algorithms, namely the computation of all prefix sums and sorting.

3.1.   Prefix sums

Every processor stores some values, and all processors compute the partial sums of these values with respect to some associative operator:

Given: a set E.

Input: n elements $v_1, v_2, \ldots, v_n$ of E.

Output: $S_k = v_1 + \ldots + v_k, k = 1, \ldots, n$.

As an example, let $v_1 = 1$, $v_2 = 2$, $v_3 = 4$, $v_4 = 3$, $v_5 = 5$, $v_6 = 1$. The resulting prefix sums are hence $S_1 = 1$, $S_2 = 3$, $S_3 = 7$, $S_4 = 10$, $S_5 = 15$, $S_6 = 16$.

Algorithm 1 is a general parallel algorithm to compute the prefix sums.

Algorithm 1: Parallel Prefix Sums

Input: n elements $v_1, v_2, ..., v_n$

Output: the partial sums of the v_i values

1 Partition: divide the sequence $A = v_1, ..., v_n$ in $A_1 = v_1, ..., v_{\frac{n}{2}}$ and $A_2 = v_{\frac{n}{2}+1}, ..., v_n$

2 Recursion: compute in parallel the prefix sums $S_1, ..., S_{\frac{n}{2}}$ for A_1 and $S'_1, ..., S'_{\frac{n}{2}}$ for A_2

3 Merge: get the missing prefix sums by computing $S_{\frac{n}{2}+j} = S'_j + S_{\frac{n}{2}}$, for $j = 1, ..., n/2$

Algorithm 1 translated in the CGM model requires at least $\log p$ communication steps. Algorithm 2, below, is a CGM algorithm for computing the prefix sums that uses less communication steps than Algorithm 1.

Algorithm 2: CGM Prefix Sums

Input: p procs, n elements $v_1, ..., v_n$, n/p elements per processor i

Output: The prefix sums $S_k = v_1 + ... + v_k$, $k = 1, ..., n$

1 Each p_i: Compute the local **total** sum $T(i) = \Sigma_{j=1}^{j=\frac{n}{p}} v_{\frac{n}{p}(i-1)+j}$

2 Communication: Each p_i sends its value $T(i)$ to all other processors

3 Each p_i: Compute the sum of the $T(j)$ of smaller indexed processors, i.e., $ST(i) = \Sigma_{j<i} T(j)$

4 Each p_i: Compute the local **prefix** sums $S_{\frac{n}{p}(i-1)+k} = ST(i) + \Sigma_{j=1}^{j=k} v_{\frac{n}{p}(i-1)+j}$, $k = 1, ..., \frac{n}{p}$

We note that for Algorithm 2 only one communication operation is required and that each processor requires at most p values from all the others in order to be able to compute its own prefix sums with respect to the global set of input values, implying the relation $\frac{n}{p} \geq p$. Note also that there are two steps with $\frac{n}{p}$ local computations each. This implies that the theoretical speedup cannot be greater than $\frac{p}{2}$. Moreover, if $p = 2$, more operations will be carried out in parallel than in sequential. Therefore, it is unlikely that running Algorithm 2 with two processors will be any faster than computing prefix sums with a sequential code.

Experimental results.　　We implemented the CGM prefix sums algorithm on a PC cluster and on a parallel machine CRAY-T3E. The cluster consisted of 12 Pentium Pro 200 MHz, each with 64 Mb of RAM, linked by a 100 Mb/s Fast Ethernet network. This cluster is called Fast-Ethernet PCC henceforth. We used the C++ language for the local computations and the PVM library to realize the communications.

The implementation of Algorithm 2 is quite straightforward. The input data are integers encoded in 4 bytes. Whatever the input, the execution of the algorithm is always the same, therefore we can use any input data for the tests. The implemented sequential algorithm is not the simple algorithm that consists in linearly summing the input elements, but it uses the instruction level of parallelism of the actual computation units. The instructions handled in parallel have to be independent. The sequential algorithm uses the same principle of the parallel algorithm (Algorithm 1) where the processors are simulated by registers. In fact, it uses the recursion and merge steps of Algorithm 1 on k input elements (k is a parameter of our algorithm) and then iterates this process on the next k elements. Of course, to have a meaningful study, Step 1 of Algorithm 2 also uses this sequential algorithm based on the instruction level of parallelism.

In the figures, the x-axis is the input size n and the y-axis the execution time in seconds.

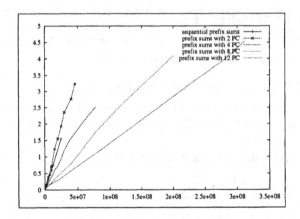

Figure 5.1.   Prefix Sums on the Fast Ethernet-PCC.

Figure 5.1 gives the execution time for 1, 2, 4, 8, and 12 PC on the Fast Ethernet-PCC. Each curve stops before the RAM overflow. First, note that the results are linear as expected. Then, note that the sequential algorithm is faster than CGM with 2 PC as expected. On the other hand, the execution is faster when we use more PC. The speedup is around 1.7 for 4 PC, 3.2 for 8 PC, and 4.2 for 12 PC. Note also that more precise measures show that the time for the communication step is negligible when compared to the time for the local computation steps. When memory swapping is required (i.e. the processor needs to use its disk to store all its dedicated input data), then the parallel computation can reduce the input size per processor, eliminating the swapping effects and thus considerably decreasing the total execution time. It takes more than 90 seconds for the sequential algorithm to compute the prefix sums of $2,800,000$ elements, whereas 12 PC can compute the prefix sums of 300 million elements in less than 5 seconds.

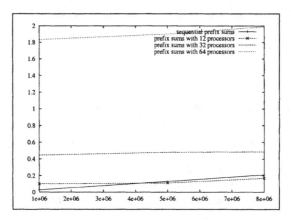

Figure 5.2.   Prefix Sums on a CRAY-T3E.

Figure 5.2 gives the execution time for $1, 12, 32$, and 64 processors on a CRAY-T3E. The input size, n, ranges from 1 million to 8 million. Beyond 8 million, the sequential program cannot be run due to memory overflow. We can see that the curves are again linear in n. The execution with 12 processors is faster than in sequential for a set of at least 4 million elements. But with 32 and 64 processors the execution time for the CGM algorithm is larger than the sequential algorithm. After some detailed measures, we noticed that the communication time is predominant compared to the local computations time for input size smaller than 8 millions. We explain this behavior by the fact that the PVM library is not efficient on the CRAY-T3E, when messages of small size are concerned. A dedicated communication library should thus give much better results. Nevertheless, this program can compute the prefix sums on very large data. Whereas one processor can handle at most 8 million elements, 64 processors can compute the prefix sum of up to 896 million integers in less than 3 seconds.

3.2.   CGM sorting

Another important and basic procedure is that of sorting $O(n)$ data in the CGM model. Recall that each processor stores $O(\frac{n}{p})$ data items.

For the PRAM model, an optimal $O(\log n)$ time algorithm with n processors exists [5]. This algorithm was adapted to the CGM setting, with the following complexity. Given $p < n^{1-\frac{1}{c}}$ $(c \geq 1)$, sorting $O(n)$ elements distributed evenly over p processors can be achieved in $O(\log n / \log(h + 1))$ communication rounds and $O(n \log n / p)$ local computation time [25], for $h = \Theta(\frac{n}{p})$, i.e. with optimal local computation and $O(1)$ communication rounds, when $\frac{n}{p} \geq p$.

Although optimal from a theoretical viewpoint, both algorithms above require a great deal of details to be implemented. Therefore, we prefer to use a rather simple CGM split-sort, as described in Algorithm 3, below. It is less scalable, but it is still optimal, with a small and constant number of rounds of communication, for $\frac{n}{p} \geq p$.

Algorithm 3: CGM Split Sort

   Input: p processors, n elements, n/p elements per processor

   Output: all elements sorted (inside each processor and by processors, i.e.
           if $i < j$ then elements in p_i are smaller than those in p_j

1   Select splitter set $S = \{s_1, s_2, \ldots, s_{p-1}\}$
2   Communication: S is sent to all processors
3   Each p_i: Partition the elements in buckets B_j^i according to S, $1 \leq i \leq$
    $p, 1 \leq j, \leq p$
4   Communication: each p_i sends B_j^i to p_j
5   Each p_i: Sort $B_i^k = B_i^0 \cup B_i^1 \cup \ldots \cup B_i^p$

It is clear that Algorithm 3 sorts any input, since we have not yet constrained the size of the buckets (and consequently of the local memories of the processors). To respect the size of $O(\frac{n}{p})$ data per processor given in the CGM model, we must define how the splitter set is chosen. Indeed, the size of the buckets depends on a good partition of the input.

3.3.   Deterministically selecting the splitters

We start by giving a deterministic way to compute the splitter set, which will need only $O(p)$ memory space per processor. This method is based on partitioning the input into p subsets of same size, as follows.

Definition 5.1 *The p-quantiles of an ordered set A of size n are the $p - 1$ elements, of rank $\frac{n}{p}, \frac{2n}{p}, \ldots, \frac{(p-1)n}{p}$, splitting A into p parts of equal size.*

The p-quantiles can be easily computed in sequential by a recursive algorithm in time $O(n \log p)$ as shown with Algorithm 4.

The corresponding CGM algorithm, from [19], is given in Algorithm 5 below.

Proposition 5.2 *The p-quantiles search algorithm on a set of n elements can be implemented in the CGM model with p processors in $O(\frac{n \log p}{p})$ time and*

Algorithm 4: Sequential p-quantiles search

Input: n data

Output: the p-quantiles set of the input data

1 Find the median of A
2 Split it into two sets
3 Recurse until $p - 1$ splitters are found

Algorithm 5: CGM p-quantiles search

Input: n data, p processors, $\frac{n}{p}$ data per processor

Output: the p-quantiles set of the input data

1 Each p_i: Extract $p - 1$ local pivots with Algorithm 4 on the local input data
2 Communication: Each p_i sends its pivots to p_0
3 p_0: Sort all pivots and choose the splitter set S
4 Communication: p_0 sends S to all $p_{i's}$

$O(1)$ *communication steps, with the scalability assumption* $\frac{n}{p} = \Omega(p^2)$ *(it can be improved to* $\frac{n}{p} = \Omega(p)$*).*

Corollary 5.3 *Sorting a set of n elements can be implemented in the CGM model with p processors in time $O(\frac{n \log n}{p})$ and $O(1)$ communication steps, with the scalability assumption $\frac{n}{p} = \Omega(p)$.*

We refer the interested reader to [19], where all the details on the algorithms, definitions, propositions and proofs concerning the p-quantiles problem can be found. Below, we describe another approach to compute the splitter set for a split sort, based on randomization.

3.4. CGM randomized sorting

This method requires less computation and is guaranteed to bound the size of the buckets with high probability, if the size of the sample set is chosen correctly (we omit here this proof and we refer the interested reader to [24]).

The whole randomized algorithm is given in Algorithm 6, below. It requires exactly three communication steps (only one of them communicates a total $O(n)$ data over the network) and $O(\frac{n \log n}{p})$ local computations per processor. Its drawback is that data may not be *exactly* equally distributed at the end of the sort. Nevertheless, a prefix sum procedure and a routing could be used to redistribute the data with a constant number of communication rounds so that each processor stores exactly $\frac{n}{p}$ data in its memory.

Algorithm 6: CGM Randomized Sorting

Input: p processors, n elements, n/p elements per processor

Output: n elements sorted (inside each processor and by processors, i.e. if $i < j$ then elements in p_i are smaller than those in p_j)

1 Each p_i: Randomly choose sample E_i, such that $|E_i| = 2\log^2 n$
2 Communication: Each p_i sends E_i to p_0
3 p_0: Sort $E = E_0 \cup E_1 \cup \ldots \cup E_p$.
4 p_0: Select splitter set $S = \{s_1, s_2, \ldots, s_{p-1}\}$
5 Communication: p_0 sends S to all processors
6 Each p_i: Partition the elements in buckets B_j^i according to s_i
7 Communication: p_i sends B_j^i to p_j, $1 \leq i \leq p, 1 \leq j \leq p$
8 Each p_i: Sort $B_i^k = B_i^0 \cup B_i^1 \cup \ldots \cup B_i^p$

Experimental results.   Algorithm 6 was implemented on the Fast Ethernet-PCC platform and the CRAY-T3E presented Section 3.1. The sorted integers are the standard 32 bit int types of the machines. The results shown are the average of ten execution times over ten different inputs generated randomly. The x-axis represents n the input size (which ranges from 1 million to 45 millions integers) and the y-axis the execution time in seconds. Measures begin at one million elements to satisfy some inequalities given by Algorithm 6 (the number of potential splitters should be smaller than $\frac{n}{p}$).

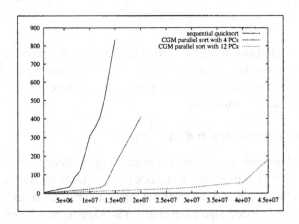

Figure 5.3.   Sorting on the Fast Ethernet-PCC.

Figure 5.3 shows the execution time on the Fast Ethernet PCC with the data balanced at the end, in comparison to the sequential performance of *quicksort* (from the standard C library). The execution times are neatly improved when we use 4 or 12 PC. For less than 7 million integers, the achieved speedup is about 2.5 for 4 processors and 6 for 12 processors. Beyond the size of 7 millions integers, the memory swapping effects increase significantly the execution time on a single processor and a super-linear speedup is obtained, with 10 for 4 processors and 35 for 12 processors. Sorting 40 millions integers takes less than one minute with 12 processors.

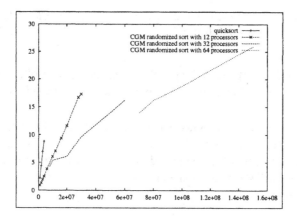

Figure 5.4. Sorting on the CRAY-T3E.

Figure 5.4 shows the results for the CRAY-T3E, that is for $1, 12, 32$, and 64 processors. Each curve stops before the local memory of each processor becomes saturated. To satisfy the inequality between the number of potential splitters and the size of the processors memory, the measures begin with 1 million integers for 12 processors, 8 million for 32 processors and 70 million elements for 64 processors. We see that the execution times improve when we use more processors. The algorithm also allows the sort of very large data since 64 processors can sort up to 200 millions integers in 32 seconds. Although the data balance at the end of the sort is implemented by a call to the CGM prefix sum algorithm (see the results on the CRAY-T3E Section 3.1), note that the CGM randomized sort is still efficient on the CRAY-T3E.

4.   Connected components

This section deals with one of the basic algorithms for graphs. The *connected components* of a graph G are the maximal connected subgraphs of G. The *connected components problem* consists of assigning to each node of G the label of the connected component that contains it. For an example, see Figure 5.5.

Several PRAM algorithms have been designed for this problem, but few algorithms for the coarse grained models have been proposed. Caceres et al. presents the first deterministic CGM algorithm in [4]. It requires $O(\log p)$ supersteps. The algorithm simulates $O(\log p)$ phases of an existing PRAM algorithm in order to reduce the original input into a *dense* graph, where $n \leq \frac{m}{p}$

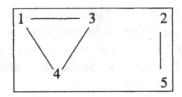

Figure 5.5. A graph with five vertices and two connected components, $label(1) = label(3) = label(4) = 1$ and $label(2) = label(5) = 2$.

(n is the number of vertices in the graph and m the number of edges). Then, it uses a specific CGM algorithm which computes the connected components for these graphs.

Section 4.1 presents this CGM part of the algorithm of Caceres et al. and shows some practical results. Then, in Section 4.2, we give a very simple algorithm that computes the connected components for interval graphs. The practical results will show that some encodings are particularly well adapted to parallel computation.

4.1.   Connected components for graphs such that $n \leq \frac{m}{p}$

Algorithm 7 computes the connected components for graphs such that $n \leq \frac{m}{p}$. Each processor has a local memory of $O(\frac{m}{p})$. At the beginning, each processor stores the n vertices of the graph and $\frac{m}{p}$ edges. At Step 1, each processor computes the connected components of the part of the graph it stores. Then each processor numbered in the upper half of the processor range sends the connected components it has just computed to a specific processor in the other half. Only processors numbered in the lower half of the range will continue in the following phase. At the end of $\lceil \log p \rceil$ steps, processor 1 stores the connected components of the whole graph. The algorithm requires $\lceil \log p \rceil$ communication steps and $O(\frac{m}{p} + \lceil \log p \rceil n)$ local computations. For a detailed analysis of this algorithm, see [26].

Theorem 5.4 *Algorithm 7 computes in the CGM model the connected components of a graph $G = (V, E)$ with n vertices and m edges and such that $n \leq \frac{m}{p}$ in $\lceil \log p \rceil$ supersteps and $O(\frac{m}{p} + \lceil \log p \rceil n)$ local computations. Each processor requires a memory of $O(\frac{m}{p})$.*   □

Experimental results.   Again, we implemented this algorithm of the Fast-Ethernet PCC platform and on the CRAY-T3E. However, for the sake of simplicity, in the remainder we will only present the results for the CRAY-T3E. Nevertheless, the results are alike with the Fast-Ethernet PCC platform, as it was the case with the results shown in the previous sections.

Algorithm 7: Connected Components for Graphs with $n \leq \frac{m}{p}$

Input: A graph $G = (V, E)$ given by the values $parent(v)$ for all vertices $v \in V$

Output: The processor 1 has the connected components of G

$active = p$;

Call a processor P_i active if $i \leq active$;

while $active > 1$ **do**

1 | **foreach** *active processor* **do**
        | compute the partial connected components of its subgraph;
           keep only edges of spanning tree of these components.

2 | **foreach** active P_i *such that* $i > \lceil active/2 \rceil$ **do**
        | send its partial connected components to processor $P_{i-\lceil active/2 \rceil}$.

3 | **foreach** active P_i *such that* $i \leq \lceil active/2 \rceil$ **do**
        | rebuild the subgraph according its partial connected components
           computed at Step 1 and those eventually received at Step 2.

    $active = \lceil active/2 \rceil$;

Our implementation uses adjacency lists to encode the graphs to reuse the sequential depth-first search algorithm for the local computation of the connected components. The test graphs are multi-graphs. The use of multi-graphs is not a drawback because the algorithm treats each edge exactly once unless it belongs to the spanning tree. We present in the following results for both $n = 1,000$ and $n = 100,000$, with varying m.

There are two parameters to vary: n and m. We decided to present results for two values of n with m varying. In all the figures, the x-axis is the number of edges in the graph and the y-axis is the execution time in seconds.

Figure 5.6 gives the results on the T3E for graphs having $1,000$ vertices. The number of edges, m, ranges from $10,000$ to $500,000$. For 16 and 32 processors, the curves begin with $m = 50,000$ because for less edges the inequality $n \leq \frac{m}{p}$ is not satisfied. If $m > 75,000$, the execution is always faster with p processors than in sequential. If $75,000 < m < 80,000$, the execution with 2 processors is the fastest. For $80,000 < m < 200,000$, it is faster with 8 processors, and for larger m it is faster with 16 processors. The execution times for 16 and 32 processors are almost alike. This can be explained by the fact that with 32 processors there is an extra step that takes more time than the time gained with the edges partition (there are less edges per processor with 32 processors than with 16 for the same graph size).

Figure 5.7 gives the results for graphs having $100,000$ vertices. The number of edges, m, ranges from $100,000$ to 48 million. Each curve stops before

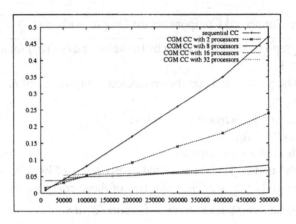

Figure 5.6.   Connected Components on the CRAY-T3E for graphs with $n = 1,000$.

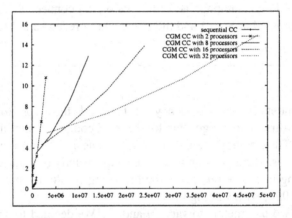

Figure 5.7.   Connected Components on the CRAY-T3E for graphs with $n = 100,000$.

the processors memory becomes saturated. The sequential execution is always faster than the CGM one. Some detailed tests show that the time for re-building the subgraph (Step 3) is the most influent. Therefore, the more processors, the longer this step, because it takes the same time whatever p, and it is called $\lceil \log p \rceil$ times. On the other hand, it is possible to compute the connected components on very large graphs. In sequential it is possible to handle graphs with at most 1.5 million edges, whereas 32 processors compute the connected components on a graph with 48 million edges in less than 16 seconds.

In the next section, we show that the encoding has an impact on the simplicity of the connected components algorithm and on its practical efficiency. We can design an algorithm for interval graphs that leads to efficient results, whatever the size of the graph.

4.2. Interval graphs

Formally, an *interval graph* is as follows: given a set n of intervals $\mathcal{I} = \{I_1, I_2, \ldots, I_n\}$ on a line, the corresponding interval graph $G = (V, E)$ has the set of nodes $V = \{v_1, \ldots, v_n\}$, and there is an edge in E between nodes v_i, v_j if and only if $I_i \cap I_j \neq \emptyset$. Figure 5.8 gives an example of an interval graph. This graph class can model a number of applications in scheduling, circuit design, traffic control, genetics and others [3, 27, 33].

Figure 5.8.   Intervals and the corresponding interval graph.

Several coarse-grained parallel algorithms to solve standard problems arising in the context of interval graphs have been designed. In this section, we concentrate on the connected components problem. For more details on the CGM algorithms for interval graphs, we refer the interested reader to [16].

Algorithm 8 computes the connected components of interval graphs in CGM.

Algorithm 8: CGM Connected Components

Input: n intervals, p processors, $\frac{n}{p}$ intervals per processor

Output: each interval has its label of the connected component that contains it

1 Sort the intervals by left endpoints distributing $2n/p$ elements to each processor

2 Assign the value 1 to each left endpoint and the value -1 to each right endpoint

3 Each p_i: Compute internal connected components. Label them $i.counter$, with $counter \geq 0$

4 Compute the prefix sums of the endpoints, on a list L

5 Each p_i: If there is a 0 in L_i then send ID= i to all processors

6 Each p_i: Let $j_1, j_2, \ldots, j_q$, $q \leq p$ be the list of ID's received. Thus, label last or only connected component with $(k + 1).0$, where $k \leq i < k + 1$.

Algorithm 8 requires a time $O(T_S(n, p))$ in Step 1 where $T_S(n, p)$ is the time for the global sort of $O(n)$ data on p processors in the CGM model and $O(\frac{n}{p})$ local computations in Steps 2, 3, 4 and 6. It also requires a constant number of supersteps.

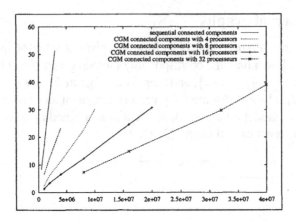

Figure 5.9. Connected Components on the CRAY-T3E for interval graphs.

Theorem 5.5 *The connected components problem in interval graphs can be solved in the CGM model in $O(T_S(n,p) + n/p)$ time, with a constant number of communication rounds.*   □

Experimental results.   The intervals are encoded with an array of structure composed of two fields. The first field is the left endpoint of the interval, the second one is its right endpoint. To generate random inputs for the tests, we use random permutations. The elements are unsigned integers of 4 bytes. The x-axis is the number of intervals and the y-axis is the execution times in seconds. We present below the results on the CRAY-T3E.

Figure 5.9 shows the execution times. The number of nodes, n, ranges from $500,000$ to 20 million. The sequential curve begins with $500,000$, whereas the CGM curves begins with 1 million (except for 32 processors that begins with 8 million). This is to respect the inequality mentioned in Section 3.2 concerning the sorting step. Each curve stops before the processors memories become saturated.

We see that the more processors, the faster the execution. The achieved speedup is approximately 2.9 for 4 processors, 5.4 for 8 processors and 9.8 for 16 processors. Also observe that the processors memories are saturated in sequential with 3 million intervals, whereas 32 processors compute the connected components on 40 million intervals in only 39 seconds.

In conclusion, to work directly on the intervals rather than on the graph simplifies the algorithm a great deal, and leads to more efficient results than those obtained with the linked lists encoding. Note also that the efficiency of Algorithm 8 is based on an efficient sort that partitions the intervals for the computation of the connected components.

These results show the impact of data encoding, allowing a good partition of the input, on practical results. In the following section, we show that it is still possible to design CGM algorithms for intricate problems, which can also lead to efficient practical results, as long as some kind of partition of the input can be used.

5.   Convex hull computation

This section describes a scalable parallel algorithm for a geometric problem, namely the two-dimensional (2D) Convex Hull. This is perhaps the most fundamental problem in computational geometry and certainly the most studied [1]. In fact, it appears to be the first problem in computational geometry for which parallel algorithms were designed. In the fine grained parallel setting, algorithms have been described for many architectures including the CRCW PRAM, the CREW PRAM, the Hypercube and the Mesh [1]. In the coarse grained setting, there are also several results [14, 7, 9].

Let ϵ be a fixed constant such that $\epsilon > 0$. The Convex Hull algorithm described in this chapter is deterministic, requires only $O(1)$ communication phases and time $O(\frac{n \log n}{p} + T_s(n, p))$ in the worst case, and is highly scalable in that it is efficient and applicable for $\frac{n}{p} \geq p^\epsilon$ ($T_s(n, p)$ is the time to sort $O(n)$ data on p processors in the CGM model). Since computing 2D convex hull requires time $T_{sequential} = \Theta(n \log n)$ [34] this algorithm runs in optimal time $\Theta(\frac{n \log n}{p})$.

We give here a succinct description of such an algorithm, and we refer the interested reader to [12] for further details. The key ideas are again to partition the data (through a sorting by x-coordinates), and to use splitters to sample local data and to perform computation (of supporting lines) that would be redundant in the sequential setting, but which reduces communication in our parallel setting. All inter-processor communication is restricted to a constant number of usages of a small set of simple communication operations. This has the desired effect of making the algorithms easy to implement.

5.1.   Outline of the algorithm

Algorithm 9 gives a CGM Convex Hull algorithm. We only show how to compute the upper hull of a set S of n points. The lower hull, and therefore the complete hull, can be computed analogously. In the remainder we assume without loss of generality that all points are in the first quadrant.

The main challenge is in performing Step 3. This step amounts to a merge algorithm in which p disjoint upper hulls are merged into a single hull. We present a simple merge procedure, described in Section 5.11, that requires $\frac{n}{p} \geq p^2$. Note that a more scalable but more complex merge procedure have

---

Algorithm 9: CGM Upper Hull(S)

Input: Each processor stores a set of $\frac{n}{p}$ points drawn arbitrarily from S

Output: A distributed representation of the upper hull of S. All points on the upper hull are identified and labeled from left to right

1 Globally sort the points in S by x-coordinate. Let S_i denote the set of $\frac{n}{p}$ sorted points now stored on processor i

2 Independently and in parallel, each processor i computes the upper hull of the set S_i. Let X_i denote the result on processor i

3 Compute for each upper hull X_i, $1 \le i \le p$, the upper common tangent lines between it and all upper hulls X_j, $i < j \le p$, and label the upper hull of S by using the upper tangent lines

---

been designed requiring only $\frac{n}{p} \ge p^\epsilon$. Both algorithms use the idea of selecting splitter sets which was introduced in the context of solving convex hulls in [31]. We refer the reader to [12] for a detailed study on the complex merge procedure, as well as on the merging presented below.

5.2.   Merging convex hulls in parallel

Some definitions.   We denote by $\overline{ab}$ the line segment connecting the points a and b and by (ab) the line passing through a and b. A point c is said to be *dominated* by the line segment $\overline{ab}$ if and only if c's x-coordinate is strictly between the x-coordinates of a and b, and c is located below the line segment $\overline{ab}$. Definitions 5.6 and 5.7, as illustrated by Figure 5.10, establish the initial condition before the merge step.

Definition 5.6 *Let* $\{S_i\}$, $1 \le i \le p$ *be a partition of S such that* $\forall x \in S_j$, $y \in S_i$, $j > i$, *the x-coordinate of x is larger than that of y (see Figure 5.10).*

Definition 5.7 *Let* $X_i = \{x_1, x_2, \ldots, x_m\}$ *be an upper hull. Then,* $pred_{X_i}(x_i)$ *denotes* x_{i-1} *and* $suc_{X_i}(x_i)$ *denotes* x_{i+1} *(see Figure 5.11).*

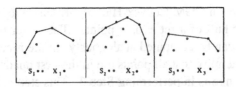

Figure 5.10.   In this example $S = \{S_1, S_2, S_3\}$ and the points in S_i that are filled in are the elements of the upper hull of S_i, namely X_i.

Definition 5.8 *Let $Q \subseteq S$. Then, $Next_Q : S \longrightarrow Q$ is a function such that $Next_Q(p) = q$ if and only if q is to the right of p and $\overline{pq}$ is above $\overline{pq'}$ for all $q' \in Q$, q' to the right of p.*

Definition 5.9 *Let $Y \subseteq S_i$. Then, $lm(Y)$ is a function such that $lm(Y) = y^*$ if and only if y^* is the leftmost point in Y such that $Next_{Y \cup S_j, j > i}(y^*) \notin S_i$.*

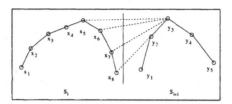

Figure 5.11. Let $S' = S_i \bigcup S_{i+1}$ then $suc_{X_i}(x_j) = x_{j+1}$, $x_3 = Next_{S'}(x_2)$, $Next_{S'}(x_3) = x_4$, $Next_{S'}(x_4) = x_5$, $Next_{S'}(x_5) = Next_{S'}(x_6) = Next_{S'}(x_7) = y_3$, $Next_{S'}(x_8) = y_2$, and $lm(S_i) = x_5$.

Let X represent the upper hull of a set of n points. Let also c be a point located to the left of this set. Algorithm 10 is a sequential algorithm to search for $q = Next_X(c)$. This binary search process takes time $O(\log |X|)$ [34].

Algorithm 10: QueryFindNext(X,c,q)

> **Input**: an upper hull $X = \{x_1, \ldots, x_m\}$ sorted by x-coordinate and a point c to the left of x_1
>
> **Output**: a point $q \in X$, $q = Next_X(c)$
>
> 1 **if** $X = \{x\}$ **then** $q \leftarrow x$ and halt
>
> 2 **if** $x_{\lceil m/2 \rceil} suc(x_{\lceil m/2 \rceil})$ *is located below the line* $(cx_{\lceil m/2 \rceil})$ **then**
> | QueryFindNext($\{x_1, \ldots, x_{\lceil m/2 \rceil}\}$,c,q)
>
> **else**
> | QueryFindNext($\{x_{\lceil m/2 \rceil}, \ldots, x_m\}$,c,q)

Characterization of the upper hull. A classical way of characterizing the upper hull of a point set S, as given in [34], is based on the observation that "a line segment $\overline{ab}$ is an edge of the upper hull of a point set S located in the first quadrant if and only if all the $n - 2$ remaining points fall below the line (ab)". We will work with a new characterization of the upper hull of S based on the same observation, but defined in terms of the partitioning of S given in Definitions 5.6 and 5.9. Consider sets S, S_i and X_i as given in Definitions 5.6 and 5.7.

Definition 5.10 *Let* $S' = \{c \in \bigcup X_i \mid c$ *is not dominated by a line segment* $x_i^* Next_{\bigcup X_j, j > i}(x_i^*), 1 \le i < p\}$, *where* $x_i^* = lm(X_i)$.

We then have the following characterization of $UH(S)$.

Theorem 5.11 $S' = UH(S)$.

Definition 5.12 *Let* G_i *a subset of* X_i *and* $g_i^* = lm(G_i)$. *Let* $R_i^- \subseteq X_i$ *be composed of the points between* $pred_{G_i}(g_i^*)$ *and* g_i^*, *and* $R_i^+ \subseteq X_i$ *be composed of the points between* g_i^* *and* $suc_{G_i}(g_i^*)$.

Lemma 5.13 *One or both of* R_i^+ *or* R_i^- *is such that all its points are under the line* $(g_i^* Next_{X_j, j > i}(g_i^*))$.

Definition 5.14 *Let* R_i *denote the set* R_i^+ *or* R_i^- *that has at least one point above the line* $(g_i^* Next_{X_j, j > i}(g_i^*))$.

Note that the size of the sets R_i is bounded by the number of points laying between two consecutive points in G_i.

Algorithm 11 computes $g_i^* = lm(G_i)$, where $G_i \subseteq X_i$. The key idea is to send the elements of G_i, at Step 2, to all larger-numbered processors so that processors receiving $G = \bigcup G_j, j < i$ can sequentially compute, at Step 3, the required points with Algorithm 10 and send the answers back at Step 4. Then, the processors can independently compute all g_i^*. When the processors are divided into groups, let q_z^i denote the z-th processor of the i-th group.

Lemma 5.15 *Algorithm 11 computes* $g^* = lm(G_i)$ *in a constant number of communication rounds. Furthermore, it requires local memory space* $\frac{n}{p} \ge p^k |G_i|$.

A merge algorithm for the case $n/p \ge p^2$.   Algorithm 12 merges p upper hulls, one per processor, into a single upper hull. This CGM algorithm uses a constant number of global communication rounds and requires that $\frac{n}{p}$ be greater than or equal to p^2.

In order to find the upper common tangent between an upper hull X_i and an upper hull X_j (to its right) Algorithm 12 computes the *Next* function, not for the whole set X_i but for a subset of p equally spaced points from X_i. We call this subset of equally spaced points a *splitter* of X_i. This approach based on splitters greatly reduces the amount of data that must be communicated between processors without greatly increasing the number of global communication rounds that are required.

Lemma 5.16 *Algorithm MergeHulls1 computes* $UH(S)$ *in* $O(\frac{n \log n}{p} + T_s(n, p))$ *time. It requires* $\frac{n}{p} \ge p^2$ *local memory space and a constant number of communication rounds.*

Algorithm 11: FindLMSubset($\Delta_i, k, w, G_i, g_i^*$)

Input: Upper hulls $\Delta_i, 1 \leq i \leq p^k$; represented each in p^w consecutively numbered processors $q_z^i, 1 \leq z \leq p^w$, and a set $G_i \subseteq \Delta_i$

Output: The point $g_i^* = lm(G_i)$. There will be a copy of g_i^* in each $q_z^i, 1 \leq z \leq p^w$

1 Gather G_i in processor q_1^i, for all i
2 Each processor q_1^i sends its G_i to all processors $qj_z, j > i, 1 \leq z \leq p^w$. Each processor q_z^i, for all i, z, receives $\mathcal{G}^i = \bigcup G_j, \forall j < i$
3 Each processor q_z^i, for all i, z, sequentially computes $Next_{\Delta_i}(g)$ for every $g \in \mathcal{G}^i$, using procedure **QueryFindNext**
4 Each processor q_z^i, for all i, z, sends back to all processors $q_1^j, j < i$, the computed $Next_{\Delta_i}(g), \forall g \in G_j$. Each processor q_1^i, for all i, receives for each $g \in G_i$ the computed $Next_{\Delta_j}(g), \forall j > i$
5 Each processor q_1^i, for all i, computes for each $g \in G_i$ the line segment with the largest slope among $\overline{gsuc_{G_i}(g)}$ and $\overline{gNext_{\Delta_j}(g)}, j > i$, finding $Next_{G_i \cup \Delta_j, j > i}(g)$. Then, it computes $g_i^* = lm(G_i)$
6 Each processor q_1^i, for all i, broadcasts g_i^* to $q_z^i, 1 \leq z \leq p^w$

Algorithm 12: MergeHulls1($X_i (1 \leq i \leq p), S, n, p, UH$)

Input: The set of p upper hulls X_i consisting of a total of at most n points from S, where X_i is stored on processor $q_i, 1 \leq i \leq p$

Output: A distributed representation of the upper hull of S. All points on the upper hull are identified and labeled from left to right

1 Each processor q_i sequentially identifies a splitter set G_i composed of p evenly spaced points from X_i
2 The processors find in parallel $g_i^* = lm(G_i)$. This is done via a call to Procedure **FindLMSubset**
3 Each processor q_i computes its own R_i^- and R_i^+ sets according to Definition 5.12, and the set R_i of points which are above the line $(g_i^* Next_{G_i \cup X_j, j > i}(g_i^*))$, according to Lemma 5.13
4 The processors find in parallel $x_i^* = lm(R_i \cup g_i^*)$, using Procedure **FindLMSubset**. Note that by definition $Next_S(x_i^*) \notin X_i$
5 Each processor q_i broadcasts its $Next_S(x_i^*)$ to all $q_j, j > i$, and computes its own S_i' according to Definition 5.10

5.3.   Experimental results

Algorithm 9 has been implemented on a CRAY-T3D with 128 processors. As for the other implementations presented in this chapter, PVM has been used for the communications. In our experimental study, we assume that the initial points are sorted in order to focus on the behavior of our algorithm, since parallel sorting algorithms are well studied and efficient (see Section 3.2). The sequential algorithm used for computing the local upper hulls is a slight modification of Procedure 8.2 from [13]. Its time complexity is $O(n \log n)$ for an arbitrary set of points. However in the case of pre-sorted data, the complexity becomes linear. The code derives from the one in [2]. Each point is implemented as a pair of two single precision floating point numbers (coordinates of the point).

We remark that if S is a set of uniformly distributed points in the plane, the upper hull of S consists in average of just a few points of S [10]. This data reduction can considerably speedup our parallel merging phase, since the amount of data remaining in this phase is, in average, significantly smaller than the size of the initial set S. Therefore, in the remainder we will also study the worst case behavior of Algorithm 9 (i.e., all the points of the initial set belong to the upper hull). We present our experimental results for random point sets and for a worst case set (a linear set of points). The execution time for the random point set is the average of the execution times on 10 different random point sets. In all the figures, the solid curves represent the total running time and the dashed curves the total communication time.

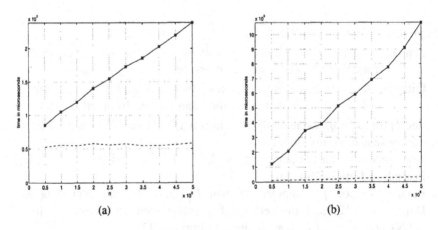

(a)   (b)

Figure 5.12.   Hupper Hull for fixed p (p = 64).

Figure 5.12 plots data for an increasing number of points when the number of processors is fixed. Figure 5.12(a) corresponds to random point sets. The total time grows proportional to n, since p is fixed, while the communication time is almost constant. As mentioned above, for random point sets, the data reduction is dramatic. The computation time curve in Figure 5.12 accounts mainly for the computation of the initial upper hull.

Figure 5.12(b) shows plotted data for the worst case. Notice that the upper hull takes almost 50 times longer to compute for worst case data than for random data.

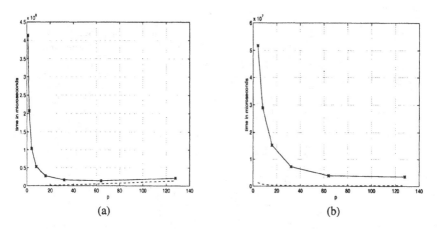

(a)   (b)

Figure 5.13.   Hupper Hull for fixed n (n = 2 000 000).

Figure 5.13 plots data for an increasing number of processors when the number of points is fixed. Figure 5.13(a) corresponds to random point sets, where a good behaviour is observed when p is small. However, when p increases, so does the total number of messages and the number of communication startups. Hence, the good behaviour is maintained until the communication time is no longer negligible compared to the local computation time. For instance, when $p \geq 64$, we observe a complete inversion of the slope of the curve of the total running time, meaning that we are using too many processors to solve a small problem.

In the worst case (Figure 5.13(b)), the observed speedup is better because the amount of local computation (computation of the local hull and the supporting tangents) is large enough compared to the communication.

6.   Conclusion

In this chapter we showed that CGM is a nice bridging model, because it allows for both theory and practice of parallel algorithms. Naturally, now, we can say that parallel discrete algorithms need experiments, and, eventually, a library for discrete applications, as it was done for parallel linear algebra procedures.

At the theoretical level, a nice result would consist of a bridging theorem translating PRAM algorithms into CGM ones. Finally, Table 5.1 shows that we have a good know-how on models for homogeneous, single user and/or single application systems. What about heterogeneous, multi user and/or multi application systems?

Table 5.1.   A non-exhaustive table of CGM deterministic solutions for discrete problems arising in different areas.

Area	Problem	Communication Rounds	Scalability Assumption
Binary	Medial Axis Transform [21]	$O(1)$	$n/p \geq p^\epsilon$
Images	Contour Tracking [20]	$O(1)$	$n/p \geq p^\epsilon$
Computational	Convex Hull/Triang. [12]	$O(1)$	$n/p \geq p^\epsilon$
Geometry	Range-search [17]	$O(1)$	$n/p \geq p^\epsilon$
	Voronoï [11]	$O(\log p)$	$n/p \geq p$
Graphs	LCA [32]	$O(1)$	$n/p \geq p$
	List Ranking [4]	$O(\log p)$	$n/p \geq p$
	List Ranking [28]	$O(\log p \log^* p)$	$n/p \geq p$
	Connected Components [4]	$O(\log p)$	$n/p \geq p$
	Ear Decomposition [4]	$O(\log p)$	$n/p \geq p$
	Coloring [23]	$O(p)$	$n/p \geq p$
	Others [4]	$O(\log p)$	$n/p \geq p$
Interval	Max. Clique [16]	$O(1)$	$n/p \geq p$
Graphs	Connected Components [16]	$O(1)$	$n/p \geq p$
	BFS & DFS Tree [16]	$O(1)$	$n/p \geq p$
	Others [16]	$O(\log p)$	$n/p \geq p$
Others	Knapsack Problem [18]	$O(1)$	$\mathcal{M} \geq 2^{n/4}$

Acknowledgments

We thank our many colleagues and co-authors for fruitful discussions on coarse-grained computing: E. Cáceres, F. Dehne, M. Diallo, P. Flocchini, C. Kenyon, K. Marcus, A. Rau-Chaplin, I. Rieping, M. Robson, A. Roncato, N. Santoro, S. Song, S. Ubéda.

References

[1] S. Akl and K. Lyons. *Parallel Computational Geometry*. Prentice Hall, 1993.

[2] J. L. Bentley, K. L. Clarkson, and D. B. Levine. Fast linear expected-time algorithms for computing maxima and convex hulls. *Algorithmica*, pages 168–183, 2993.

[3] A.A. Bertossi and M.A. Bonuccelli. Some parallel algorithms on interval graphs. *Discrete Applied Mathematics*, 16:101–111, 1987.

[4] E. Caceres, F. Dehne, A. Ferreira, P. Flocchini, I. Rieping, A. Roncato, N. Santoro, and S. Song. Efficient parallel graph algorithms for coarse grained multicomputers and bsp. In P. Degano, R. Gorrieri, and A. Marchetti-Spaccamela, editors, *Proceedings of ICALP'97*, Lecture Notes in Computer Science, pages 390–400, 1997.

[5] R. Cole. Parallel merge sort. *SIAM Journal on Computing*, pages 770–785, 1988.

[6] D. Culler, R. Karp, D. Patterson, A. Sahay, K. Schauser, E. Santos, R. Subramonian, and T. von Eicken. LogP: Towards a realistic model of parallel computation. In *Fifth ACM SIGPLAN Symposium on the Principles and Practice of Parallel Programming*, 1993.

[7] F. Dehne, A. Fabri, and C. Kenyon. Scalable and archtecture independent parallel geometric algorithms with high probability optimal time. In *Proceedings of the 6th IEEE SPDP*, pages 586–593. IEEE Press, 1994.

[8] F. Dehne, A. Fabri, and A. Rau-Chaplin. Scalable parallel geometric algorithms for coarse grained multicomputers. In *Proc. 9th ACM Symp. on Computational Geometry*, pages 298–307, 1993.

[9] X. Deng and N. Gu. Good algorithm design style for multiprocessors. In *Proc. of the 6th IEEE Symposium on Parallel and Distributed Processing*, pages 538–543, Dallas, USA, 1994.

[10] L. Devroye and G.T. Toussaint. A note on linear expected time algorithms for finding convex hulls. *Computing*, 26:361–366, 1981.

[11] M. Diallo, A. Ferreira, and A. Rau-Chaplin. Communication-efficient deterministic parallel algorithms for planar point location and 2d Voronoi diagram. In Springer Verlag, editor, *Proceedings of the 15th Symposium on Theoretical Aspects of Computer Science – STACS'98*, volume 1373 of *Lecture Notes in Computer Science*, pages 399–409, Paris, France, 1998.

[12] M. Diallo, A. Ferreira, A. Rau-Chaplin, and S. Ubeda. Scalable 2d convex hull and triangulation algorithms for coarse-grained multicomputers. *Journal of Parallel and Distributed Computing*, 56(1):47–70, 1999.

[13] H. Edelsbrunner. *Algorithms in Combinatorial Geometry*. Springer-Verlag, 1987.

[14] A. Fabri, F. Dehne, and A. Rau-Chaplin. Scalable parallel geometric algorithms for coarse grained multicomputers. In *Proc. of the 9th ACM Symposium on Computational Geometry*, pages 298–307, 1993.

[15] A. Ferreira. *Handbook of Parallel and Distributed Computing*, chapter Parallel and communication algorithms for hypercube multiprocessors, pages 568–589. McGraw-Hill, A. Zomaya edition, 1996.

[16] A. Ferreira, Isabelle Guérin-Lassous, Karina Marcus, and A. Rau-Chaplin. Parallel computation on interval graphs using pc clusters: Algorithms and

experiments. In D. Pritchard and J. Reeves, editors, *Proceedings of Europar'98*, volume 1470 of *Lecture Notes in Computer Science*, pages 875–886, Southampton, UK, September 1998. Springer Verlag. Distinguished Paper.

[17] A. Ferreira, C. Kenyon, A. Rau-Chaplin, ánd S. Ubéda. d-dimensional range search on multicomputers. *Algorithmica*, 24(3/4):195–208, 1999. Special Issue on Coarse Grained Parallel Algorithms.

[18] A. Ferreira and J.M. Robson. Fast and scalable parallel algorithms for knapsack and similar problems. *Journal of Parallel and Distributed Computing*, 39(1):1–13, November 1996.

[19] A. Ferreira and N. Schabanel. A randomized BSP algorithm for the maximal independent set problem. In *Proceedings of the 4th Int. Symp. on Parallel Architectures, Algorithms and Networks*, pages 284–289, 1999.

[20] A. Ferreira and S. Ubéda. Ultra-fast parallel contour tracking, with applications to thinning. *Pattern Recognition*, 27(7):867–878, 1994.

[21] A. Ferreira and S. Ubéda. Computing the medial axis transform in parallel with 8 scan operations. *IEEE Transactions on Pattern Analysis & Machine Intelligence (PAMI)*, 21(3):277–282, March 1999.

[22] S. Fortune and J. Wyllie. Parallelism in Random Access Machines. In *10-th ACM Symposium on Theory of Computing*, pages 114–118, 1970.

[23] A.H. Gebremedhin, Guérin Lassous I., J. Gustedt, and J.A. Telle. Graph Coloring on a Coarse Grained Multiprocessor. In *Proceedings of Workshop on Graph-Theoretic Concepts in Computer Science (WG'00)*, Lecture Notes in Computer Science. Springer Verlag, 2000. To appear.

[24] A.V Gerbessiotis and L.G Valiant. Direct bulk-synchronous parallel algorithms. *Journal of Parallel and Distributed Computing*, pages 251–267, 1994.

[25] M. Goodrich. Communication-efficient parallel sorting. In *Proc. of the 28th annual ACM Symposium on Theory of Computing*, Philadephia, USA, May 1996.

[26] I. Guérin-Lassous, J. Gustedt, and M. Morvan. Feasability, Portabiblity, Predictability and Efficiency: Four Ambitious Goals for the Design and Implementation of Parallel Coarse Grained Graph Algorithms. Technical Report 3885, INRIA, 2000.

[27] U.I. Gupta, D.T. Lee, and J.Y.-T. Leung. An optimal solution for the channel-assignment problem. *IEEE Transaction on Computers*, C-28:807–810, 1979.

[28] Guérin Lassous I. and J. Gustedt. List Ranking on PC clusters. In *Proceedings of Workshop on Algorithm Engineering (WAE'00)*, Lecture Notes in Computer Science. Springer Verlag, 2000. To appear.

[29] R. M. Karp and V. Ramachandran. *Handbook of Theoretical Computer Science, volume A: Algorithms and Complexity*, chapter A survey of parallel algorithms for shared-memory machines. Elsevier/MIT Press, J. van Leeuwen edition, 1990.

[30] F.T. Leighton. *Introduction to Parallel Algorithms and Architectures: Arrays, Trees, Hypercubes*. Morgan Kaufmann Publishers, San Mateo, CA, 1992.

[31] R. Miller and Q. Stout. Efficent parallel convex hull algorithms. *IEEE Transcations on Computers*, 37(12):1605–1618, 1988.

[32] H. Mongelli and S. Song. A range minima parallel algorithm for coarse grained multicomputers. In *Proc. of Irregular'99*, volume 1586 of *Lecture Notes in Computer Science*, pages 1075–1084. Springer-Verlag, 1999.

[33] S. Olariu. *Handbook of Parallel and Distributed Computing*, chapter Parallel graph algorithms, pages 355–403. McGraw-Hill, A. Zomaya edition, 1996.

[34] F. Preparata and M. Shamos. *Computational Geometry: An Introduction*. Springer Verlag, 1985.

[35] L. Valiant. A bridging model for parallel computation. *Communication of ACM*, 38(8):103–111, 1990.

REFERENCES

II

PARALLEL APPLICATIONS

Chapter 6

PARALLEL GRAPH ALGORITHMS FOR COARSE-GRAINED MULTICOMPUTERS

S. W. Song

*Universidade de São Paulo**

song@ime.usp.br

Abstract   Current applications in parallel machines are restricted to trivially parallelizable problems. In real machines communication time is often much greater than computation time. Therefore for many non-trivial graph problems, many theoretically efficient parallel algorithms for PRAM or fine grained network models often give disappointing speedups when implemented on real machines. The CGM (Coarse Grained Multicomputer) model was proposed by F. Dehne to be an adequate model of parallelism sufficiently close to existing parallel machines. It is a simple model but nevertheless intends to give a reasonable prediction of performance when parallel algorithms on this model are implemented. CGM algorithms are expected to have theoretical complexity analyses close to actual times observed in real implementations. In this chapter we present the CGM model and discuss several CGM scalable parallel algorithms to solve some basic graph problems, including connected components and list ranking. It is important to have very efficient algorithms for such basic problems because, as shown by J. H. Reif, many important graph problems are based on these subproblems.

Keywords: Coarse grained multicomputer, parallel algorithms, scalable algorithms, graph algorithms, list ranking, connected components, range minima.

1.   Introduction

Many applications in parallel machines that appeared in the literature are restricted to trivially parallelizable problems with low communication requirements. In real machines communication time is often much greater than compu-

---

*IME/USP, Rua do Matão, 1010. CEP 05508-900 São Paulo, SP, Brazil. This author is supported by FAPESP (Fundação de Amparo à Pesquisa do Estado de São Paulo) Proc. No. 98/06138-2, CNPq Proc. No. 52 3778/96-1, and CNPq/NSF Collaborative Research Program Proc. No. 680037/99-3.

R. Corrêa et al. (eds.), Models for Parallel and Distributed Computation. Theory, Algorithmic Techniques and Applications, 147–178.

tation time. Therefore for non-trivial problems, such as many graph problems, theoretically efficient parallel algorithms for PRAM or fine grained network models often give disappointing speedups when implemented on real machines.

The CGM (Coarse Grained Multicomputer) model was proposed [16, 15, 13] to be an adequate model of parallelism sufficiently close to existing parallel machines with distributed memory. It is a simple model but nevertheless intends to give a reasonable prediction of performance when parallel algorithms on this model are implemented. CGM algorithms are expected to have theoretical complexity analyses close to actual times observed in real implementations.

In [30], Reif presents the *synthesis approach*Synthesis approach to algorithm design. With this approach, a family of related algorithms can be synthesized from some very basic algorithms. He presents a family of graph problems for which sophisticated parallel algorithms are synthesized from parallel algorithms for simpler basic problems. For instance, to solve the open ear decomposition problem, we can first solve the lowest common ancestor problem, which in turn can be based on the range minima problem and the list ranking problem. It is our purpose to consider some of the very basic problems such as list ranking, determination of the connected components of a graph, and range minima. All the presented algorithms in [30] however are based on the PRAM model. This text considers the CGM model for which there exist, to the best of our knowledge, only a few results on parallel graph algorithms.

In Section 2 we present the CGM model. In Section 3 we consider the basic list ranking problem and give two algorithms: a random algorithm and a deterministic algorithm. In Section 4 we present the problem of determining the connected components of a graph. In Section 5 we present a very efficient algorithm for the range minima problem. At the end of each of these sections we give simulation or experimental results obtained on real parallel machines. In Section 6 we mention the lowest common ancestor problem and observe that it can be based on the range minima problem. Finally in Section 7 we conclude with a summary of the presented results.

The current text is based on results presented in several recent conferences and journal articles. The probabilistic algorithm for list ranking has appeared in ASIAN'96 [11] and its deterministic version in ICALP'97 [12]. The connected components algorithm has been presented in ICALP'97. The parallel algorithm for range minima has appeared in the proceedings of IRREGULAR'99 [25]. This text is therefore the result of many researchers.

2.   The coarse grained multicomputer (CGM)

The CGM (Coarse Grained Multicomputer) model [16, 15, 13] is similar to Valiant's BSP (Bulk-Synchronous Parallel) model [14]. However it uses only two parameters, n and p, where n is the size of the input and p the number of processors each with $O(n/p)$ local memory. The term *coarse grained* means the local memory size is considerably greater than $O(1)$. Usually we require $n/p > p$.

All processors are connected by a router that can deliver messages in a point to point fashion. A CGM algorithm consists of an alternating sequence of *computation rounds* and *communication rounds* separated by barrier synchronizations.

A computation round is equivalent to a computation superstep in the BSP model. In this computation round, we usually use the best possible sequential algorithm in each processor to process its data locally. A communication round consists of a single h-relation with $h \leq n/p$, that is, each processor can exchange a total of $O(n/p)$ data with other processors in one communication round.

In the CGM model the communication cost of a parallel algorithm is modeled by the number of communication rounds. The objective is to design algorithms that require a small number of communication rounds. Many algorithms for graph and geometric problems [12, 16, 15, 13] require only a constant or $O(\log p)$ rounds. As p is usually small compared to n the resulting CGM algorithm is expected to be efficient when implemented in practice. Observe that PRAM algorithms frequently are designed for $p = O(n)$ and each processor receives a small amount of input data. The CGM model considers a more realistic case of $p << n$.

Notice that the CGM model is particularly suitable in cases where the overall computation speed is considerably larger than the overall communication speed, and the problem size is considerably larger than the number of processors, which is usually the case in practice.

3.   List ranking

List ranking

Consider a linear linked list consisting of a set S of n nodes and, for each node $x \in S$, a pointer $(x \rightarrow next(x))$ to its successor, $next(x)$, in the list. Let $\lambda \in S$ be the last list element and $next(\lambda) = \lambda$. The list ranking problem consist of computing for each $x \in S$ the distance of x to λ, referred to as $dist(x)$. We assume that, initially, every processor stores n/p nodes and, for each of these nodes the pointer $(x \rightarrow next(x))$ to the next list element. See Figure 6.1. As output we require that every processor stores for each of its n/p nodes $x \in S$ the value $dist(x)$.

A trivial sequential algorithm solves the list ranking problem in optimal linear time by traversing the list. Several PRAM list ranking algorithms have been proposed [21, 23]. The first optimal $O(\log n)$ EREW PRAM algorithm is due to Cole and Vishkin [10]. Another optimal deterministic algorithm is given by

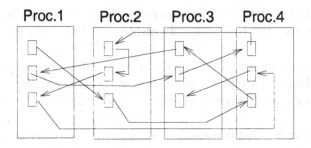

Figure 6.1.   A Linear linked list stored in a distributed memory multiprocessor.

Anderson and Miller [4]. Parallel list ranking algorithms using randomization were proposed by Miller and Reif [24, 18]. The algorithms use $O(n)$ processors. The optimal algorithm by Anderson and Miller [5] improves this by using an optimal number of processors. A $O(\sqrt{(n)})$ time mesh algorithm is described in [6].

3.1.   A randomized parallel algorithm for list ranking

Consider a linear linked list with a set S of n nodes. In this section we will show that if we select $\frac{n}{p}$ random elements (pivots) of S then, with high probability, these pivots will split S into sublists whose maximum size is bound by $3p\ln(n)$.

Lemma 6.1 $xk \leq n$ *randomly chosen elements of S (pivots) partition list S into sublists S_i such that the size of the largest sublist is at most $\frac{n}{x}$ with probability at least $1 - 2x(1 - \frac{1}{2x})^{xk}$.*

Proof: (Analogous to [17]) Assume that the nodes of S are sorted by their rank. This sorted list can be viewed as $2x$ segments of size $\frac{n}{2x}$. If every segment contains at least one pivot (chosen element), then $\max_{1 \leq j \leq xk} |S_j| \leq \frac{n}{x}$. Consider one segment. Since the pivots are chosen randomly, the probability that a specific pivot is not in the segment is $(1 - \frac{1}{2x})$. Since xk pivots are selected independently, the probability that none of the pivots are in the segment is $(1 - \frac{1}{2x})^{xk}$. Therefore, even assuming mutual exclusion, the probability that there exists a segment which contains no pivot is at most $2x(1 - \frac{1}{2x})^{xk}$. Hence, every segment contains at least one pivot with probability at least $1 - 2x(1 - \frac{1}{2x})^{xk}$.
□

Corollary 6.2 $xk \leq n$ *randomly chosen pivots partition list S into $xk + 1$ sublists S_i such that there exists a sublist S_i of size larger than $c\frac{n}{x}$ with probability at most $\frac{2x}{c}(1 - \frac{c}{2x})^{xk} \leq \frac{2x}{c}e^{-\frac{1}{2}ck}$.*

Lemma 6.3 *Consider* $xk \leq n$ *randomly chosen pivots which partition S into* $xk + 1$ *sublists S_i, and let* $m = \max_{0 \leq i \leq xk} |S_i|$. *If* $k \geq \ln(x) + 2\ln(n)$ *then* $Prob\{m > c\frac{n}{x}\} \leq \frac{1}{n^c}$, $c > 2$.

Proof: Corollary 6.2 implies that $Prob\{m > c\frac{n}{x}\} \leq \frac{2x}{c} e^{-\frac{1}{2}ck}$.

We observe that, for $c > 2$, $\ln(x) + 2\ln(n) \leq k \Rightarrow \frac{2}{c}\ln(\frac{2x}{c}) + 2\ln(n) \leq k$

$\Rightarrow \ln(\frac{2x}{c}) + c\ln(n) \leq \frac{ck}{2} \Rightarrow \frac{2x}{c} n^c \leq e^{\frac{ck}{2}} \Rightarrow Prob\{m > c\frac{n}{x}\} \leq n^{-c}$ □

Theorem 6.4 $\frac{n}{p}$ *randomly chosen pivots partition S into* $\frac{n}{p} + 1$ *sublists S_j with* $m = \max_{0 \leq j \leq p} |S_j|$ *such that* $Prob\{m \geq c3p\ln(n)\} \leq \frac{1}{n^c}$, $c > 2$

Proof: Let $x = \frac{n}{3p\ln(n)}$, $k = \ln(x) + 2\ln(n) = 3\ln(n) - \ln(3p\ln(n))$.

Then $xk = \frac{n}{p} \frac{3\ln(n) - \ln(3p\ln(n))}{3\ln(n)} \leq \frac{n}{p}$, and Theorem 6.4 follows from Lemma 6.3. □

3.1.1 A simple algorithm using a single random sample.

We present a simple list ranking algorithm which requires, with high probability, at most $\log(3p) + \log\ln(n) = \tilde{O}(\log p + \log\log n)$ communication rounds. This algorithm is based on a single random sample of nodes.

($\tilde{O}(n)$ denotes $O(n)$ "with high probability". More precisely, $X = \tilde{O}(f(n))$, if and only if $(\forall c > c_0 > 1)$ $Prob\{X \geq cf(n)\} \leq \frac{1}{n^{g(c)}}$ where c_0 is a fixed constant and $g(c)$ is a polynomial in c with $g(c) \rightarrow \infty$ for $c \rightarrow \infty$ [26].)

Consider a random set $S' \subset S$ of pivots. For each $x \in S$ let $nextPivot(x, S')$ refer to the closest pivot following x in the list S. (Without loss of generality, assume that the last element, λ, of S is selected as a pivot and let $nextPivot(\lambda, S') = \lambda$. Note that for $x \neq \lambda$, $nextPivot(x, S') \neq x$.) Let $distToPivot(x, S')$ be the distance between x and $nextPivot(x, S')$ in list S. Furthermore, let $m(S, S') = \max_{x \in S} distToPivot(x, S')$.

The *modified list ranking problem* for S with respect to S' refers to the problem of determining for each $x \in S$ its next pivot $nextPivot(x, S')$ as well as the distance $distToPivot(x, S')$. The input/output structure for the modified list ranking problem is the same as for the list ranking problem.

(1) *Select a set $S' \subset S$ of $\tilde{O}(\frac{n}{p})$ random pivots as follows: Every processor P_i makes for each $x \in S$ stored at P_i an independent biased coin flip which selects x as a pivot with probability $\frac{1}{p}$.*

(2) *All processors solve collectively the* modified list ranking problem *for S with respect to S'.*

(3) *Using an all-to-all broadcast, the values $nextPivot(x, S')$ and $distToPivot(x, S')$ for all pivots $x \in S'$ are broadcast to all processors.*

(4) Using the data received in Step 3, each processor P_i can solve the list ranking problem for the nodes stored at P_i sequentially in time $\tilde{O}(\frac{n}{p})$.

For the correctness of Step 1, we recall the following

Lemma 6.5 *[26] Consider a random variable X with binomial distribution. Let n be the number of trials, each of which is successful with probability q. The expectation of X is $E(X) = nq$, $Prob\{X > cnq\} \leq e^{-\frac{1}{2}(c-1)^2 nq}$, for any $c > 1$.*

In order to implement Step 2, we simply simulate the standard recursive doubling (also called pointer jumping) technique. >From Theorem 6.4 it follows that, with high probability, $m(S, S') \leq 3p \ln(n)$. Hence, Step 2 requires, with high probability, at most $\log(3p \ln(n)) = \log(3p) + \log \ln(n)$ communication rounds. Step 3 requires 1 communication round, and Step 4 is straightforward. In summary, we obtain

Theorem 6.6 *Algorithm 3.1.1 solves the list ranking problem using, with high probability, at most $1 + \log(3p) + \log \ln(n)$ communication rounds and $\tilde{O}(\frac{n}{p})$ local computation.*

3.1.2   Improving the maximum sublist size.

We now present an improved algorithm that solves the list ranking problem by using, with high probability, only $r \leq (4k + 6) \log(\frac{2}{3}p) + 8$ communication rounds and $\tilde{O}(\frac{n}{p})$ local computation where $k := \min\{i \geq 0 | \ln^{(i+1)} n \leq (\frac{2}{3}p)^{2i+1}\}$.

Note that $k < \ln^*(n)$ is an extremely small number (see Table 6.1).

The basic idea of the algorithm is that any two pivots should not be closer than $O(p)$ because this creates large "gaps" elsewhere in the list. If two pivots are closer than $O(p)$, then one of them is "useless" and should be "relocated". The non-trivial part is to perform the "relocation" without too much overhead and such that the new set of pivots has a considerably better distribution. The algorithm uses three colors to mark nodes: *black* (pivot), *red* (a node close to a pivot), and *white* (all other nodes).

Table 6.1. Values of k and $R := (4k + 6) \log(\frac{2}{3}p) + 8$ [upper bound on r] for various combinations of n and p.

$p =$	4	8	16	32	64	128	256	512	1024	2048
n	$k; R$	$k; R$	$k; R$	$k; R$	$k; R$	$k; R$	$k; R$	$k; R$	$k; R$	$k; R$
10^{10}	1;18	0;26	0;32	0;38	0;44	0;50	0;56	0;62	0;68	0;74
10^{100}	1;18	1;38	0;32	0;38	0;44	0;50	0;56	0;62	0;68	0;74
10^{1000}	1;18	1;38	1;48	0;38	0;44	0;50	0;56	0;62	0;68	0;74
$10^{(10^4)}$	1;18	1;38	1;48	1;58	0;44	0;50	0;56	0;62	0;68	0;74
$10^{(10^5)}$	1;18	1;38	1;48	1;58	1;68	0;50	0;56	0;62	0;68	0;74
$10^{(10^6)}$	1;18	1;38	1;48	1;58	1;68	1;78	0;56	0;62	0;68	0;74
$10^{(10^7)}$	1;18	1;38	1;48	1;58	1;68	1;78	1;88	0;62	0;68	0;74
$10^{(10^8)}$	1;18	1;38	1;48	1;58	1;68	1;78	1;88	1;98	0;68	0;74
$10^{(10^9)}$	1;18	1;38	1;48	1;58	1;68	1;78	1;88	1;98	1;108	0;74
$10^{(10^{10})}$	1;18	1;38	1;48	1;58	1;68	1;78	1;88	1;98	1;108	1;118
$10^{(10^{11})}$	1;18	1;38	1;48	1;58	1;68	1;78	1;88	1;98	1;108	1;118
$10^{(10^{12})}$	1;18	1;38	1;48	1;58	1;68	1;78	1;88	1;98	1;108	1;118
$10^{(10^{14})}$	2;22	1;38	1;48	1;58	1;68	1;78	1;88	1;98	1;108	1;118
$10^{(10^{16})}$	2;22	1;38	1;48	1;58	1;68	1;78	1;88	1;98	1;108	1;118
$10^{(10^{18})}$	2;22	1;38	1;48	1;58	1;68	1;78	1;88	1;98	1;108	1;118
$10^{(10^{20})}$	2;22	1;38	1;48	1;58	1;68	1;78	1;88	1;98	1;108	1;118
$10^{(10^{30})}$	2;22	1;38	1;48	1;58	1;68	1;78	1;88	1;98	1;108	1;118
$10^{(10^{40})}$	2;22	1;38	1;48	1;58	1;68	1;78	1;88	1;98	1;108	1;118
$10^{(10^{50})}$	2;22	1;38	1;48	1;58	1;68	1;78	1;88	1;98	1;108	1;118
$10^{(10^{60})}$	2;22	1;38	1;48	1;58	1;68	1;78	1;88	1;98	1;108	1;118
$10^{(10^{70})}$	2;22	1;38	1;48	1;58	1;68	1;78	1;88	1;98	1;108	1;118
$10^{(10^{80})}$	2;22	1;38	1;48	1;58	1;68	1;78	1;88	1;98	1;108	1;118
$10^{(10^{90})}$	2;22	1;38	1;48	1;58	1;68	1;78	1;88	1;98	1;108	1;118
$10^{(10^{100})}$	2;22	1;38	1;48	1;58	1;68	1;78	1;88	1;98	1;108	1;118

(1) Perform Step 1 of Algorithm 3.1.1. Mark all selected pivots black *and all other nodes* white.

(2) For $i = 1, \ldots, k$ do

(2a) For each black *node x, all nodes which are to the right of x (in list S) and have distance at most $\frac{2}{3}p$ are marked* red. *Note: previously* black *nodes (pivots) that are now marked* red *are no longer considered pivots.*

(2b) For each black *node x, all nodes which are to the left of x (in list S) and have distance at most $\frac{2}{3}p$ are marked* red.

(2c) *Every processor P_i makes for each* white *node $x \in S$ stored at P_i an independent biased coin flip which selects x as a new pivot, and marks it* black, *with probability $\frac{1}{p}$.*

(2d) *Every processor P_i marks* white *every* red *node $x \in S$ stored at P_i.*

(3) *Let $S' \in S$ be the subset of* black *nodes obtained after Step 2. Continue with Steps 2 – 4 of Algorithm   3.1.1.*

Observe that Steps 2a and 2b have to be performed in a left-to-right scan, respectively, as if executed sequentially. We can simulate this sequential scanning process in the parallel setting because the number of pivots is bounded by n/p. For Step 2a, we build linked lists of pivots by computing for each of them a pointer to the next pivot of distance at most $2\,p/3$, if any, and the distance. These linked lists of pivots are compressed into one processor and we run on these lists a sequential left-to-right scan to mark pivots red. We return the pivots to their original location and mark every non-pivot red for which there exists a non-red pivot that attempts to mark it red. Step 2b is performed analogously.

Let r be the number of communication rounds required by Algorithm 6.1. We will now show that, with high probability, $r \leq (4k + 6) \log(\frac{2}{3}p) + 8 = \tilde{O}(k \log p)$.

Let n_i be the maximum length of a contiguous sequence of *white* nodes after the ith execution of Step 2b, and define $n_0 = n$. Let S_i be the set of *black* nodes after the ith execution of Step 2c, $1 \leq i \leq k$, and let S_0 be the set of *black* nodes after the execution of Step 1. Note that, in Step 3, $S' = S_k$. Define $m_i = m(S_i)$ for $0 \leq i \leq k$.

Lemma 6.7 *With high probability, the following holds:*
(a) $n_0 = n$ and $n_i \leq 3p \ln(n_{i-1}), 1 \leq i \leq k$
(b) $m_i \leq 3p \ln(n_i), 0 \leq i \leq k$

Proof: It follows from Theorem 6.4 that, with high probability, $n_0 = n$ and $m_0 \leq 3p \ln(n)$ and, for a fixed $1 \leq i \leq k$ $n_i \leq m_{i-1}$ and $m_i \leq 3p \ln(n_i)$.
Since $k \leq \ln^*(n)$ and $\log^*(n)\frac{1}{n^c} \leq \frac{1}{n^{c-\epsilon}}, \epsilon > 0$, the above bounds for n_i and m_i hold, with high probability, for all $1 \leq i \leq k$.   □

Lemma 6.8 *With high probability, for all $1 \leq i \leq k$,*
(a) $n_i \leq 3p(2 \ln(3p) + \ln^{(i)}(n))$
(b) $m_i \leq 6p \ln(3p) + 3p \ln^{(i+1)}(n)$

Proof:
(a) Applying Lemma 6.7 we observe that

$$n_1 \quad \leq \quad 3p \ln(n)$$

$$n_2 \leq 3p\ln(3p\ln(n)) = 3p(\ln(3p) + \ln\ln(n))$$
$$n_3 \leq 3p\ln(n_2) \leq 3p(\ln(3p) + \ln(\ln(3p) + \ln\ln(n)))$$
$$\leq 3p(\ln(3p) + \ln\ln(3p) + \ln\ln\ln(n))$$
$$n_4 \leq 3p\ln(n_3) \leq 3p(\ln(3p) + \ln\ln(3p) + \ln\ln\ln(3p) + \ln\ln\ln\ln(n))$$
$$\vdots$$
$$n_i \leq 3p(2\ln(3p) + \ln^{(i)}(n))$$

(b) It follows from Lemma 6.7 that $m_i \leq 3p\ln(n_i) \leq 3p\ln(3p(2\ln(3p) + \ln^{(i)}(n))) \leq 3p(\ln(3p) + \ln(2) + \ln^{(2)}(3p) + \ln^{(i+1)}(n)) \leq 6p\ln(3p) + 3p\ln^{(i+1)}(n)$. □

Theorem 6.9 *With high probability, Algorithm 6.1 solves the list ranking problem with $r \leq (4k + 6)\log(\frac{2}{3}p) + 8 = \tilde{O}(k\log p)$ communication rounds and $\tilde{O}(\frac{n}{p})$ local computation.*

Proof: With high probability, the total number of communication rounds in Algorithm 6.1 is bounded by $2k\log(\frac{2}{3}p) + \log(m_k) + 1$
$$\leq 2k\log(\tfrac{2}{3}p) + \log(6p) + \log\ln(3p) + \log(3p) + \log\ln^{(k+1)}(n) + 1$$
$$\leq (2k + 3)\log(\tfrac{2}{3}p) + \log 9 + \log 4.5 + \log\ln^{(k+1)}(n) + 1$$
$$\leq (2k + 3)\log(\tfrac{2}{3}p) + \log\ln^{(k+1)}(n) + 8$$
$$\leq \log((\tfrac{2}{3}p)^{2k+3}) + \log\ln^{(k+1)}(n) + 8 \leq 2\log((\tfrac{2}{3}p)^{2k+3}) + 8$$
$$[\text{ if } (*)\ \ln^{(k+1)}(n) \leq (\tfrac{2}{3}p)^{2k+3}] \leq (4k + 6)\log(\tfrac{2}{3}p) + 8 = \tilde{O}(k\log p)$$
Condition $(*)$ is true because we selected $k = \min\{i \geq 0|\ \ln^{(i+1)}n \leq (\tfrac{2}{3}p)^{2i+1}\}$. Note that this bound is not tight. □

3.1.3 Simulation and experimental results.
We simulated the behavior of Algorithm 6.1. In particular, we simulated how the above method improves the sample by reducing the maximum distance, m_i, between subsequent pivots. We examined the range of $4 \leq p \leq 2048$ and $100,000 \leq n \leq 1,500,000$ as shown in Table 6.2 and applied Algorithm 6.1 for each n, p combination shown 100 times with different random samples. Table 6.2 shows the values of k and the upper bound R on the number of communication rounds required according to Theorem 6.9. We then measured the maximum distance, m_k^{obs}, observed between two subsequent pivots in the sample chosen at the end of the algorithm, as well as the number, r^{obs}, of communication rounds actually required. Each of the numbers shown is the worst case observed in the respective 100 test runs.

According to Theorem 6.9, for the range of test data used, the number of communication rounds in our algorithm should not exceed 78. This is an upper bound, though. The actual number of communication rounds observed in Table 6.2 is 25 in the worst case. The number of rounds observed is usually around 30% of the upper bound according to Theorem 6.9. We also observe that for a given p (i.e. in a vertical column), the values of m_k^{obs} and r^{obs} are essentially stable and show no monotone increase or decrease with increasing n.

Table 6.2. k, $R := (4k+6)\log(\frac{2}{3}p)+8$, m_k^{obs} and r^{obs} For various combinations of n and p, where m_k^{obs} and r^{obs} are the observed *worst case* values of m_k and r, respectively. (For each shown combination of n and p, the m_k^{obs} and r^{obs} shown are the worst case values observed during 100 test runs.)

$p =$	4	8	16	32	64	128	256	512	1024
n	k	k	k	k	k	k	k	k	k
	R	R	R	R	R	R	R	R	R
	m_k^{obs}	m_k^{obs}	m_k^{obs}	m_k^{obs}	m_k^{obs}	m_k^{obs}	m_k^{obs}	m_k^{obs}	m_k^{obs}
	r^{obs}	r^{obs}	r^{obs}	r^{obs}	r^{obs}	r^{obs}	r^{obs}	r^{obs}	r^{obs}
100,000	1	1	1	1	1	0	0	0	0
	18	38	48	58	68	50	56	62	68
	28	59	119	238	409	1400	2421	5900	9136
	8	13	16	19	22	12	13	14	15
500,000	1	1	1	1	1	1	0	0	0
	18	38	48	58	68	78	56	62	68
	32	72	117	264	474	860	3150	6144	11179
	8	14	16	20	22	25	13	14	15
1,000,000	1	1	1	1	1	1	0	0	0
	18	38	48	58	68	78	56	62	68
	40	69	127	264	440	851	3406	7924	11861
	9	14	16	20	22	25	13	14	15
1,500,000	1	1	1	1	1	1	0	0	0
	18	38	48	58	68	78	56	62	68
	33	89	172	270	551	903	3893	6120	11938
	9	14	17	20	23	25	13	14	15

Table 6.3.   No. of communication rounds on the PowerXplorer (worst values in 5 runs.)

NProc/n	64	128	256	512	1024	2048	4096	16384	65536	131072
4	7	8	8	8	9	9	9	9	10	9
8	8	9	10	10	10	10	10	10	10	10

In Table 6.3 we show the actual number of communication rounds needed by Algorithm 1 on the Parsytec PowerXplorer machine, with 16 nodes (each with a PowerPC601 processor and a T805 transputer).

3.1.4   Applications.   The problem of list ranking is a special case of computing the suffix sums of the elements of a linked list. The above algorithm can obviously be generalized to compute prefix or suffix sums for associative operators. List ranking is a very popular tool for obtaining numerous parallel tree and graph algorithms [6, 7]. An important application outlined in [6] is to use list ranking for applying Euler tourEuler tour techniques to tree problems: for an undirected forest of trees, rooting every tree at a given vertex chosen as root, determining the parent of each vertex in the rooted forest, computing the preorder (or postorder) traversal of the forest, computing the level of each vertex, and computing the number of descendants of each vertex. All these problems can be easily solved with one or a small constant number of list ranking operations.

3.1.5   Conclusion.   We presented a randomized parallel list ranking algorithm for distributed memory multiprocessors, using the coarse grained multicomputer model. The algorithm requires, with high probability, $r \leq (4k + 6) \log(\frac{2}{3}p) + 8 = \tilde{O}(k \log p)$ communication rounds. For all practical purposes, $k \leq 2$. Therefore, we expect that our result will have considerable practical relevance.

3.2.   A deterministic algorithm for list ranking

We revisit the important list ranking problem. Reid-Miller's [29] presented an empirical study of parallel list ranking for the Cray C-90. The paper followed essentially the CGM/BSP model and claimed that this was the fastest list ranking implementation so far. The algorithm in [29] required $O(\log n)$ communication rounds. In [11], an improved algorithm was presented which required, with high probability, only $O(k \log p)$ rounds, where $k \leq \log^* n$.

We improve these results by giving the first *deterministic* algorithm for list ranking using $O(\log p)$ rounds. This improvement is an important step towards the ultimate goal, a deterministic algorithm with only $O(1)$ communication rounds. In fact, it is an open problem whether this is possible for these graph problems. In contrast to the previous deterministic results, the improved number of communication rounds obtained, $O(\log p)$, is *independent* of n and grows only very slowly with respect to p. Hence, for most practical purposes, the number of communication rounds can be considered as constant. We expect that this will be of considerable practical relevance.

As in [29] we will, in general, assume that $n >> p$ (coarse grained), because this is usually the case in practice. Note, however, that our results for the list ranking problems hold for arbitrary ratios $\frac{n}{p}$.

3.2.1   The basic idea.   Let L be a list represented by a vector s such that $s[i]$ is the node following i in the list L. The last element l of the list L is the one with $s[l] = l$. The distance between i and j, $d_L(i,j)$, is the number of nodes between i and j plus 1 (i.e. the distance is 0 iff $i = j$, and it is one if and only if one node follows the other). The list ranking problem consists of computing for each $i \in L$ the distance between i and l, referred to as $rank_L(i) = d_L(i,l)$.

For our algorithm, we need the following definitions. A *r-ruling set* is defined as a subset of selected list elements that has the following properties: (1) No two neighboring elements are selected. (2) The distance of any unselected element to the next selected element is at most r.

An overview of our CGM list ranking algorithm is as follows. First, we compute a $O(p^2)$-ruling set R with $|R| = O(n/p)$ and broadcast R to all processors. More precisely, the $O(p^2)$-ruling set R is represented as a linked list where each element i is assigned a pointer to the next element j of R with respect to the order implied by L as well as the distance between i and j in L. Then, every processor sequentially performs a list ranking of R, computing for each $i \in R$ its distance to the last element of L. All other list elements have at most distance $O(p^2)$ from the next element of R in the list. Their distance is determined by simulating standard PRAM pointer jumping until the next element of R is reached.

All steps, except for the computation of the $O(p^2)$-ruling set R, can be easily implemented in $O(\log p)$ communication rounds.

In the remainder of this section we introduce a new technique, called *deterministic list compression*, which will allows us to compute a $O(p^2)$-ruling set in $O(\log p)$ communication rounds.

The basic idea behind *deterministic list compression* is to have an alternating sequence of *compress* and *concatenate* phases. In a compress phase, we select a subset of list elements, and in a *concatenate* phase we use pointer jumping to work our way towards building a linked list of selected elements.

For the compress phase, we apply the *deterministic coin tossing* technique of [10] but with a different set of labels. Instead of the memory address used in [10], we use the number of the processor storing list item i as its label $l(i)$. During the computation, we select sequentially the elements of R in the sublists of subsequent nodes in L which are stored at the same processor. The term "subsequent" refers to successor with respect to the *current* value of s.

Note that there are at most p different labels, and subsequent nodes in those parts of L that are not processed sequentially have different labels. We call list element $s[i]$ a *local maximum* if $l(i) < l(s[i]) > l(s[s[i]])$. We apply deterministic coin tossing to those parts of L that are not processed sequentially.

The naïve approach of applying this procedure $O(\log p)$ times would yield a $O(p^2)$-ruling set, but unfortunately it would require more than $O(\log p)$ communication rounds. Note that, when we want to apply it for a second, third, etc. time, the elements selected previously need to be linked by pointers. Since two subsequent elements selected by deterministic coin tossing can have distance

$O(p)$, this may require $O(\log p)$ communication rounds, each. Hence, this straightforward approach requires a total of $O(\log^2 p)$ communication rounds.

Notice, however, that if two selected elements are at distance $\Theta(p)$ at a given moment, then it is unnecessary to further apply deterministic coin tossing in order to reduce the number of selected elements. The basic approach of our algorithm is therefore to interleave pointer jumping and deterministic coin tossing operations with respect to our new labeling scheme. More precisely, we will have only one pointer jumping step between subsequent deterministic coin tossing steps, and such pointer jumping operations will not be applied to those list elements that are pointing to selected elements.

This concludes the high level overview of our *deterministic list compression* techniques. The following describes the algorithm in detail.

CGM Algorithm for computing a p^2-ruling set.

Input: *A linked list L and a vector s where $s[i]$ is the node following i in the list L. L and s are stored on a p processor CGM with total $O(n)$ memory.*

Output: *A set of selected nodes of L (which is a p^2-ruling set).*

(1) Mark all list elements as not selected.

(2) FOR EVERY list element i IN PARALLEL:
IF $l(i) < l(s[i]) > l(s[s[i]])$ THEN mark $s[i]$ as selected.

(3) Sequentially, at each processor, process the sublists of subsequent list elements which are stored at the same processor. For each such sublist, mark every second element as selected. If a sublist has only two elements, and not both neighbors have a smaller label, then mark both elements of the sublist as not selected.

(4) FOR $k = 1 \ldots \log p$ DO

   (4.1) FOR EVERY list element i IN PARALLEL:
   IF $s[i]$ is not selected THEN set $s[i] := s[s[i]]$.

   (4.2) FOR EVERY list element i IN PARALLEL:
   IF $(i, s[i]$ and $s[s[i]]$ are selected) AND NOT $(l(i) < l(s[i]) > l(s[s[i]]))$ AND $(l(i) \neq l(s[i]))$ AND $(l(s[i]) \neq l(s[s[i]]))$ THEN mark $s[i]$ as not selected.

   (4.3) Sequentially, at each processor, process the sublists of subsequent selected list elements which are stored at the same processor. For each such sublist, mark every second selected element as not selected. If a sublist has only two elements, and not both neighbors

have a smaller label, then mark both elements of the sublist as not
selected.

(5) Select the last element of L.

We first prove that the set of elements selected at the end of Algorithm 3.2.1
is of size at most $O(n/p)$.

Lemma 6.10 *After the kth iteration in Step 4, there are no more than two
selected elements among any 2^k subsequent elements of the original list L.*

We now prove that subsequent elements selected at the end of Algorithm 3.2.1
have distance at most $O(p^2)$. We first need the following.

Lemma 6.11 *After every execution of Step 4.3, the distance of two subsequent
selected elements with respect to the current pointers (represented by vector s)
is at most $O(p)$.*

Lemma 6.12 *After the k-th execution of Step 4.3, two subsequent elements with
respect to the current pointers (represented by vector s) have distance $O(2^k)$
with respect to the original list L.*

Lemma 6.13 *No two subsequent selected elements have a distance of more
than $O(p^2)$ with respect to the original list L.*

In summary, we obtain

Theorem 6.14 *The list ranking problem for a linked list with n vertices can
be solved on a CGM with p processors and $O(\frac{n}{p})$ local memory per processor
using $O(\log p)$ communication rounds and $O(\frac{n}{p})$ local computation per round.*

The proofs can be found in [12] and will be omitted here.

4.   Connected components and spanning forest

Consider an undirected graph $G = (V, E)$ with n vertices and m edges.
Each vertex $v \in V$ has a unique label between 1 and n. Two vertices u and v
are connected if there is an undirected path of edges from u to v. A *connected
subset*Connected subset of vertices is a subset of vertices where each pair of
vertices is connected. A *connected component*Connected component of G is
defined as a maximal connected subset.

Bäumker and Dittrich [8] present a randomized connected components algorithm for planar graphs using $O(\log p)$ communication rounds. They suggest
an extension of this algorithm for general graphs with the same number of
communication rounds.

In this section, we study the problem of computing the connected components of G on a CGM with p processors and $O(\frac{n+m}{p})$ local memory per processor. We improve the previous result by giving the first *deterministic* algorithm for connected components problem using $O(\log p)$ rounds. We introduce a new technique, called *clipping*, which refers to the idea of taking a PRAM algorithm for the same problem but running it for only $O(\log p)$ rounds and then finishing the computation with some other $O(\log p)$ rounds CGM algorithm. (See also JáJá's *accelerated cascading* technique for the PRAM [21].)

Steps 1 and 2 of Algorithm 4 simulate Shiloch and Vishkin's PRAM algorithm [32], but for $\log p$ phases only. Each vertex v has a pointer to a vertex *parent(v)* such that the *parent(v)* pointers always form trees. The trees are also referred to as a *supervertices*. A tree of height one is called a star. An edge (u, v) is *live if parent(u)* $\neq$ *parent(v)*. Shiloch and Vishkin's PRAM algorithm merges supervertices along live edges until they equal the connected components. When simulated on a CGM or BSP computer, Shiloch and Vishkin's PRAM algorithm results in $\log n$ communication rounds or supersteps, respectively.

Our CGM algorithm requires $O(\log p)$ rounds only. It simulates only the first $\log p$ iterations of the main loop in the PRAM algorithm by Shiloch and Vishkin and then completes the computation in another $\log p$ communication rounds (Steps 3 - 7).

CGM Algorithm for Connected Component Computation

Input: *An undirected graph $G = (V, E)$ with n vertices and m edges stored on a p processor CGM with total $O(n + m)$ memory.*

Output: *The connected components of G represented by the the values parent(v) for all vertices $v \in V$.*

(1) FOR all $v \in V$ IN PARALLEL DO parent(v) := v.

(2) FOR $k := 1$ to $\log p$ DO

    (2.1) FOR all $v \in V$ IN PARALLEL DO parent(v) := parent(parent(v)).

    (2.2) FOR every live edge (u, v) IN PARALLEL DO (simulating concurrent write)

        (a) IF parent(parent(v)) = parent(v) AND parent(parent(u)) = parent(u) THEN { IF parent(u) > parent(v) THEN parent(parent(u)) := parent(v) ELSE parent(parent(v)) := parent(u) }

(b) *IF parent(u) = parent(parent(u)) AND parent(u) did not get new links in steps 2.1 and 2.2(a) THEN parent(parent(u)) := parent(v)*

(c) *IF parent(v) = parent(parent(v)) AND parent(v) did not get new links in steps 2.1 and 2.2.1 THEN parent(parent(v)) := parent(u)*

(2.3) *FOR all $v \in V$ IN PARALLEL DO parent(v) := parent(parent (v)).*

(3) *Use the Euler TourEuler tour algorithm in [12] to convert all trees into stars. For each $v \in V$, set parent(v) to be the root of the star containing v. Let $G' = (V', E')$ be the graph consisting of the supervertices and live edges obtained. Distribute G' such that each processor stores the entire set V' and a subset of $\frac{m}{p}$ edges of E'. Let E_i be the edges stored at processor i, $0 \le i \le p - 1$.*

(4) *Mark all processors as* active.

(5) *FOR $k := 1$ to $\log p$ DO*

(5.1) *Partition the active processors into groups of size two.*

(5.2) *FOR each group P_i, P_j of active processors, $i < j$ IN PARALLEL DO*

(a) *processor P_j sends it's edge set E_j to processor P_i.*

(b) *processor P_j is marked as* passive.

(c) *processor P_i computes the spanning forest (V', E_s) of the graph $SF = (V', E_i \cup E_j)$ and sets $E_i := E_s$.*

(6) *Mark all processors as* active *and broadcast E_0.*

(7) *Each processor i computes sequentially the connected components of the graph $G'' = (V', E_0)$. For each vertex v of V' let parent'(v) be the smallest label parent(w) of a vertex $w \in V'$ which is in the same connected component with respect to $G'' = (V', E_0)$. For each vertex $u \in V$ stored at processor P_i set parent(u) := parent'(parent(u)). (Note that parent(u) $\in V'$.)*

Lemma 6.15 [32] *The number of different trees after iteration k of Step 2 is bounded by $(\frac{2}{3})^k n$.*

We obtain

Theorem 6.16 *Algorithm 4 computes the connected components and spanning forest of a graph $G = (V, E)$ with n vertices and m edges on a CGM with p processors and $O(\frac{n+m}{p})$ local memory per processor, $\frac{n+m}{p} \geq p^\epsilon$ ($\epsilon > 0$), using $O(\log p)$ communication rounds and $O(\frac{n+m}{p})$ local computation per round.*

Proof: Steps 1 and 2 simulate the PRAM algorithm by Shiloch and Vishkin [32] but with only $\log p$ iterations of the main loop in Step 2 instead of the $O(\log n)$ iterations in Shiloch and Vishkin's original algorithm. Step 3 converts all trees into stars. It follows from Lemma 6.15 that the graph $G' = (V', E')$ has at most $O(\frac{n}{p})$ vertices. Hence, V' can be broadcast to all processors. E' can still be of size $O(m)$ and is distributed over the p processors. Recall that E_i refers to the edges in E' stored at processor i. We note that the spanning forest of $(V', E_i \cup E_j)$ for two sets E_i, E_j is of size $O(|V'|) = O(\frac{n}{p})$. Hence, each of the spanning forests computed in Step 5 can be stored in the local memory of a single processor. Steps 3-6 merge pairs of spanning forests until, after $\log p$ rounds, the spanning forest of G' is stored at processor P_0. In Step 7, all processors update the partial connected component information obtained in Step 2.

The PRAM simulations in Steps 2.1 - 2.3 require $O(1)$ rounds, each.

Hence, the entire algorithm requires $O(\log p)$ communication rounds. The local computation per round is obviously bounded by $O(\frac{n+m}{p})$.

We observe that the algorithm is easily extended to report the entire spanning forest for G within the same time bounds. $\qquad \square$

5. Parallel range minima

Given an array of n real numbers $A = (a_1, a_2, \ldots, a_n)$, define $MIN(i, j) = \min\{a_i, \ldots, a_j\}$. The *range minima problem*Range minima consists of preprocessing array A such that queries $MIN(i, j)$, for any $1 \leq i \leq j \leq n$, can be answered in constant time.

We present an algorithm, under the $CGM(n, p)$ model, that solves the problem requiring a constant number of communication rounds and $O(\frac{n}{p})$ computation time. In the PRAM model, there is an optimal algorithm of $O(\log n)$ time [21]. This algorithm, however, is not immediately transformable to the CGM model. Berkman et al. [27] describe an optimal algorithm of $O(\log \log n)$ time for a PRAM CRCW. This algorithm motivates the CGM algorithm presented in this paper.

Range minima is a basic problem and appears in other problems. In [21], range minima is used in the design of a PRAM algorithm for the problem of Lowest Common Ancestor (LCA). This algorithm uses twice the Euler-tourEuler tour problem. The idea of this algorithm can be used in the CGM model, by using the same sequence of steps. In [12], the Euler-tourEuler tour problem in the CGM model is discussed. Its implementation reduces to the list ranking problem, that uses $O(\log p)$ communication rounds in the CGM model [12]. Applications of the LCA problem, on the other hand, are found in graph problems (see [30]) and in the solution of some other problems. Besides the LCA problem, the range minima problem is used in several other applications [1, 19, 28].

Definition 6.17 *Consider an array $A = (a_1, a_2, \ldots, a_n)$ of n real numbers. The prefix minimum array is the array*

$$P = (\min\{a_1\}, \min\{a_1, a_2\}, \ldots, \min\{a_1, a_2, \ldots, a_n\})$$

The suffix minimum array is the array

$$S = (\min\{a_1, a_2, \ldots, a_n\}, \min\{a_2, \ldots, a_n\}, \ldots, \min\{a_{n-1}, a_n\}, \min\{a_n\})$$

Definition 6.18 *The lowest common ancestor of two vertices u and v of a rooted tree is the vertex w that is an ancestor of u and v and that is farthest from the root. We denote $w = LCA(u, v)$.*

Definition 6.19 *The lowest common ancestor problem consists of preprocessing a rooted tree such that queries of the type $LCA(u, v)$, for any vertices u and v of the tree, can be answered in constant sequential time.*

5.1.   Sequential algorithms

The algorithm for the range minima problem, in the CGM model, uses two sequential algorithms that are executed using local data in each processor. These algorithms are described in the following.

5.1.1   Algorithm of Gabow et al..

This sequential algorithm was designed by Gabow et al. [19] and uses the *Cartesian tree* data structure, introduced in [33]. An example of Cartesian tree is given in figure 6.2. In this algorithm we need also an algorithm for the lowest common ancestor problem.

Definition 6.20 *Given an array $A = (a_1, a_2, \ldots, a_n)$ of n distinct real numbers, the Cartesian tree of A is a binary tree whose nodes have as label the values of array A. The root of the tree has as label $a_m = \min\{a_1, a_2, \ldots, a_n\}$. Its left subtree is a Cartesian tree for $A_{1,m-1} = (a_1, a_2, \ldots, a_{m-1})$, and its right subtree is a Cartesian tree for $A_{m+1,n} = (a_{m+1}, \ldots, a_n)$. The Cartesian tree for an empty array is the empty tree.*

Range Minima (Gabow et al.)

1 Build a Cartesian tree for A.

2 Use a linear sequential algorithm for the LCA problem using the Cartesian tree.

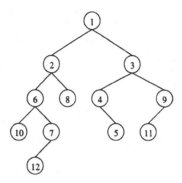

Figure 6.2.   Cartesian tree corresponding to the array $(10, 6, 12, 7, 2, 8, 1, 4, 5, 3, 11, 9)$.

The construction of the Cartesian tree takes linear time [19]. Linear sequential algorithms for the LCA problem can be found in [9, 20, 31]. Thus, any query $MIN(i, j)$ is answered as follows. From the recursive definition of the Cartesian tree, the value of $MIN(i, j)$ is the value of the LCA of a_i and a_j. Thus, each range minima query can be answered in constant time through a query of LCA in the Cartesian tree. Therefore, the range minima problem is solved in linear time.

5.1.2   Sequential algorithm of Alon and Schieber.   In this section we describe the algorithm of Alon and Schieber [3] of $O(n \log n)$ time. In spite of its non linear complexity, this algorithm is important in the description of the CGM algorithm, as will be seen in section 5.2. Without loss of generality, we consider n to be a power of 2.

  Range Minima (Alon and Schieber)

  1 Construct a complete binary tree T with n leaves.

  2 Associate the elements of A to the leaves of T.

  3 For each vertex v of T we compute the arrays P_v and S_v, the prefix minimum and the suffix minimum arrays, respectively, of the elements of the leaves of the subtree with root v.

Tree T constructed by algorithm 5.1.2 will be called *PS-tree*. Figure 6.3 illustrates the execution of this algorithm. Queries of type $MIN(i, j)$ are answered as follows. To determine $MIN(i, j)$, $1 \leq i \leq j \leq n$, we find $w = LCA(a_i, a_j)$ in T. Let v and u be the left and right sons of w, respectively. Then, $MIN(i, j)$ is the minimum between the value of S_v in the position corresponding to a_i and the value of P_u in the position corresponding to a_j. *PS*-tree is a complete binary tree. In [21, 30] it is shown that a LCA query in complete binary trees can be answered in constant time.

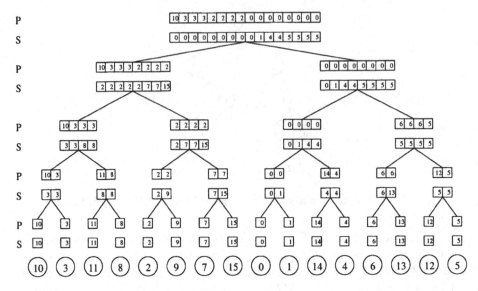

Figure 6.3. PS-tree generated by the algorithm of Alon and Schieber for the array $(10, 3, 11, 8, 2, 9, 7, 15, 0, 1, 14, 4, 6, 13, 12, 5)$.

5.2.   The CGM algorithm for the range minima problem

>From the sequential algorithms seen in the previous section, we can design a CGM algorithm for the range minima problem. This algorithm is executed in $O(\frac{n}{p})$ time and uses $O(1)$ communication rounds. The major difficulty is how to store the required data structure among the processors so that the queries can be done in constant time, without violating the limit of $O(\frac{n}{p})$ memory, required by the CGM model.

Given an array $A = (a_1, a_2, \ldots, a_n)$, we write $A[i] = a_i$, $1 \leq i \leq n$, and $A[i \ldots j] =$ subarray $(a_i, \ldots, a_j)$, $1 \leq i \leq j \leq n$.

The idea of the algorithm is based on how queries $MIN(i, j)$ can be answered. Each processor stores $\frac{n}{p}$ contiguous positions of the input array. Thus, given $A = (a_1, a_2, \ldots, a_n)$, we have subarrays $A_i = (a_{i\frac{n}{p}+1}, \ldots, a_{(i+1)\frac{n}{p}})$, for $0 \leq i \leq p - 1$. Depending on the location of a_i and a_j in the processors, we have the following cases:

1 if a_i and a_j are in a same processor, the problem domain reduces to the subarray stored in that processor. Thus, we need a data structure to answer this type of queries in constant time. This data structure is obtained in each processor by Algorithm 6.2 (section 5.1.1).

2 if a_i and a_j are in distinct processors $p_{\bar{i}}$ e $p_{\bar{j}}$ (without loss of generality, $\bar{i} < \bar{j}$), respectively, we have two subcases:

(a) if $\bar{i} = \bar{j} - 1$, a_i and a_j are in neighbor processors, $MIN(i,j)$ corresponds to the minimum between the minimum of a_i through the end of array $A_{\bar{i}}$ and the minimum of the beginning of $A_{\bar{j}}$ to a_j. These minima can be determined by Algorithm 6.2. To determine the minimum of the minima we need one communication round.

(b) if $\bar{i} < \bar{j} - 1$, $MIN(i,j)$ corresponds to the minimum among the minimum of subarray $A_{\bar{i}}[i - \bar{i}\frac{n}{p} \ldots (i+1)\frac{n}{p}]$, the minimum of subarray $A_{\bar{j}}[\bar{j}\frac{n}{p} + 1 \ldots j - \bar{j}\frac{n}{p}]$ and the minima of subarrays $A_{\bar{i}+1}, \ldots, A_{\bar{j}-1}$. The first two minima are obtained as in the previous subcase. The minima of subarrays $A_{\bar{i}+1}, \ldots, A_{\bar{j}-1}$ are easily obtained by using the Cartesian tree. The minimum among them corresponds to the range minima problem restricted to the array of minima of the data in the processors. Thus we need a data structure to answer these queries in constant time. As the array of minima contains only p values, this data structure can be obtained by Algorithm 5.1.2 (section 5.1.2).

The difficulty of case 2.b is that we cannot construct the PS-tree explicitly in each processor as described in section 5.1.2, since this would require a memory of size $O(p \log p)$, larger than the memory requirement in the CGM model, which is $O(\frac{n}{p})$, with $\frac{n}{p} \geq p$. To overcome this difficulty we construct arrays P' and S' of $\log p + 1$ positions each, described as follows, to store some partial information of the PS-tree T in each processor. Let us describe this construction in processor i, with $0 \leq i \leq p - 1$. Let b_i be the value of the minimum of array A_i. Let v be any vertex of T such that the subtree with root v has b_i as leaf and let d_v be the *depth of v* in T, that is the path length from the root to v, as defined in [2]; and l_v be the *level of v*, which is the height of the tree minus the depth of v, as defined in [2] ($l_v = \log p - d_v$, because tree T has height $\log p$). Array P' (respectively, S') contains in position l_v the value of array P_v (respectively, S_v), of the level l_v of T, in the position corresponding to the leaf b_i. In other words, we have $P'[l_v] = P_v[i \bmod 2^{l_v} + 1]$.

Figure 6.4 illustrates the correspondence between arrays P' and S' stored in each processor and the PS-tree constructed by the algorithm of Alon and Schieber [3]. In Figure 6.4.b, the bold face positions in arrays S in each level of the tree correspond to the suffix minimum of $b_0 = 3$ in each level. In this way, we obtain array S' in P_0 (figure 6.4.c).

Range Minima (CGM model)

{Each processor receives $\frac{n}{p}$ contiguous positions of array A, partitioned into subarrays $A_0, A_1, \ldots, A_{p-1}$.}

1 Each processor i executes sequentially Algorithm 6.2 (section 5.1.1).

2 {*Each processor constructs an array* $B = (b_i)$ *of size* p, *that contains the minimum of the data stored in each processor.*}

  2.1. *Each processor* i *calculates* $b_i = \min A_i = \min\{a_{i\frac{n}{p}+1}, \ldots,$
  $a_{(i+1)\frac{n}{p}}\}$.

  2.2. *Each processor* i *sends* b_i *to the other processors.*

  2.3. *Each processor* i *puts in* b_k *the value received from processor* k,
  $k \in \{0, \ldots, p-1\} \setminus \{i\}$.

3 *Each processor* i *executes procedures* Construct_P' *and* Construct_S'
  (see below).

  {*By observing the description of Algorithm   5.1.2 (section   5.1.2),* $P'[k]$
  contains position i *of the prefix minimum array in level* k, $0 \le k \le \log p$.}

Given array B, the following procedure constructs array P' of $\log p + 1$ postitions in processor i, for $0 \le i \le p - 1$. This procedure constructs P' in $O(p)$ $(= O(\frac{n}{p}))$ time using only local data. The construction of array S' is done in a symmetric way, considering array B in reverse order.

Procedure 6.21 *PROCEDURE* Construct_P'.

1. $P'[0] \leftarrow b_i$
2. pointer $\leftarrow i$
3. inorder $\leftarrow 2 * i + 1$
4. for $k \leftarrow 1$ *until* $\log p$ *do*
5. $P'[k] \leftarrow P'[k-1]$
6.   *if* $\lfloor inorder/2^k \rfloor$ *is odd*
7.   *then for* $l \leftarrow 1$ *until* 2^{k-1} *do*
8.   *pointer* $\leftarrow pointer - 1$
9.   *if* $P'[k] > B[pointer]$
10.   *then* $P'[k] \leftarrow B[pointer]$

To simplify the correctness proof of this procedure, we consider p to be a power of 2 and that, in each processor i, array B contains leaves of a complete binary tree, as in the description of the sequential algorithm of section 5.1.2. We do not store an entire array in each internal node of this tree, but only the $\log p + 1$ positions of the prefix minimum array in each level corresponding to position i of array B.

The value of the variable *inorder* of line 3 is the value of the in-order number of the leaf containing b_i, obtained in an in-order traversal of the tree nodes. It can easily be seen that these values, from the left to the right, are the odd numbers in the interval $[1, 2p - 1]$.

Theorem 6.22 *Procedure 6.21 correctly calculates array P', for each processor i, $0 \le i \le p - 1$.*

Proof: We give an idea of the proof. For a processor i, $0 \le i \le p - 1$, we prove the following invariant at each iteration of the for loop of line 4: At each iteration k, let T_k be subtree that contains b_i and has root at level k. $P'[k]$ stores the position i of the minimum prefix array of the subarray of B corresponding to the leaves of subtree T_k and the variable *pointer* contains the index, in B, of the leftmost leaf of T_k. □

To determine S' we have a similar theorem with a similar proof.

Lemma 6.23 *Execution of Procedure 6.21 in each processor takes $O(\frac{n}{p})$ sequential time.*

Proof: The maximum number of iterations is $\log p$. In each iteration, the value of *pointer* is decremented or remains the same. The worst case is when $i = p - 1$. In this case, at each iteration the value of the variable *pointer* is updated. As the number of elements of B is p, the algorithm updates the value of *pointer* at most p times. Therefore, procedure 6.21 is executed in $O(p) = O(\frac{n}{p})$ sequential time. □

Theorem 6.24 *Algorithm 5.2 solves the range minima problem in $O(\frac{n}{p})$ time using $O(1)$ communication rounds.*

Proof: Step 1 is executed in $O(\frac{n}{p})$ sequential time and does not use communication. Step 2 runs in $O(\frac{n}{p})$ sequential time and uses one communication round. By theorem 6.23, step 3 is executed in $O(\frac{n}{p})$ sequential time and does not require communication. Therefore, algorithm 5.2 solves the range minima problem in $O(\frac{n}{p})$ time using $O(1)$ communication rounds. □

5.3.   Query processing

In this section, we show how to use the output of algorithm 5.2 to answer queries $MIN(i, j)$ in constant time. A query $MIN(i, j)$ is handled as follows. Assume that i and j are known by all the processors and the result will be given by processor 0. If a_i and a_j are in a same processor, then $MIN(i, j)$ can be determined by step 1 of algorithm 5.2. Otherwise, suppose that a_i and a_j are in distinct processors $p_{\bar{i}}$ and $p_{\bar{j}}$, respectively, with $\bar{i} < \bar{j}$. Let $right(\bar{i})$ the index in A of the rightmost element in array $A_{\bar{i}}$, and $left(\bar{j})$ the index in A of the leftmost element in array $A_{\bar{j}}$. Calculate $MIN(i, right(\bar{i}))$ and $MIN(left(\bar{j}), j)$, using step 1. We have two cases:

1 If $\bar{j} = \bar{i}+1$, then $MIN(i, j) = \min\{MIN(i, right(\bar{i})), MIN(left(\bar{j}), j)\}$.

2 If $\bar{i} + 1 < \bar{j}$, then calculate $MIN(right(\bar{i}) + 1, left(\bar{j}) - 1)$, using step 3 of the algorithm. Notice that $MIN(right(\bar{i}) + 1, left(\bar{j}) - 1)$ corresponds to $\min\{b_{\bar{i}+1}, \ldots, b_{\bar{j}-1}\}$. Thus, $MIN(i, j) = \min\{MIN(i, right(\bar{i})), MIN(right(\bar{i}) + 1, left(\bar{j}) - 1), MIN(left(\bar{j}), j)\}$.

The value of $MIN(right(\bar{i}) + 1, left(\bar{j}) - 1)$ is obtained using step 3, as follows. Each processor calculates $w = LCA(b_{\bar{i}+1}, b_{\bar{j}-1})$ in constant time. Then determines the level l_w; determines v and u, the left and right sons of w, respectively; determines l, the index of the leftmost leaf in the subtree of u. Processor $\bar{i} + 1$ calculates $S_v(2^{l_w - 1} - l + \bar{i})$ and sends this value to processor 0. Processor $\bar{j} - 1$ calculates $P_u(\bar{j} - l)$ and sends this value to processor 0. Processor 0, finally, calculates the minimum of the received minima. In both cases, processor 0 receives the minima of processors $\bar{i}$ and $\bar{j}$.

Notice finally that the correctness of Algorithm 5.2 results from the observations presented in this section.

5.4.   Experimental results

The proposed algorithm has been implemented on a distributed memory parallel machine, the Parsytec PowerXplorer 16/32 Parallel Computing System, with 16 nodes under a two-dimensional topology. Each node consists of one PowerPC601 application processor and one T805 communication processor, and 32 MBytes of local memory. The implementation is done in PVM (parallel virtual machine). We use the interface PowerPVM that implements PVM 3.2. Compared to the implementation with the native system called PARIX, we observe very similar performance results. PVM is chosen for portability.

For the experiments we use the input size $n = 2^k$, with $k = 5, \ldots, 19$, and $p = 2^j$ processors, with $j = 0, \ldots, 4$. The inputs consist of integers randomly distributed among the processors. The input distribution is not important, since the number of operations required in the execution of the algorithm does not depend on how the input data is distributed among the processors.

There is only one communication round in which the minima of the subarrays in each processor are broadcast to the others. Therefore, the amount of data exchanged depends on p and does not depend on n. Communication thus becomes less significant for large n.

We use the term *speedup*Speedup as the quotient between the sequential running time and the parallel running time. For our experiments, the sequential time is obtained by using an optimized sequential algorithm (Gabow et al. [19]) running on a single processor, instead of running the parallel algorithm run on one processor.

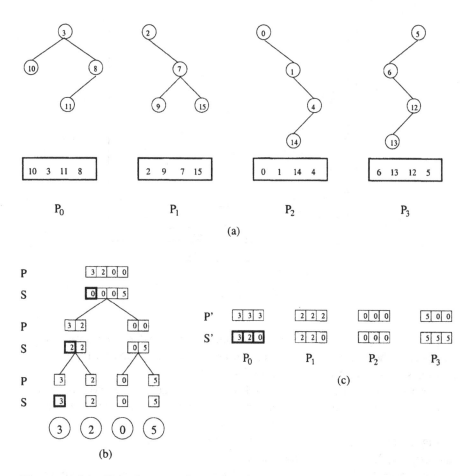

Figure 6.4. Execution of algorithm 5.2 using the array $(10, 3, 11, 8, 2, 9, 7, 15, 0, 1, 14, 4, 6, 13, 12, 5)$. (a) The data distributed in the processors and the corresponding Cartesian trees. (b) PS-trees constructed by the algorithm of Alon and Schieber [3] of section 5.1.2 for the array $(3, 2, 0, 5)$ of the minima of processors. (c) Arrays P' and S' constructed by step 3 of algorithm 5.2 corresponding to arrays P and S of T.

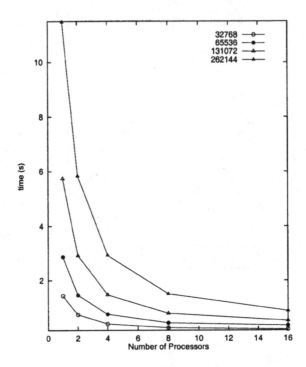

Figure 6.5.   Maximum times.

For small n ($n < 2^{10}$), no significant gain is obtained due to communication overhead. The effect of communication becomes less influential for $n > 2^{10}$ and the speedup becomes significant for $n > 2^{15}$.

For a given experiment, with fixed n and p, we do not observe significant difference in the running times in different executions. Therefore, we run the experiments only once each to get the results.

Fig. 6.5 shows the running times (in seconds) for $n = 2^k$ with $k = 15, \ldots, 18$ and using 1, 2, 4, 8 and 16 processors.

Fig. 6.6 shows the speedup obtained. Notice the speedup curves approximate the optimal curve for large n.

5.5.   Conclusion

In this work we present an algorithm, in the CGM(n, p) model, for the range minima problem that runs in $O(n/p)$ time using $O(1)$ communication rounds. This algorithm can be used to design a CGM algorithm for the LCA problem, based on the PRAM algorithm of JáJá [21], instead of other PRAM algorithms as in [22, 31], which would give less efficient implementations. Many graph algorithms in turn are based on the LCA problem.

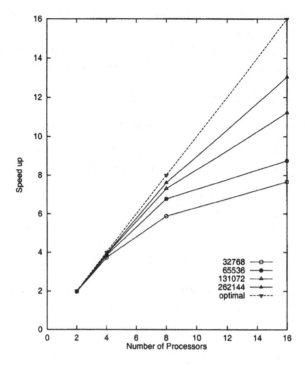

Figure 6.6.   Speed up.

In the design of the CGM algorithm, we chose to use the PS-tree to handle case 2(b) instead of merely constructing a Cartesian tree of an array of p numbers. This second approach, though feasible, would increase the local processing time. The small communication overhead, together with efficient local computing, enable us to obtain the final CGM algorithm, which is able to beat the very efficient sequential algorithm, as shown in the experimental results.

The experimental results are very promising, with speedup curves getting closer to the optimal speedup curve for large n, indicating the efficiency and scalability of the proposed algorithm. The efficiency of the proposed algorithm results from the constant number of communication rounds with low communication overhead and $O(n/p)$ computing time.

6.   Lowest common ancestor

The *lowest common ancestor*Lowest common ancestor,$LCA(u,v)$, of two vertices u and v of a rooted tree $T = (V, E)$ is the vertex w that is an ancestor to both u and v, and is farthest from the root. We address the problem of preprocessing T in order to answer a query $LCA(u, v)$ quickly for any pair (u, v). We refer to this problem as the *lowest-common-ancestor* (LCA) *problem.*Lowest common ancestor problem

We apply the approach in [21] that uses Euler tourEuler tour and range-minimum calculation. It consists of the following operations:

(1) compute an Euler tourEuler tour for T;

(2) find the levels, in T, for all vertices of the Euler tourEuler tour;

(3) for each vertex v find $l(v)$ and $r(v)$ which denote the leftmost and right-most) appearances, respectively, of v in the Euler tour;

(4) solve the range-minima problem defined as follows: given a list of numbers $\{b_1, b_2, \ldots, b_n\}$ and an interval $[i, j]$, with $1 \leq i \leq j \leq n$, find the minimum of $\{b_i, \ldots, b_j\}$.

Operation 4 is important because of the following:

Lemma 6.25 *Consider a rooted tree* $T = (V, E)$. *Let* u *and* v *be two arbitrary distinct vertices of* T. *Then, the following statements hold:*

- u *is an ancestor of* v *if and only if* $l(u) < l(v) < r(u)$.

- u *and* v *are not related (that is,* u *is not a descendant of* v *and* v *is not a descendant of* u*) if and only if either* $r(u) < l(v)$ *or* $r(v) < l(u)$.

- *If* $r(u) < l(v)$ *then* $LCA(u, v)$ *is the vertex with minimum level contained in the interval* $[r(u), l(v)]$.

Operation 1 can be performed in $O(\log p)$ communication rounds as shown in [12]. The same holds for Operation 2 because it can also be reduced to Euler tourEuler tour computation.

Consider Operation 3. Given an Euler tour of vertices $a_1, a_2, \ldots, a_n, a_1$. The element $a_i = v$ is the leftmost (rightmost) appearance of v if and only if $level(a_{i-1}) = level(v) - 1$ ($level(a_{i+1}) = level(v) - 1$, respectively) [21]. This requires the use of indices of the vertices in the Euler tour. Our Euler tour is not given as an array of vertices but rather by pointers to successor vertices in the tour. This is easily solved by considering the index as the rank obtained by list ranking from Section 3.2.1. The rank can be viewed as an index going backwards from the list. After list ranking (in $O(\log p)$ communication rounds), Operation 3 can be completed in $O(1)$ communication rounds.

Operation 4 also uses indices. Likewise, instead of indices, we utilize the ranks of the vertices of the Euler tour. Given two vertices u and v of T. In order to find the minimum level over the interval $[r(u), l(v)]$, let $rank(r(u)) = i$ and $rank(l(v) = j$. To find the required minimum, each of the p processors considers vertices in its local memory with ranks between j and i and finds the minimum level ($O(n/p$ time). The minimum of the resulting p numbers can be found in O(1) communication rounds.

We obtain

Theorem 6.26 *Consider a rooted tree $T = (V, E)$ with n vertices. The LCA problem can be solved on a CGM with p processors and $O(\frac{n}{p})$ local memory per processor using $O(\log p)$ communication rounds and $O(\frac{n}{p})$ local computation per round.*

7. Concluding remarks

In the CGM(n, p) model, we are interested in designing parallel algorithms with linear local executing time and that utilize few communication rounds – preferably a constant number. This is because the communication between processors is more expensive than the computing time of the local data.

Several results are presented in this text.

We present a randomized parallel list ranking algorithm for distributed memory multiprocessors. A simple version requires, with high probability, $\log(3p) + \log \ln(n) = \tilde{O}(\log p + \log \log n)$ communication rounds (h-relations with $h = \tilde{O}(\frac{n}{p})$) and $\tilde{O}(\frac{n}{p})$ local computation. An improved version requires, with high probability, only $r \leq (4k + 6)\log(\frac{2}{3}p) + 8 = \tilde{O}(k \log p)$ communication rounds where $k = \min\{i \geq 0 | \ln^{(i+1)} n \leq (\frac{2}{3}p)^{2i+1}\}$. Note that $k < \ln^*(n)$ is an extremely small number. For $n \leq 10^{10^{100}}$ and $p \geq 4$, the value of k is at most 2. For a given number of processors, p, the number of communication rounds required is, for all practical purposes, independent of n. For $n \leq 10^{10^{100}}$ and $4 \leq p \leq 2048$, the number of communication rounds in our algorithm is bounded, with high probability, by 118. We conjecture that the actual number of communications rounds will not exceed 50.

We then present *deterministic* parallel algorithms for the coarse grained multicomputer (CGM) which solve the list ranking and determination of the connected components of a graph. The algorithms require $O(\log p)$ communication rounds and linear sequential work per round. We view the algorithms presented as an important step towards the final goal of $O(1)$ communication rounds. Note that the number of communication rounds obtained in this paper is independent of n, and grows only very slowly with respect to p. Hence, for most practical purposes, the number of communication rounds can be considered as constant. The result for these two problems constitutes a considerable improvement over those previously reported.

Given an array of n real numbers $A = (a_1, a_2, \ldots, a_n)$, define $MIN(i, j) = \min\{a_i, \ldots, a_j\}$. The range minima problem consists of preprocessing array A such that queries $MIN(i, j)$, for any $1 \leq i \leq j \leq n$, can be answered in constant time. Range minima is a basic problem that appears in many other important problems such as lowest common ancestor, Euler tourEuler tour, pattern matching with scaling etc. We present a parallel algorithm under the CGM model (Coarse Grained Multicomputer), that solves the range minima problem in $O(\frac{n}{p})$ time and constant number of communication rounds.

Acknowledgments

This text is based on the results of many researchers. In particular I would like to thank my co-authors E. Cáceres, A. Chan, F. Dehne, A. Ferreira, P. Flocchini, H. Mongelli, I. Rieping, A. Roncato, N. Santoro, and E. Saukas.

References

[1] G. M. Landau A. Amir and U. Vishkin. Efficient pattern matching with scaling. *Journal of Algorithms*, 13:2–32, 1992.

[2] J. E. Hopcroft A. V. Aho and J. D. Ullman, editors. *The Design and Analysis of Computer Algorithms*. Addison-Wesley Publishing Company, 1995.

[3] N. Alon and B. Schieber. Optimal preprocessing for answering on-line product queries. Technical Report TR 71/87, The Moise and Frida Eskenasy Inst. of Computer Science, Tel Aviv University, 1987.

[4] J. R. Anderson and G. L. Miller. Deterministic parallel list ranking. In J. H. Reif, editor, *VLSI Algorithms and Architectures: 3rd Aegean Workshop on Computing*, volume 319, pages 81–90. Springer-Verlag, Lecture Notes in Computer Science, 1988.

[5] J. R. Anderson and G. L. Miller. A simple randomized parallel algorithm for list ranking. *Information Processing Letters*, 33(5):269–273, 1990.

[6] M. J. Atallah and S. E. Hambrusch. Solving tree problems on a mesh-connected processor array. *Information and Control*, 69:168–187, 1986.

[7] S. Baase. Introduction to parallel connectivity, list ranking, and euler tour techniques. In J. H. Reif, editor, *Synthesis of Parallel Algorithms*. Morgan Kaufmann Publisher, 1993.

[8] A. Bäumker and W. Dittrich. Parallel algorithms for image processing: Practical algorithms with experiments. In *International Parallel Processing Symposium*, pages 429–433. IEEE Computer Society, 1996.

[9] O. Berkman and U. Vishkin. Recursive star-tree parallel data structure. *SIAM J. Comput.*, 22(2):221–242, 1993.

[10] R. Cole and U. Vishkin. Approximate parallel scheduling, part i: the basic technique with applications to optimal parallel list ranking in logarithmic time. *SIAM J. Computing*, 17(1):128–142, 1988.

[11] F. Dehne and S. W. Song. Randomized parallel list ranking for distributed memory multiprocessors. In *Second Asian Computing Science Conference*, volume 1179, pages 1–10. Springer-Verlag, Lecture Notes in Computer Science, 1996.

[12] E. Caceres et al. Efficient parallel graph algorithms for coarse grained multicomputers and bsp. In A. Marchetti-Spaccamela P. Degano, R. Gorrieri, editor, *Proceedings ICALP'97 - 24th International Colloquium on Automata, Languages, and Programming*, volume 1256, pages 390–400. Springer-Verlag, Lecture Notes in Computer Science, 1997.

[13] F. Dehne et al. A randomized parallel 3d convex hull algorithm for coarse grained multicomputers. In *Proc. ACM Symposium on Parallel Algorithms and Architectures (SPAA'95)*, pages 27–33. ACM, 1995.

[14] L. G. Valiant et al. General purpose parallel architectures. In J. van Leeuwen, editor, *Handbook of Theoretical Computer Science*, pages 943–972. MIT Press/Elsevier, 1990.

[15] A. Fabri F. Dehne and C. Kenyon. Scalable and architecture independent parallel geometric algorithms with high probability optimal time. In *Proc. 6th IEEE Symposium on Parallel and Distributed Processing*, pages 586–593. IEEE Computer Society, 1994.

[16] A. Fabri A. F. Dehne and Rau-Chaplin. Scalable parallel geometric algorithms for coarse grained multicomputers. In *Proc. ACM 9th Annual Computational Geometry*, pages 298–307. ACM, 1993.

[17] B. M. Maggs G. E. Blelloch, C. E. Leiserson and C. G. Plaxton. A comparison of sorting algorithms for the connection machine cm-2. In *Proc. ACM Symp. on Parallel Algorithms and Architectures*, pages 3–16. ACM, 1991.

[18] G. L. Miller R. H. and Reif. Parallel tree contraction part 1: Further applications. *SIAM J. Computing*, 20(6):1128 – 1147, December 1991.

[19] J. L. Bentley H. N. Gabow and R. E. Tarjan. Scaling and related techniques for geometry problems. In *Proc. 16th ACM Symp. On Theory of Computing*, pages 135–143. ACM, 1984.

[20] D. Harel and R. E. Tarjan. Fast algorithms for finding nearest common ancestors. *SIAM J. Comput.*, 13(2):338–345, 1984.

[21] J. JáJá, editor. *An introduction to parallel algorithms*. Addison-Wesley Publishing Company, 1992.

[22] R. Lin and S. Olariu. A simple optimal parallel algorithm to solve the lowest common ancestor problem. In *Proc. of International Conference on Computing and Information*, volume 497, pages 455–461. Springer-Verlag, Lecture Notes in Computer Science, 1991.

[23] C. L. Miller M. Reid-Miller and F. Modugno. List ranking and parallel tree compaction. In J. H. Reif, editor, *Synthesis of Parallel Algorithms*. Morgan Kaufmann Publisher, 1993.

[24] G. L. Miller and J. H. Reif. Parallel tree contraction part 1: Fundamentals. *Advances in Computing Research*, 5:47–72, 1989.

[25] H. Mongelli and S. W. Song. A range minima parallel algorithm for coarse grained multicomputers. In José Rolim et al., editor, *IPPS99/Irregular - Sixth International Workshop on Solving Irregularly Structured Problems in Parallel*, volume 1586, pages 1075–1084. Springer-Verlag, Lecture Notes in Computer Science, 1999.

[26] K. Mulmuley, editor. *Computational Geometry: An Introduction Through Randomized Algorithms*. Prentice Hall, 1993.

[27] B. Schieber O. Berkman and U. Vishkin. Optimal doubly logarithmic parallel algorithms based on finding all nearest smaller values. *Journal of Algorithms*, 14:344–370, 1993.

[28] V. L. Ramachandran and U. Vishkin. Efficient parallel triconectivity in logarithmic parallel time. In *Proceedings of AWOC'88*, volume 319, pages 33–42. Springer-Verlag, Lecture Notes in Computer Science, 1988.

[29] M. Reid-Miller. List ranking and list scan on the cray c-90. In *ACM Symp. on Parallel Algorithms and Architectures*, pages 104–113. ACM, 1994.

[30] J. H. Reif, editor. *Synthesis of Parallel Algorithms*. Morgan Kaufmann Publishers, 1993.

[31] B. Schieber and U. Vishkin. On finding lowest common ancestors: simplification and parallelization. *SIAM J. Comput.*, 17:1253–1262, 1988.

[32] Y. Shiloch and U. Vishkin. An $o(\log n)$ parallel connectivity algorithm. *Journal of Algorithms*, 3(1):57–67, 1983.

[33] J. Vuillemin. A unified look at data structures. *Comm. of the ACM*, 23:229–239, 1980.

Chapter 7

PARALLEL METAHEURISTICS FOR COMBINATORIAL OPTIMIZATION

Sandra Duni Ekşioğlu
*University of Florida**
duni@cao.ise.ufl.edu

Panos M. Pardalos
*University of Florida**
pardalos@ufl.edu

Mauricio G. C. Resende
AT&T Labs Research[†]
mgcr@research.att.com

Abstract   In this chapter, we review parallel metaheuristics for approximating the global optimal solution of combinatorial optimization problems. Recent developments on parallel implementation of genetic algorithms, simulated annealing, tabu search, variable neighborhood search, and greedy randomized adaptive search procedures (GRASP) are discussed.

Keywords: Parallel local search, parallel GRASP, parallel genetic algorithms, parallel simulated annealing, parallel tabu search, variable neighborhood descent, parallel computing environments, heuristics, combinatorial optimization.

---
[*]Center for Applied Optimization, Department of Industrial and Systems Engineering, University of Florida, Gainesville, FL 32611 USA.
[†]Information Sciences Research Center, AT&T Labs Research, 180 Park Avenue, Room C241, Florham Park, NJ 07932 USA.

179

R. Corrêa et al. (eds.), Models for Parallel and Distributed Computation. Theory, Algorithmic Techniques and Applications, 179–206.
© 2002 *Kluwer Academic Publishers.*

1.   Introduction

Search techniques are fundamental problem-solving methods in computer science and operations research. Search algorithms have been used to solve many classes of problems, including path-finding problems, two-player games, and constraint satisfaction problems. Classical examples of path-finding problems include many combinatorial optimization problems (e.g. integer programming) and puzzles (e.g. Rubic's cube, Eight Puzzle). Chess, backgammon, and Othello belong to the class of two player games, while a classic example of a constraint satisfaction problem is the eight-queens problem.

In this chapter, we focus on NP-hard combinatorial optimization problems. A combinatorial optimization problem P in its minimization form can be formulated as follows. Given a finite set $E = \{1, 2, \ldots, n\}$, the set of feasible solutions $F \subseteq 2^E$ of P, and an objective function $f : 2^E \to \mathbb{R}$, one wishes to find $S^* \in F$ such that $f(S^*) \leq f(S), \forall S \in F$, where $f(S)$ is the objective function defined as $f(S) = \sum_{e \in S} c(e)$, where $c(e)$ is the cost of including $e \in S$ in the solution S. For a specific optimization problem, the sets E and F, as well as the function $c(e)$ need to be defined. For combinatorial optimization problems, exact search algorithms, such as branch and bound or dynamic programming, may degenerate to complete enumeration. Because of this, exact approaches limit us to solve only moderately sized problem instances, due to the exponential increase in CPU time and memory when problem size increases. Therefore, in practice, heuristic search algorithms are necessary to find (not necessarily optimal) solutions to these problems. Heuristics explore a small portion of the solution space, where one believes good solutions can be found. Most heuristics for combinatorial optimization make use of a basic local search method called iterative improvement which can be described as in Figure 7.1 in its minimization form ([59, 74]).

The initial solution s_0 can be produced in many different ways. A solution s' is "near" another solution s if it is in the set of neighborhood solutions $N(s)$ of s. Usually, a solution is in the neighborhood of another if it can be obtained from the other solution by an elementary perturbation or move. The neighborhood is usually constructed as the set $M_N(s)$ of all elementary moves from s. Often the set $M_N(s)$ is too large for a practical algorithm, and one searches in a subset $C_N(s) \subset M_N(s)$ called the candidate set. Iterative improvement proceeds until a locally optimal solution is found. Once such a solution is found, no further improvement is possible.

One limitation of local search is getting trapped in local optima. To get around this limitation, several general purpose heuristic strategies have been developed in the last two decades. For large-scale problems, another limitation is the exponential computational complexity of iterative improvement [60]. Parallel local search algorithms are one way to cope with this problem. Parallel

Layout of Iterative Improvement

Input: *A problem instance P, and a feasible solution $s_0 \in F$*

Output: *A (sub-optimal) solution $s \in F$*

1. Initialize $s = s_0$;

2. **while** (there exists $s' \in N(s)$ such that $f(s') < f(s)$) **do**
   (a) Select $s' \in C_N(s)$ such that $f(s') < f(s)$;
   (b) Let $s = s'$;

3. Output s as the (sub-optimal) solution;

Figure 7.1. Iterative Improvement

implementation can significantly increase the size of the problems that can be solved. While there is a large body of work on search algorithms, work on parallel search algorithms is relatively sparse.

A survey of parallel local search algorithmsis given by Verhoeven and Aarts [118]. They give a review of some concepts that can be used to incorporate parallelism into local search. They distinguish between single walk and multiple walk parallel local search and between asynchronous and synchronous parallelism. In a single walk algorithm only a single walk in the solution space is carried out, whereas in a multiple-walk algorithm several walks are performed simultaneously. In the class of single walk algorithms they distinguish between single step and multiple step parallel local search. The idea of single step parallelism is to evaluate neighbors simultaneously and subsequently make a single step. In an algorithm with multiple steps parallelism, several consecutive steps through the solution space are made simultaneously. In the class of multiple-walks they distinguish between algorithms that perform interacting walk and algorithms that perform multiple independent walks. Finally, both in single walk and multiple walk parallel local search, they distinguish between synchronous and asynchronous algorithms. In a synchronously parallel algorithm one or more steps of the algorithm are performed simultaneously by all processors, while synchronous parallelism requires a global clocking scheme that guarantees that communication occurs at fixed points of time. They observed that multiple-walk parallelism offers promising results to parallel tabu search and simulated annealing algorithms for a wide range of problems, and single-step parallelism can be used for most of the tabu search algorithms.

For a comprehensive overview of parallel search algorithms the reader is referred to [22, 80, 31, 32, 90, 16].

Macready et al. [71] report on some interesting investigations of the effect of local search on combinatorial optimization. Although they accept that local search methods constitute one of the most successful approaches to combinatorial optimization problems, they demonstrate that for a wide class of search techniques, increasing parallelism leads to better solution faster, but only to a certain point. At some degree of parallelism, the quality of solution abruptly degrades to that of random sampling.

In this chapter, we explore different approaches of parallel heuristic search for solving combinatorial optimization problems. We focus on issues of parallelization of genetic algorithms, simulated annealing, tabu (or taboo) search, variable neighborhood search, and GRASP (greedy randomized adaptive search procedures). These heuristic methods are often called metaheuristic since they are general purpose optimization schemes that can be adapted to address specific optimization problems. These metaheuristics have been used to approximately solve a wide spectrum of combinatorial optimization problems [98].

Portions of this chapter are based on Pardalos, Mavridou, Pitsoulis, and Resende [94].

2.   Parallel variable neighborhood descent

Variable neighborhood descent (VND) is a metaheuristic recently proposed by Hansen and Mladenović [45, 46, 47, 48, 49, 50, 51]. The approach is a local improvement heuristic as described in the previous section, with the difference that instead of utilizing a single neighborhood structure, several neighborhood structures are used. These extended neighborhoods allow searching for improving solutions that are "further" from the current solution, thus allowing the method to escape local optima with respect to a smaller neighborhood.

Let $N_1, N_2, \ldots, N_p$ be p neighborhood structures such that $N_k(s)$ is the set of solutions in the k-th neighborhood of s. One usually assumes that neighborhood N_{k+1} is larger than neighborhood N_k. Figure 7.2 illustrates variable neighborhood descent. The while loop (2) is repeated while $k < p$. During each iteration of the loop, a random starting solution s' in the k-th neighborhood of s is generated and local search is applied using neighborhood N_1. Let s'' be the local optimal solution. If an improvement is found with local search, then $s = s''$ and the search restarts with $k = 1$. If no improvement is found in the local search, the search proceeds with $k = k + 1$.

Since VND is relatively new, it has not been investigated much from a parallelization point of view. Martins [75] outlined how parallel implementations of VND can be accomplished. In a low-level parallelization, each search of the neighborhood $N_k(s)$ could be done in parallel. $N_k(s)$ would be divided

Layout of Variable Neighborhood Descent

Input:　*A problem instance P, a feasible solution $s_0 \in F$, and neighborhoods $N_1, N_2, \ldots, N_p$.*
Output:　*A (sub-optimal) solution $s \in F$*

1. Initialize $s = s_0$; Improve = **true**;

2. **while** (Improve == **true**) **do**
　(a) Improve = **false**;
　(b) $k = 1$;
　(c) **while** $k \leq p$ **do**
　　(i) Generate s' at random $N_k(s)$;

　　(ii) Apply local search using N_1 and s' as the
　　initial solution. Let s'' be the local optimum;
　　(iii) **if** $f(s'') < f(s)$ **then do**
　　　Let $s = s''$;
　　　Improve = **true**;
　　　break;
　　else $k = k + 1$;
3. Output s as the (sub-optimal) solution;

Figure 7.2.　Variable Neighborhood Descent

between the processors and each would return an improving neighbor in its partition. The best neighbor would be chosen as the current solution. In another strategy, $N_k(s)$ would again be partitioned and each processor would search for the best neighbor in its partition. Again, the best neighbor of the best would be chosen as the current solution. In a multi-search thread strategy, the processors would execute independent sequential VNDs and in the end, the best solution found would be output.

3.　Parallel genetic algorithms

In the 1960's, biologists began to use digital computers to perform simulations of genetic systems. Although these studies were aimed at understanding natural phenomena, some were not too distant from the modern notion of a genetic algorithm (GA). Genetic algorithms, as they are used today, were first introduced by Holland [52]. Genetic algorithms try to imitate the development of new and better populations among different species during evolution, just as their early biological predecessors. Unlike most standard heuristic algorithms, GAs use information of a population of individuals (solutions) when they conduct their search for better solutions as opposed to only information from a single individual. GAs have been applied to a number of problems in combinatorial optimization. In particular, the development of parallel computers has made this an interesting approach.

A GA aims at computing sub-optimal solutions by letting a set of random solutions undergo a sequence of unary and binary transformations governed by a selection scheme biased towards high-quality solutions. Solutions to optimization problems can often be encoded to strings of finite length. GAs work on these strings [52]. The encoding is done through the structure named *chromosomes*, where each chromosome is made up of units called *genes*. The values of each gene are binary, and are sometimes called *alleles*. The problem is encoded by representing all its variables in binary form and placing them together in a single chromosome. A fitness function evaluates the quality of a solution represented by a chromosome.

There are several critical parts which strongly affect the success of genetic algorithms:

- Representation of the solutions.

- Generation of the initial population.

- Selection of the individuals in an old population that will be allowed to affect the individuals of a new population. In terms of evolution, this relates to the selection of suitable parents in the new population.

- Genetic operators, such as crossover and mutation. That is, how to recombine the genetic heritage from the parents in the previous generation.

If $P(t)$ denotes the population at time t, the GA can be described as in Figure (7.3).

$P(0)$ is usually generated at random. The evaluation of a population $P(t)$ involves computing the fitness of the individuals and checking if the current population satisfies certain termination conditions. Types of termination rules include:

- A given time limit which is exceeded.

- The population is dominated by a few individuals.

- The best objective function value of the populations is constant over a given number of generations.

Due to their inherent parallel properties, GAs have been successfully implemented on parallel computers, introducing this way a new group of GAs, *Parallel Genetic Algorithms* (PGAs). In a PGA the population is divided into subpopulations and an independent GA is performed on each of these subpopulations. Furthermore, the best individuals of a local population are transferred

Layout of Genetic Algorithm

Input: *A problem instance*

Output: *A (sub-optimal) solution*

1. $t = 0$, initialize $P(t)$, evaluate fitness of individuals in $P(t)$;

2. **while** (termination condition is not satisfied) **do**
   (a) $t = t + 1$;
   (b) Select $P(t)$, recombine $P(t)$ and evaluate $P(t)$;

3. Output the best solution among all the population as
   the (sub-optimal) solution;

Figure 7.3. Layout of Genetic Algorithm

to the other subpopulations. Communication among the subpopulations is established to facilitate the operations of selection and recombination. There are two types of communication [86]: (1) *among all nodes*, where the best string in each subpopulation is broadcast to all the other subpopulations, and (2) *among the neighboring nodes*, i.e. only the neighboring subpopulations receive the best strings.

The most important aspects of PGAs, which result in a considerable speedup relative to sequential GAs, are the following [58]:

- *Local selection*, i.e. a selection of an individual in a neighborhood is introduced, in contrast with the selection in original GAs which is performed by considering the whole population.

- *Asynchronous behavior* which allows the evolution of different population structures at different speeds, possibly resulting in an overall improvement of the algorithm in terms of CPU time.

- *Reliability* in computation performance, i.e. the performance of one processor does not affect the performance of the other processors.

Jog et al. [58] consider two basic categories of *parallel genetic algorithms*:

1 *Coarse-grained* PGAs, where subpopulations are allocated to each processor of the parallel machine. The selection and recombination steps are performed within a subpopulation.

2 *Fine-grained* PGAs, where a single individual or a small number of individuals are allocated to each processor.

In the same reference, a review of parallel genetic algorithms applied to the *traveling salesman problem* (TSP) is presented. A PGA developed by Suh and Gucht [113], has been applied to TSPs of growing size (100-1000 cities). PGAs without selection and crossover, so called *independent strategies*, were used. These algorithms consist of running an "unlimited" number of independent sequential local searches in parallel. PGAs with *low amount of local improvement* were used and performed better in terms of quality solution than the independent strategies. In terms of computational time, the PGA showed nearly a linear-speedup for various TSPs, using up to 90 processors. The algorithms were run on a BBN Butterfly.

Another implementation of a PGA applied to the TSP can be found in [41], where an asynchronous PGA, called ASPARAGOS, is presented in detail. An application of ASPARAGOS was also presented by Muhlenbein [85] for the *quadratic assignment problem* (QAP) using a polysexual voting recombination operator. The PGA was implemented on QAPs of size 30 and 36 and for TSPs of size 40 and 50 with known solutions. The algorithm found a new optimum for the Steinberg problem (QAP of size 36). The numbers of processors used to run this problem were 16, 32, and 64. The 64 processor implementation (on a system with distributed memory) gave by far the best results in terms of computational time.

Battiti and Tecchiolli [9] presented parallelization schemes of genetic algorithms and tabu search for combinatorial optimization problems and in particular for *quadratic assignment* and *N-k problems* giving indicative experimental results.

Tanese [115] presents a parallel genetic algorithm implemented on a 64-processor NCUBE/six hypercube. Each processor runs the algorithm on each own subpopulation (coarse-grained PGA), and sends in an adjustable frequency a portion of good individuals to one of its neighboring processors. Each exchange takes place along a different dimension of the hypercube. The PGA is applied to a function optimization problem and its performance is compared with the corresponding GA, which is found to be similar. In addition, the PGA achieves comparable results with near linear speedup.

Pettey et al. [101] proposed an "island model" that restricts the communication among the subpopulations to some adjacent interconnected subpopulations [69]. The PGA was tested on four of DeJong's testbed of functions [24]. Population sizes of 50 to 800 were used, with the best answer in terms of quality solution, corresponding to size 50, which is approximately the theoretical optimal population size. The algorithm was implemented on an Intel *iPSC*, a message-based multiprocessor system with a binary *n-cube* interconnection network.

More recently, Lee and Kim [69] developed PGAs, based on the island model, for solving job scheduling problems with generally weighted earliness and tardiness penalties (GWET), satisfying specific properties, such as the *V-shaped schedule* around the due date. A binary representation scheme is used to code the job schedules into chromosomes. A GA is developed by parallelizing the population into subgroups, each of which keeps its distinct feasible schedules. The initial population is constructed so that the resulting genotype sequence satisfies the *V-shaped schedule*. Two different reproduction methods are employed, the *roulette wheel selection* and the *N-best selection method*. Also, the *crossover* and *mutation* operators are implemented in parallel. Several instances of problems with subpopulations of 30 chromosomes (jobs) and total population size of 100 to 200 where used to evaluate the efficiency of the PGA. The authors report that the roulette wheel selection scheme performs far better than the N-best selection in terms of quality solution, though the N-best selection gives good results in terms of CPU time. The mutation operator seems to improve the performance of the PGA. The chapter concludes, showing the superior performance of the parallel algorithm when compared to the sequential GA. In terms of CPU time, the parallelization of the population into several groups speeds up the convergence of the GAs as the size of the problem increases.

Parallel genetic algorithms have been applied to the graph partition problem [68], scheduling problems [19], and global optimization problems [111].

4. Parallel simulated annealing

The simulated annealing method (SA) was proposed in 1983 by Kirkpatrick et al. [62], based on the pioneering work of Metropolis et al. [79]. Since then, much research has been accomplished regarding its implementation (sequential and parallel) to solve a variety of difficult combinatorial optimization problems. Simulated annealing is based on the analogy between statistical mechanics and combinatorial optimization. The term *annealing* derives from the roughly analogous thermal process of melting at high temperatures and then lowering the temperature slowly based on an annealing schedule, until the vicinity of the solidification temperature is reached, where the system is allowed to reach the *ground state* (the lowest energy state of the system). Simulated annealing is a simple Monte Carlo approach to simulate the behavior of this system to achieve thermal equilibrium at a given temperature in a given annealing schedule. This analogy has been applied in solving combinatorial optimization problems. Given an objective function $f(x)$ over a feasible domain D, a generic simulated annealing algorithm for finding the global minimum of $f(x)$ is given in Figure 7.4.

An introduction to general concepts of parallel simulated annealing techniques can be found in Aarts and Korst [1]. Several parallel algorithms based on SA have been implemented for a variety of combinatorial optimization problems [3, 70, 63]. In the context of annealing, a parallel implementation can be

Layout of Simulated Annealing

Input: *A problem instance*

Output: *A (sub-optimal) solution*

1. Generate initial solution at random and initialize temperature T;

2. **while** $(T > 0)$ **do**
    (a) **while** (thermal equilibrium not reached) **do**
        (i) Generate a neighbor state at random and evaluate
        the change in energy level ΔE;

        (ii) If $\Delta E < 0$ update current state with new state;

        (iii) If $\Delta E \geq 0$ update current state with new state
        with probability $e^{\frac{-\Delta E}{K_B T}}$;

    (b) Decrease temperature T according to annealing schedule;

3. Output the solution having the lowest energy;

Figure 7.4. Layout of Simulated Annealing

presented in two forms [108]: (1) *functional parallelism*, which uses multiple processors to evaluate different phases of a single move, (2) *data parallelism*, which uses different processors or group of processors to propose and evaluate moves independently. The second form has the advantage of "easily scaling the algorithm to large ensembles of processors."

The *standard-cell approach* is a semi-custom designing method in which functional building blocks called *cells* are used to construct a part of, or an entire, VLSI chip [108]. Two basic approaches have been applied to the placement problem [108] – *constructive* methods and *iterative* methods. The SA technique belongs to the second group, as an approach that uses probabilistic hill climbing to avoid local minima. The TimberWolf program has been proposed as a version of SA implemented for the cell placement problem. It has been shown to provide enormous chip area savings compared to the already existed standard methods of cell layout. Jones and Banerjec [61] developed a parallel algorithm based on TimberWolf, where multiple cell moves are proposed and evaluated by using pairs of processors. The algorithm was implemented on a iPSC/1 Hypercube. Rose et al. [106] presented two parallel algorithms for the

same problem. The *Heuristic Spanning* approach replaces the high temperature portion of simulated annealing, assigning cells to fixed sub-areas of the
chip. The *Section Annealing* approach is used to speed-up the low temperature
portion of simulated annealing. The placement here is geographically divided
and the pieces are assigned to separate processors. Other parallel algorithms
based on SA and implemented for the cell placement problem can be found in
[15, 108, 119].

Greening [42] examines the *asynchronous parallel* SA algorithm in relation
with the effects of calculation errors resulting from the parallel implementation
or an approximate cost function. More analytically, the relative work analyzes
the effects of *instantaneous* and *accumulated* errors. The first category contains
the errors which result as the difference between the true and inaccurate costs
computed at a given time. An accumulated error is the sum of a stream of instantaneous errors. The author proves a direct connection between the accumulated
errors measured in previous research work, and annealing properties.

Most recently, Boissin and Lutton [10] proposed a new SA algorithm that
can be efficiently implemented on a massively parallel computer. A comparison to the sequential algorithm has been presented by testing both algorithms on two classical combinatorial optimization problems: (1) the *quadratic
sum assignment problem* and (2) the *minimization of an unconstrained* $0 - 1$
quadratic function. The numerical results showed that the parallel algorithm
converges to high-quality suboptimal solutions. For the problem of minimization of quadratic functions with 1000 variables, the parallel implementation on
a 16K Connection Machine was 3.3 times faster than the sequential algorithm
implemented on a SPARC 2 workstation, for the same quality of solution.

SA general purpose software packages are provided for public use. Ingber
provides an ASA-package (adaptive SA) [57] and Carter provides a straightforward, simulated annealing algorithm as a public available package [14]. A
parallel simulated annealing library *parSA* can be found on [64].

5. Parallel tabu search

Tabu search (TS), first introduced by Glover [37, 38, 40], is a heuristic procedure to find good solutions to combinatorial optimization problems. A *tabu
list* is a set of solutions determined by historical information from the last t iterations of the algorithm, where t is fixed or is a variable that depends on the state
of the search, or a particular problem. At each iteration, given the current solution x and its corresponding neighborhood $N(x)$, the procedure moves to the
solution in the neighborhood $N(x)$ that most improves the objective function.
However, moves that lead to solutions on the tabu list are forbidden, or are tabu.
If there are no improving moves, TS chooses the move which least changes the
objective function value. The tabu list avoids returning to the local optimum

from which the procedure has recently escaped. A basic element of tabu search is the *aspiration* criterion, which determines when a move is admissible despite being on the tabu list. One *termination* criterion for the tabu procedure is a limit in the number of consecutive moves for which no improvement occurs. A more detailed description of the main features, aspects, applications and extensions of tabu search, can be found in [39]. Several implementations of parallel algorithms based on TS have been developed for classical optimization problems, such as TSP and QAP, which will be discussed below. Given an objective function $f(x)$ over a feasible domain D, a generic tabu search for finding an approximation of the global minimum of $f(x)$ is given in Figure 7.5.

Layout of Tabu Search

Input: *A problem instance*

Output: *A (sub-optimal) solution*

1. Initialization:
   (a) Generate an initial solution x and set $x^* = x$;
   (b) Initialize the tabu list $T = \emptyset$;
   (c) Set iteration counters $k = 0$ and $l = 0$;

2. **while** $(N(x) \setminus T \neq \emptyset)$ **do**
   (a) $k = k + 1; l = l + 1$;
   (b) Select x as the best solution from set $N(x) \setminus T$;
   (c) If $f(x) < f(x^*)$ then update $x^* = x$ and set $l = 0$;
   (d) If $k = \bar{k}$ or if $l = \bar{l}$ go to step 3;

3. Output the best solution found x^*;

Figure 7.5.   Layout of Tabu Search

Taillard [114] presents two implementations of parallel TS for the quadratic assignment problem. Computational results on instances of size up to 64 are reported. The form of TS considered is claimed to be more robust than earlier implementations, since it requires less computational effort and uses only the basic elements of TS (moves to neighboring solutions, tabu list, aspiration function). The neighborhood used is a *2-exchange neighborhood*. The tabu list is made up of pairs (i, j) of interchanged units both of which are placed at locations they had occupied within the last s iterations, where s is the size of tabu list. The size s changes its value randomly in an interval during the search.

The *aspiration criterion* is introduced to allow the tabu moves to be chosen, if both interchanged units are assigned to locations they have not occupied within the t most recent iterations. In the first method, the neighborhood is divided into p parts of approximately the same size. Each part is distributed for evaluation to one of p different processors. Using a number of processors proportional to the size of the problem (precisely $p = n/10$), the complexity is reduced by a factor of n, where n is the size of the problem. Moreover, the computational results showed improvement to the quality of solution for many large problems and the best published solutions of other problems have been found. The second method performs independent searches, each of which starts with a different initial solution. The parallel algorithm was implemented on a ring of 10 transputers (T800C-G20S). The computational complexity of this algorithm is less than that of earlier TS implementations for QAP by a factor n.

Another parallel tabu search algorithm for solving the quadratic assignment problem has been developed by Chakrapani and Skorin-Kapov [17]. The algorithm includes elements of TS, such as aspiration criterion, dynamically changing tabu list sizes and long-term memory [37]. A new intensification strategy is proposed, based on the intermediate term memory, restricting the searches in a neighborhood. This results in less amount of computational effort during one iteration since the procedure does not examine the entire neighborhood. A massively parallel implementation was tested on the Connection Machine CM-2 for large QAPs of size ranging from $n = 42$ to 100, using n^2 processors. The new tabu strategy gave good results in terms of quality of solutions. For problems up to size 90, it obtained the best known solutions or close to those. For size $n = 100$, every previously best solution found, was improved upon.

Battiti and Tecchiolli [9] describe a parallelization scheme for tabu search called *reactive* tabu scheme, and apply it to the quadratic assignment problem and the N-k problem with the objective of achieving a speedup in the order of the number of processors. In the reactive tabu search, each processor executes an independent search. Furthermore, the same problems are solved by a parallel genetic algorithm in which the interaction between different search processes is strong because the generation of new candidate points depends on the consideration of many members of the population.

A tabu search for the traveling salesman problem has been proposed by Fiechter [33], and was tested on instances having 500 to 10000 vertices. A new estimation of the asymptotic normalized length of the shortest tour through points uniformly distributed in the unit square is given. Numerical results and speedups obtained by the implementation of the algorithm on a network of transputers show the efficiency of the parallel algorithm.

6.   Parallel GRASP

A GRASP [28, 30] is an iterative process for finding approximate solutions to combinatorial optimization. Let, as before, $f(x)$ denote the objective function to be minimized over a feasible domain D. A generic layout of GRASP is given in Figure 7.6. The GRASP iterations terminate when some stopping criterion,

---

Layout of GRASP

Input: *A problem instance*

Output: *A (sub-optimal) solution*

1. Initialization: set $x^* = \infty$;
2. **while** (stopping criterion not satisfied) **do**
       (a) Construct a greedy randomized solution x;
       (b) Find local minimum $\tilde{x}$ in neighborhood $N(x)$ of x;
       (c) If $f(\tilde{x}) < f(x^*)$ then update $x^* = \tilde{x}$;
3. Output the best solution found x^*;

Figure 7.6.   Layout of GRASP

such as maximum number of iterations, is satisfied. Each iteration consists of a construction phase, and a local search phase. If an improved solution is found, then the incumbent is updated. A high-level description of these two phases is given next.

In the construction phase, a feasible solution is built up, one element at a time. The choice of the next element to be added is determined by ordering all elements in a candidate list with respect to a greedy function. An element from the candidate list is randomly chosen from among the best candidates in the list, but is not necessarily the top candidate. Figure 7.7 displays a generic layout for the construction phase of a GRASP. The solution set S is initially made empty, and the construction iterations are repeated until the solution is built. To do this, the restricted candidate list is setup. This candidate list contains a subset of candidates that are well ordered with respect to the greedy function. A candidate selected from the list at random is added to the solution. The greedy function is adapted to take into account the selected element.

The solutions generated in the construction are not guaranteed to be locally optimal and therefore local search is applied to produce a locally optimal solution. Figure 7.8 illustrates a generic local search procedure.

Parallel implementation of GRASP is straightforward. Two general strategies have been proposed. In search space decomposition, the search space is

Layout of GRASP Construction Phase

Input: *A problem instance and pseudo random number stream*

Output: *A (sub-optimal) solution*

1. Initialization: set solution $S = \emptyset$;
2. **while** (solution construction not done) **do**
   (a) Using greedy function, make restricted candidate list (RCL);
   (b) At random, select element s from RCL;
   (c) Place s in solution, i.e. $S = S \cup \{s\}$;
   (d) Change greedy function to take into account updated S;
3. Output the solution x corresponding to set S;

Figure 7.7. Layout of GRASP Construction Phase

Layout of GRASP Local Search Phase

Input: *A problem instance, solution x, neighborhood $N(x)$*

Output: *A locally-optimal solution $\tilde{x}$*

1. **while** (x not locally optimal) **do**
   (a) Find $\tilde{x} \in N(x)$ with $f(\tilde{x}) < f(x)$;
   (b) Update $x = \tilde{x}$;
3. Output the locally optimal solution $\tilde{x}$;

Figure 7.8. Layout of GRASP Local Search Phase

partitioned into several regions and GRASP is applied to each in parallel. An example of this is the GRASP for maximum independent set [29, 104] where the search space is decomposed by fixing two vertices to be in the independent set. In iteration parallelization, the GRASP iterations are partitioned and each partition is assigned to a processor. See [87, 94, 95, 96, 103] for examples of parallel implementations of GRASP. Some care is needed so that different random number generator seeds are assigned to the different iterations. This can be done by running the random number generator through an entire cycle, recording all N_g seeds in a seed array. Iteration i is started with seed(i). GRASP has been implemented on distributed architectures. In [96] a PVM-based implementation is described. Two MPI-based implementations are given in [6, 76]. Alvim [6] proposes a general scheme for MPI implementations. A master process manages seeds for slave processors. It passes blocks of seeds to each slave processor and awaits the slaves to indicate that they have finished processing the block and need another block. Slaves also pass back to the master the best solution found for each block of iterations.

Aiex et al. [2] study the probability distribution of solution time to a sub-optimal target value for GRASP. They conclude that the solution time to a sub-optimal target value fits a two-parameter exponential distribution, which shows that it is possible to approximately achieve linear speed-up by implementing GRASP in parallel. The same behavior is observed in a number of metaheuristics such as SA [26, 89], Iterative Improvement for the TSP [116], and Tabu search [9, 114].

7.   Parallel computing environments

The development of efficient and user friendly software tools, together with the availability of a wide range of parallel machines with large processing capacities, high reliability, and low costs, have brought parallel processing into reality as an efficient way for implementing techniques to solve large scale optimization problems. A good choice of the programming environment is essential to the development of a parallel program.

Parallel environments are composed of programming tools, performance evaluation tools, debuggers, and optimization libraries. Parallel programs consist of a number of processes working together. The processes are executed on parallel machines, based on different memory organization methods, as shared memory or distributed memory. In shared memory machines, all processors are able to address the whole memory space. The computers communicate through operations performed by the parallel tasks on the shared memory. Each task shares a common address space, which they can asynchronously access. The advantage of this approach is that the communication can be easy and fast. However, there is a limitation to this approach, system stability is limited by the number of paths between the memory and the processors. To overcome this disadvantage, local memory is added to each processor. An alternative to the shared memory organization is the distributed memory organization. In the framework of the distributed memory organization, the memory is physically distributed among the processors. Each processor can only access its own memory, and communication between processors is performed by messages passed through the communication network. Networks of workstations are an example of this architecture. A drawback of this architecture is that the control of the memory access that is required to maintain the consistency of the data has to be done by the programmer.

The parallel programming tools used in shared or distributed memory organization, are usually sequential languages augmented by a set of special system calls. The calls provide primitive message passing, process synchronization, process creation, mutual exclusion, and other functions.

In developing parallel programs, two types of parallelism can be used: data parallelism and functional parallelism [34, 67, 82]. Data parallelism programs apply the same instruction set to different items of a data structure. Programming languages called data-parallel programming have been developed to help with data parallelism. One of these languages is HPF. Functional parallelism is based on partitioning the program into cooperative tasks. Each task executes a different set of functions/codes and the tasks run asynchronously. The application of each technique depends on the characteristics of the problem to be solved. However, functional and data parallelism are complementary techniques and sometimes can be applied together to different parts of the same problem.

Many programming tools are available to implement parallel programs, each of them being suitable for specific problem types. Examples of programming tools are:

- **PVM** (Parallel Virtual Machine) [36] is an extensively used message passing library that supports the development of parallel programs on interconnected heterogeneous machines.

- **MPI** (Message Passing Interface) [34, 83, 84, 43, 44] is a proposal for the standardization of a message passing interface for distributed memory parallel machines, to enable program portability among different parallel machines.

- **Linda** [12, 13] is a language that offers a collection of tools for process creation and communication, based on the concept of an associative shared memory.

- **HPF** (High Performance FORTRAN) [53, 65, 66, 78] extends FORTRAN 90 with parallel construct and data placement features.

- **Threads** are a general operating systems concept, not specifically related to parallel programming. However, their understanding is essential to a parallel programmer, because of the support they provide in concurrent programming. In 1995, IEEE defined a standard Application Program Interface (API) for programming with threads, known as *POSIX threads* or *Pthreads* [56].

In the process of evaluating the performance of a parallel implementation, it is important to collect suitable and reliable data. There are three basic categories of data collectors: profilers, counters, and event traces.

Profilers record the amount of time spent in different parts of a program, usually in subroutines. This information is often obtained by sampling the program counter and generating a histogram of the execution frequencies. These frequencies are used to estimate the amount of time spent in different program components. *Gprof* is a commercial tool provided by IBM to profile parallel programs developed using the IBM MPI library. *Gprof* shows execution times of the subroutines executed by each processor.

Counters record the time a specific event occurs. *Counters* are used to record global events of the program, such as the number of procedure calls, the number of messages exchanged between pairs of processors, etc. This information cannot be obtained by profilers.

Event traces record each occurrence of specific events. They can be used to investigate relationships between communications and determine sources of idle time and communication bottlenecks. Some of the public domain and commercial packages available for *event traces* are:

- *AIMS* (Automated Instrumentation and Monitoring System) [120, 121] is a public domain package that can be downloaded from *http://science.nas .nasa.gov/Software/AIMS/*.

- *Paradyn* [81] handles the interactive run-time analysis for parallel programs using message passing libraries such as PVM and MPI. *Paradyn* can be downloaded from *http://www.cs.wisc.edu/paradyn/*.

- The *Pablo* performance analysis environment [25, 105] is a public domain package that consists of several components for instrumenting and tracing parallel MPI programs. It can be downloaded from *http://vibes.cs. uiuc.edu/*.

- *VAMPIR* (Visualization and Analysis of MPI resource) [35, 88] is a commercial trace visualization tool from Pallas GmbH. Information about this tool is available at *http://www.pallas.de/pages/vampir.htm*.

- *VT* (Visualization Tool) can be used for post-mortem trace visualization or for on-line performance monitoring of parallel programs developed with the IBM implementation of MPI. It is a commercial tool provided by IBM and can only be used on the IBM SP parallel environment.

- *XPVM* is a public domain software that provides a graphical console and monitor for PVM. It is available from *http://www.netlib.org/pvm3/xpvm*.

A common way to debug a parallel program is by inserting instructions in the code to print relevant information. However, often the output data lacks temporal information, since the information is not necessarily printed in the same order as generated by the parallel program, due to data buffering in the processors.

PVM and LAM public domain implementation of MPI allow the programmer to launch debuggers on remote nodes. Sometimes this kind of technique may interfere with the execution of the parallel program, making it behave differently from the way it runs. Total View is a commercial debugger developed by Etnus Inc. [27] that can be used to debug MPI, PVM, HPF, and multi-threaded programs. Tuplescope [109] is a debugging tool for Linda programs.

Optimization libraries consist of a set of subroutines that can be used to solve large optimization problems. These libraries include functions that solve linear and mixed-integer programming problems, provide important features such as control variables, debugging capabilities, file reading and writing etc. CPLEX [21] and OSL provide parallel implementations of their libraries. For an extensive review of parallel computing environments refer to [77, 8].

8.   Concluding remarks

Since the 1980's, we have witnessed an explosion and availability of parallel, multiprocessing computers. At the same time, we have also seen the development of new, powerful, metaheuristics for combinatorial optimization. Since most of the interesting combinatorial optimization problems are NP-hard, it is quite natural to consider implementing algorithms on such environments. In this brief survey, we summarized recent developments in parallel implementations of several classes of new heuristics for combinatorial optimization.

References

[1] E. Aarts and J. Korst. *Simulated Annealing and Boltzmann Machines – A Stochastic Approach to Combinatorial Optimization and Neural Computing*, John Wiley and Sons, 1989.

[2] R. M. Aiex and M.G.C. Resende and C.C. Ribeiro. *Probability Distribution of Solution Time in GRASP: an Experimental Investigation*, AT&T Labs Research Technical Report: 00.7.2 (2000).

[3] I. Azencott. *Simulated Annealing: Parallelization Techniques*, Wiley (1992).

[4] S.G. Akl, D.T. Barnard and R.J. Doran. Design, Analysis, and Implementation of a Parallel Tree Search Algorithm, *IEEE Transactions on Pattern Analysis and Machine Intelligence*, Vol. PAMI-4 (1982), pp. 192–203.

[5] I. Alth. A Parallel Game Tree Search Algorithm with a Linear Speedup, *Journal of Algorithms*, 15:175–198, 1993.

[6] A.C.F. Alvim. Parallelization strategies for the metaheuristic GRASP. Master's thesis, Department of Computer Science, Catholic University

of Rio de Janeiro, Rio de Janeiro, RJ 22453-900 Brazil, April 1998. In Portuguese.

[7] G.Y. Ananth, V. Kumar and P. M. Pardalos. Parallel Processing of Discrete Optimization Problems, In *Encyclopedia of Microcomputers* Vol. 13 (1993), pp. 129–147, Marcel Dekker Inc., New York.

[8] D. Bader. http://computer.org/parascope (2000).

[9] R. Battiti and G. Tecchiolli. Parallel Biased Search for Combinatorial Optimization: Genetic Algorithms and TABU, *Microprocessors and Microsystems*, 16:351–367, 1992.

[10] N. Boissin and J.-L. Lutton. A Parallel Simulated Annealing Algorithm, *Parallel Computing*, 19:859–872, 1993.

[11] R.J. Brouwer, P. Banerjee. A Parallel Simulated Annealing Algorithm for Channel Routing on a Hypercube Multiprocessor, *Proceedings of 1988 IEEE International Conference on Computer Design*, pp. 4–7.

[12] N. Carriero and D. Gelernter and T. Mattson. Linda in Context, *Communications of the ACM*, 32:444–458, 1989.

[13] N. Carriero and D. Gelernter and T. Mattson. The Linda Alternative to Message-Passing Systems, *Parallel Computing*, 20:458–633, 1994.

[14] E. Carter. (2000), http://www.taygeta.com/annealing/simanneal.html.

[15] A. Casotto and A. Sanngiovanni-Vincentelli. Placement of Standard Cells Using Simulated Annealing on the Connection Machine *Proceedings ICCAD*, 1987, pp. 350–352.

[16] Y. Censor and S.A. Zenios. *Parallel Optimization: Theory, Algorithms, and Applications*, Oxford University Press (1997).

[17] J. Chakrapani and J. Skorin-Kapov. Massively Parallel Tabu Search for the Quadratic Assignment Problem, *Annals of Operation Research*, 41:327–341, 1993.

[18] R.D. Chamberlain, M.N. Edelman, M.A. Franklin, E.E. Witte. Simulated Annealing on a Multiprocessor, *Proceedings of 1988 IEEE International Conference on Computer Design*, pp. 540–544.

[19] G.A. Cleveland and S.F. Smith. Using Genetic Algorithms to Schedule Flow Shop Releases, *Proceedings of the Third International Conference on Genetic Algorithms*, (1990), Morgan Kaufmann, Los Altos, CA.

[20] J.P. Cohoon, S.U. Hegde, W.N. Martin and D. Richards. Punctuated Equilibria: A Parallel Genetic Algorithm, *Proceedings of the Second International Conference on Genetic Algorithms and their Applications*, J.J. Grefenstette (editor), July 1987, pp. 148–154.

[21] CPLEX. *Home Page of CPLEX, a division of ILOG* (2000),

[22] T.G. Crainic and M. Toulouse. *Parallel Metaheuristics*, Technical report, Centre de recherche sur les transports, Université de Montr'eal, November 1997.

[23] D. Cvijovic and J. Klinowski. Taboo Search: An Approach to the Multiple Minima Problem, *Science*, 267:664–666, 1995.

[24] K.A. De Jong. *An Analysis of the Behavior of a Class of Genetic Adaptive Systems*, Doctoral dissertation, Department of Computer and Communication Sciences, University of Michigan, 1975.

[25] L. DeRose and Y. Zhang and D.A. Reed. SvPablo: A Multi-language Performance Analysis System, *10th International Conference on Computer Performance Evaluation - Modeling Techniques and Tools*, (1998), pp. 352–355.

[26] N. Dodd. Slow Annealing Versus Multiple Fast Annealing Runs: An Empirical Investigation, *Parallel Computing*, 16:269–272, 1990.

[27] Etnus Com. *Home page for Etnus Com.* (1999),

[28] T.A. Feo and M.G.C. Resende. A probabilistic heuristic for a computationally difficult set covering problem, *Operations Research Letters*, 8:67–71, 1989.

[29] T.A. Feo and M.G.C. Resende and S.H. Smith. A Greedy Randomized Adaptive Search Procedure for Maximum Independent Set, *Operations Research*, 42:860–878, 1994.

[30] T.A. Feo and M.G.C. Resende. Greedy Randomized Adaptive Search Procedures, *Journal of Global Optimization*, 6:109–133, 1995.

[31] A. Ferreira and P.M. Pardalos (Eds.), Solving Combinatorial Optimization Problems in Parallel, Methods and Techniques, *Lecture Notes in Computer Science* (1995), 1054.

[32] A. Ferreira and J.D.P. Rolim (Eds.), *Parallel Algorithms for Irregular Problems: State of the Art*, Kluwer Academic Publishers (1995).

[33] C.-N. Fiechter. A Parallel Tabu Search Algorithm for Large Traveling Salesman Problems, *Discrete Applied Mathematics*, 51:243–267, 1994.

[34] I. Foster. *Designing and Building Parallel Programs: Concepts and Tools for Parallel Software Engineering* Addison-Wesley (1995).

[35] J. Galarowicz and B. Mohr. Analyzing Message Passing Programs on the Cray T3 with PAT and VAMPIR, *Proceedings of Fourth European CRAY-SGI MPP workshop*, H. Lederer and F. Hertwick, eds. IPP-Report des MPI für Plasmaphysik, IPP R/46 (1998), pp. 29–49,

[36] A. Geist and A. Beguelin and J. Dongarra and W. Jiang and R. Manchek and V. Sunderman. *PVM: Parallel Virtual Machine - A user's*

guide and tutorial for networked parallel computing, MIT Press (1994), (http://www.netlib.org/pvm3/book/pvm-book.html).

[37] F. Glover. Tabu Search. Part I, *ORSA J. Comput.*, 1:190–206, 1989.

[38] F. Glover. Tabu Search. Part II, *ORSA J. Comput.*, 2:4–32, 1990.

[39] F. Glover, E. Taillard and D. de Werra. A User's Guide to Tabu Search, *Annals of Operation Research*, 41:3–28, 1993.

[40] F. Glover and M. Laguna. *Tabu Search*, Kluwer Academic Publishers (1997).

[41] M. Gorges-Schleuter. ASPARAGOS: A Parallel Genetic Algorithm and Population Genetics, *Lecture Notes on Computer Science*, 565:407–518, 1989.

[42] D.R. Greening. Asynchronous Parallel Simulated Annealing, *Lectures in Complex Systems*, 3:497–505, 1990.

[43] W. Gropp and E. Lusk and A. Skjellum. *Using MPI: Portable Parallel Programming with the Message Passing Interface*, MIT Press (1994).

[44] W. Gropp and S. Huss-Lederman and A. Lumsdaine and E. Lusk and B. Nitzberg and W. Saphir and M. Snir. *MPI: The Complete Reference*, In The MPI Extensions, Vol. 2, MIT Press (1998).

[45] P. Hansen and B. Jaumard and N. Mladenović and A. Parreira. *Variable Neighborhood Search for Weighted Maximum Satisfiability*. In preparation (2000).

[46] P. Hansen and N. Mladenović. Variable Neighborhood Search for the *p*-Median. *Location Science*, 5:207–226, 1998.

[47] P. Hansen and N. Mladenović. An Introduction to Variable Neighborhood Search. In *Metaheuristics, Advances and Trends in Local Search Paradigms for Optimization*, S. Voss et al. (Eds.), pp. 433–458, 1999, Kluwer, Dordrecht.

[48] P. Hansen and N. Mladenović. Variable Neighborhood Search: Principles and Applications. *European Journal of Operational Research* (to appear).

[49] P. Hansen and N. Mladenović. Variable Neighborhood Search. In *Handbook of Applied Optimization*, P.M. Pardalos and M.G.C. Resende (Eds.), Oxford University Press (2001).

[50] P. Hansen and N. Mladenović and N. Labbe and M. Gendreau. *Variable Neighborhood Search for the Traveling Salesman Problem*. In preparation (2000).

[51] P. Hansen and N. Mladenović and N. Perez-Brito. Variable Neighborhood Decomposition Search. *Journal of Heuristics* (to appear).

[52] J.H. Holland. *Adaptation in Natural and Artificial Systems*, University of Michigan Press, Ann Arbor, MI, 1975.

[53] High Performance FORTRAN Forum. *High Performance FORTRAN Language Specification Version 2.0*,

[54] S. R. Huang and L.S. Davis. Speedup Analysis of Centralized Parallel Heuristic Search Algorithms, *Proceedings of the International Conference on Parallel Processing*, Vol. 3. Algorithms and Applications (1990), pp. 18–21.

[55] IBM. *Parallel ESSL Version 2 Release 1.2 Guide and Reference (SA22-7273)*, (1999),

[56] IEEE (Institute of Electric and Electronic Engineering). *Information Technology - Portable Operating Systems Interface (POSIX) - Part 1 - Amendment 2: Threads Extension*, (1995).

[57] L. Ingber. *ASA-package* (2000), http://www.ingber.com.

[58] P. Jog, J.Y. Suh and D. Van Gucht. Parallel Genetic Algorithms Applied to the Traveling Salesman Problem, *SIAM Journal of Optimization*, 1:515–529, 1991.

[59] D.S. Johnson. Local Optimization and the Traveling Salesman Problem, *Proceedings of the Seventh Colloquium on Automata, Languages, and Programming*, Lecture Notes in Computer Science, 447:446–461, 1990.

[60] D. S. Johnson, C. H. Papadimitriou and M. Yannakakis. How Easy is Local Search?, *Journal of Computer and System Sciences*, 17:(1)79-100, 1988.

[61] M. Jones and P. Banerjee. Performance of a Parallel Algorithm for Standard Cell Placement on the Intel Hypercube, *Proceedings of the 24th Design Automation Conference*, 1987, pp. 807–813.

[62] S. Kirkpatrick, C.D. Gellat Jr. and M.P. Vecchi. Optimization by Simulated Annealing, *Science*, 220:671–680, 1983.

[63] G. Kliewer and S. Tschöke. A General Parallel Simulated Annealing Library and its Application in Airline Industry, *Proceedings of the 14th International Parallel and Distributed Processing Symposium*, IPDPS (2000), pp. 55–61.

[64] G. Kliewer and K. Klohs and S. Tschöke. *Parallel Simulated Annealing Library (parSA): User manual*, Technical report, University of Paderborn (1999), http://www.upb.de/ parsa.

[65] A. Knies and M. O'Keefe and T. MacDonald. High Performance FORTRAN: A Practical Analysis, *Scientific Programming*, 3:187–199, 1994.

[66] C. Koelbel and D. Loveman and R. Schreiber and G. Steele Jr. and M. Zosel. *The High Performance FORTRAN Handbook*, MIT Press (1994).

[67] V. Kumar and A. Grama and A. Gupta and G. Karypis. *Introduction to Parallel Computing Design and Analysis of Parallel Algorithms*, Benjaming/Cummings (1994).

[68] G. von Laszewski. Intelligent Structural Operators for K-Way Graph Partitioning Problem, *Proceedings of the Fourth International Conference on Genetic Algorithms*, (1991), Morgan Kaufmann, San Mateo, CA.

[69] C.Y. Lee and S.J. Kim. Parallel Genetic Algorithms for the Earliness-Tardiness Job Scheduling Problem with General Penalty Weights, *Computers & Ind. Eng.*, 28:231–243, 1995.

[70] F. Lee. *Parallel Simulated Annealing on a Message-Passing Multi-Computer*, Ph.D. thesis, Utah State University (1995).

[71] W.G. Macready and A.G. Siapas and S.A. Kauffman. Criticality and Parallelism in Combinatorial Optimization, *Science*, 271:56–59, 1996.

[72] A. Mahanti and C.J. Daniels. A SIMD Approach to Parallel Heuristic Search, *Artificial Intelligence*, 10:243–282, 1993.

[73] T.A. Marsland and F. Popowich. Parallel Game-Tree Search, *IEEE Transactions on Pattern Analysis and Machine Intelligence*, Vol. PAMI-7 (1985), pp. 442–452.

[74] O. Martin and S.W. Otto and E.W. Felten. Large-Steps Markov Chains for the Travelling Salesman Problem, *Complex Systems*, 5:299–326, 1991.

[75] S.L. Martins. *Parallelization strategies for metaheuristics in distributed memory environments*, Ph.D. thesis, Department of Computer Sciences, Catholic University of Rio de Janeiro, Rio de Janeiro, Brazil, 1999.

[76] S.L. Martins and C.C. Ribeiro. A parallel GRASP for the Steiner problem in graphs. In A. Ferreira and J. Rolim (Editors), *Proceedings of IRREGULAR'98 – 5th International Symposium on Solving Irregularly Structured Problems in Parallel*, Lecture Notes in Computer Science, Springer-Verlag, 1457:285–297, 1998.

[77] S.L. Martins and C.C. Ribeiro and N. Rodriguez. Parallel Computing Environments. In *Handbook of Applied Optimization*, P.M. Pardalos and M.G.C. Resende (Editors), Oxford University Press, 2001.

[78] J. Merlin and A. Hey. *An Introduction to High Performance FORTRAN*, Scientific Programming, 4:87–113, 1995.

[79] N. Metropolis and A. Rosenbluth and M. Rosenbluth and A. Teller and E. Teller. Equation of State Calculations by Fast Computing Machines, *Journal of Chemical Physics*, 21:1087–1092, 1953.

[80] A. Migdalas and P.M. Pardalos (Eds.), *Parallel Computing in Optimization*, Kluwer Academic Publishers, 1997.

[81] B.P. Miller and M.D. Callagham and J.M. Cargille and J.K. Hollingsworth and R.B. Irvin and K.L. Karavanic and K. Kunchithapadam and T. Newhall. The Paradyn Parallel Performance Tools, *IEEE Computer*, 28:37–46, 1995.

[82] *Practical Parallel Computing.* AP Professional (1994).

[83] MPI Forum. A Message-Passing Interface Standard, *The International Journal of Supercomputing Applications and High Performance Computing*, 8:138–161, 1994.

[84] MPI Forum. *A Message-Passing Interface Standard,* (1995),

[85] H. Muhlenbein. Parallel Genetic Algorithms, Population Genetics and Combinatorial Optimization, *Lecture Notes in Computer Science*, 565:398–406, 1989.

[86] H. Muhlenbein, M. Schomisch and J. Born. The Parallel Genetic Algorithm as Function Optimizer, *Proceedings on an International Conference on Genetic Algorithms,* (1991).

[87] R.A. Murphey, P.M. Pardalos, and L.S. Pitsoulis. A parallel GRASP for the data association multidimensional assignment problem, In P.M. Pardalos (Editor), *Parallel processing of discrete problems*, volume 106 of *The IMA Volumes in Mathematics and Its Applications*, pages 159–180. Springer-Verlag, 1998.

[88] W. Nagel and A. Arnold and M. Weber and H. Hoppe and K. Solchenbach. VAMPIR: Visualization and Analysis of MPI Resources, *Supercomputer*, 63:69–80, 1996.

[89] L.J. Osborne and B.E. Gillett. A Comparison of two Simulated Annealing Algorithms Applied to the Directed Steiner Problem in Networks, *ORSA J. on Computing*, 3:213–225, 1991.

[90] P. M. Pardalos (Eds), *Parallel Processing of Discrete Problems*, The IMA Volumes in Mathematics and its Applications, Vol. 106 (1999).

[91] P. M. Pardalos, Y. Li and K. A. Murthy. Computational Experience with Parallel Algorithms for Solving the Quadratic Assignment Problem, In *Computer Science and Operations Research: New Developments in their Interface*, O. Balci, R. Sharda, S.A. Zenios (eds.), Pergamon Press, pp. 267–278 (1992).

[92] P. M. Pardalos, and G. Guisewite. Parallel Computing in Nonconvex Programming, *Annals of Operations Research*, 43:87–107, 1993.

[93] P. M. Pardalos, A. T. Phillips and J. B. Rosen. *Topics in Parallel Computing in Mathematical Programming*, Science Press, 1993.

[94] P. M. Pardalos, L. S. Pitsoulis, T. Mavridou, and M.G.C. Resende. Parallel search for combinatorial optimization: Genetic algorithms, simulated annealing and GRASP, In *Parallel Algorithms for Irregularly Structured Problems, Proceedings of the Second International Workshop – Irregular'95*, A. Ferreira and J. Rolim (Editors), *Lecture Notes in Computer Science*, Springer-Verlag, 980:317–331, 1995.

[95] P. M. Pardalos, L. S. Pitsoulis and M.G.C. Resende. A Parallel GRASP Implementation for the Quadratic Assignment Problem, In *Proceedings of Parallel Algorithms for Irregularly Structured Problems (Irregular '94)*, Kluwer Academic Publishers (1995).

[96] P.M. Pardalos, L.S. Pitsoulis, and M.G.C. Resende. A parallel GRASP for MAX-SAT problems, *Lecture Notes in Computer Science*, 1180:575–585, 1996.

[97] P.M. Pardalos, M.G.C. Resende, and K.G. Ramakrishnan (Editors), *Parallel Processing of Discrete Optimization Problems*, DIMACS Series Vol. 22, American Mathematical Society, (1995).

[98] P. M. Pardalos, M. G. C. Resende (Editors), *Handbook of Applied Optimization*, Oxford University Press, (2000).

[99] P.M. Pardalos and H. Wolkowicz (Editors), *Quadratic Assignment and Related Problems*, DIMACS Series Vol. 16, American Mathematical Society (1994).

[100] C. Peterson. Parallel Distributed Approaches to Combinatorial Optimization: Benchmark Studies on Traveling Salesman Problem, *Neural Computation*, 2:261–269, 1990.

[101] C.B. Pettey, M.R. Leuze and J.J. Grefenstette. A Parallel Genetic Algorithm, *Proceedings of the Second International Conference on Genetic Algorithms and their Applications*, J.J. Grefenstette (editor), July 1987, pp. 155–161.

[102] C. Powler, C. Ferguson and R. E. Korf. Parallel Heuristic Search: Two Approaches, In *Parallel Algorithms for Machine Intelligence and Vision*, V. Kumar, P.S. Gopalakrishnan and L.N. (eds.), Springer-Verlag, pp. 42–65 (1990).

[103] M.G.C. Resende. Computing approximate solutions of the maximum covering problem using GRASP, *J. of Heuristics*, 4:161–171, 1998.

[104] M.G.C. Resende, T.A. Feo, and S.H. Smith. Algorithm 787: Fortran subroutines for approximate solution of maximum independent set problems using GRASP, *ACM Trans. Math. Software*, 24:386–394, 1998.

[105] D.A. Reed and R.A. Aydt and R.J. Noe and P.C. Roth and K.A. Shields and B. Schwartz and L.F. Tavera. Scalable Performance Analysis: The Pablo Performance Analysis Environment, *Proceedings of the Scalable Parallel Libraries Conference*, IEEE Computer Society (1993), pp. 104–113.

[106] J.S. Rose, W.M. Snelgrove and Z.G. Vranesic. Parallel Standard Cell Placement Algorithms with Quality Equivalent to Simulated Annealing, *IEEE Transactions on Computer-Aided Design*, 7:387–396, 1988.

[107] A.V. Sannier and E.D. Goodman. Genetic Learning Procedures in Distributed Environments, *Proceedings of the Second International Conference on Genetic Algorithms and their Applications*, J.J. Grefenstette (editor), July 1987, pp. 162–169.

[108] J.S. Sargent. *A Parallel Row-Based Algorithm with Error Control for Standard-Cell Placement on a Hypercube Multiprocessor*, Thesis, University of Illinois, Urbana-Illinois, 1988.

[109] Scientific Computing Associates. *Linda User's Guide and Reference Manual*, Version 3.0 (1995).

[110] B. Shirazi, M. Wang and G. Pathak. Analysis and Evaluation of Heuristic Methods for Static Task Scheduling, *Journal of Parallel and Distributed Computing*, 10:222–232, 1990.

[111] P.S. de Souza. *Asynchronous Organizations for Multi-Algorithm Problems*, Ph.D. Thesis, Department of Electrical and Computer Engineering, Carnegie Mellon University, 1993.

[112] R. Shonkwiler and E.V. Vleck. Parallel Speed-Up of Monte Carlo methods for Global Optimization, *Journal of Complexity*, 10:64–95, 1994.

[113] J. Suh and D. Van Gucht. *Distributed genetic Algorithms*, Tech. Report 225, Computer Science Department, Indiana University, Bloomington, IN, July 1987.

[114] E. Taillard. Robust Taboo Search for the Quadratic Assignment Problem, *Parallel Computing*, 17:443–445, 1991.

[115] R. Tanese. Parallel Genetic Algorithm for a Hypercube, *Proceedings of the Second International Conference on Genetic Algorithms and their Applications*, J.J. Grefenstette (editor), July 1987, pp. 177–183.

[116] H.M.M. Ten Eikelder and M.G.A. Verhoeven and T.W.M. Vossen and E.H.L. Aarts. A Probabilistic Analysis of Local Search, In I.H. Osman and J.P. Kelly (Eds.) *Metaheuristics: Theory & Applications*, pp. 605–618, Kluwer Academic Publishers (1996).

[117] N.L.J. Ulder, E.H.L. Aarts, H. -J. Bandelt, P.J.M. van Laarhoven and E. Pesch. Genetic Local Search Algorithms for the Traveling Salesman Problem, In *Lecture Notes in Computer Science*, Parallel Problem Solving from Nature-Proceedings of 1st Workshop, PPSN 1, 496:109–116, 1991.

[118] M.G.A. Verhoeven and E.H.L. Aarts. Parallel Local Search, *Journal of Heuristics*, 1:43–65, 1995.

[119] C.-P. Wong and R.-D. Fiebrich. Simulated Annealing-Based Circuit Placement on the Connection Machine System, *Proceedings of International Conference on Computer Design (ICCD '87)*, 1987, pp. 78–82.

[120] J. Yan and S. Sarukhai and P. Mehra. Performance Measurement, Visualization and Modeling of Parallel and Distributed Programs Using the AIMS Toolkit, *Software Practice and Experience*, 25:429–461, 1995.

[121] J. Yan and S. Sarukhai. Analyzing Parallel Program Performance Using Normalized Performance Indices and Trace Transformation Techniques, *Parallel Computing*, 22:1215–1237, 1996.

Chapter 8

PARALLELISM IN LOGIC PROGRAMMING AND SCHEDULING ISSUES

Cláudio F. R. Geyer
Universidade Federal do Rio Grande do Sul[*]
geyer@inf.ufrgs.br

Patrícia Kayser Vargas
Universidade Federal do Rio Grande do Sul[*]
kayser@inf.ufrgs.br

Inês de Castro Dutra
Universidade Federal do Rio de Janeiro[†]
ines@cos.ufrj.br

Abstract   This text summarises the main research work being carried out on the implementation of parallel logic programming systems. It concentrates on describing techniques for exploiting AND-parallelism and OR-parallelism while showing important aspects of some systems on shared-memory, distributed-memory and distributed-shared memory architectures. It also presents some important pointers to journals, conferences and sites with related information.

Keywords: Logic programming, parallel execution models, scheduling.

1.   Introduction

Parallel processing is a very important alternative to obtain good performance from programs, either applications or basic software. Parallelism can be

---

[*] Instituto de Informática/UFRGS, Porto Alegre-RS, Brasil.
[†] Department of Systems Engineering and Computer Science, COPPE/UFRJ, Rio De Janeiro-RJ, Brasil.

207

R. Corrêa et al. (eds.), Models for Parallel and Distributed Computation. Theory, Algorithmic Techniques and Applications, 207–241.
© 2002 *Kluwer Academic Publishers.*

exploited implicitly or explicitly. Most well-known parallel systems are based on the *imperative* programming model and parallelism is exploited explicitly, making the programming task hard. Contrasting with imperative programming, *declarative* programming presents a higher-level programming model where the programmer needs to worry on **what** to solve and not on **how** to solve a problem. Because of this declarative characteristic, logic programming offers some opportunities to exploit implicit parallelism. This clearly reduces the cost of software development, as the user writes only one application that will run on one processor just as well as on several processors.

While logic programming provides a high-level programming paradigm that releases the programmer from "programming in parallel", systems that exploit parallelism from logic programs are usually written "in parallel" using an imperative paradigm in order to obtain maximum performance from existing parallel architectures.

Our text addresses the most recent issues in exploiting parallelism from logic programs concentrating on techniques to **efficiently** implement parallel logic programming systems on several parallel architectures. This text has the following structure: section 2 briefly introduces logic programming concepts. Section 3 presents main aspects of parallel logic programming concerning mainly the implicit control parallelism. Section 4 presents some scheduling issues. Last section draws our conclusions and presents open problems for future work.

2.   Logic programming

Logic programming [65] has its basis in studies of mathematical logic, in particular, Predicate Calculus. Therefore all its basic constructs are inherited from logic. A logic program is a set of axioms, or rules, defining relations between objects. A computation of a logic program is a deduction of consequences, which is its meaning.

Kowalski [66] defines algorithm through the equation *Algorithm = Logic + Control*. In this context, many authors consider that a programmer can concentrate in Logic when using logic programming since the control is implicit in the execution environment. Because the programmers can concentrate more on what to solve and not on how to solve a problem, logic programming languages are called *declarative languages*.

Logic programming has been largely used in many research centers as an alternative to the imperative programming paradigm. Logic programming is suitable to the development of many applications such as natural language processing, theorem proving, databases, expert systems, optimisation, and symbolic problems in general. The Journal of Logic Programming has one entire volume dedicated to applications developed using logic programming languages such as intelligent information management and electronic CAD [1]. Besides there are conferences dedicated to practical applications such as PACLP (The Practical Application of Constraint Technologies and Logic Programming) and PAP (Practical Applications of Prolog).

There are several logic programming languages but Prolog [95, 8] was the first practical logic programming language implemented and it is still the most widely known and used. Prolog was created in the early 1970s by Alain Colmerauer, Robert Kowalski, Phillipe Roussel and colleagues. Modern implementation began with a virtual machine called the *Warren Abstract Machine* [109, 3] implemented by David H. D. Warren at Edinburgh, which defines a stack-based abstract architecture. Since initial Prolog interpreters were not optimised, a myth of "logic programming = slow implementations" developed and still survives [79]. Only when Prolog compilers appeared, this myth started declining. Nowadays there are extremely fast implementations available [100, 7, 29]. Almost all compiled Prolog implementations use the WAM with some extensions and/or optimisations.

The Prolog language is now defined by an ISO (1995) Standard [34]. However, much current theoretical value comes from important extensions, modifications and experimental aspects [87]. These include: Constraint Logic Programming (CLP), Parallelism and Concurrency, Linear Logic Programming (LLP), Inductive logic Programming (ILP), Committed Choice Languages (CCL), Higher-Order Logic (HOL), and Object-Oriented Logic Programming (OOLP).

In next paragraphs, basic terminology employed in this text will be presented. For a more detailed discussion on the Prolog syntax and its origin, the reader is reported to the following references [25, 95, 8].

The data structures manipulated in Prolog are first order *terms*. A term can be either a *variable*, started with an upper case letter, a *constant*, started with a lower case letter, or a *structure* in the format $f(t_1, \ldots, t_n)$ where f is a *functor* and t_i is a term.

Variables in Prolog have some important features, i.e., variables: (1) are logical variables which can be instantiated only once; (2) are untyped until instantiation; and (3) are instantiated via *unification*, a pattern matching operation finding the most general common instance of two data objects.

Logic Programming considers that each *clause* expresses a number of (maybe zero) conditions, connected by conjunctive operators, that imply a number of (maybe zero) conclusions. The set of clauses with at most one conclusion are called *Horn clauses*. Horn clauses are the basis of many logic languages including Prolog.

Prolog programs are sets of Horn clauses (assertives). In Prolog, the conditions of a clause are called the *body* of the clause. The conclusion is called the *head* of the clause. A clause can be either a fact or a rule. *Facts* are simple and unconditional, i.e., they express a tautology. Syntactically, a fact can be considered a clause without a body. *Rules* express conditional relations. All rules present head and body separated by the symbol ":-". The head has one single term representing a possible conclusion and the body is composed by a conjunction of terms also called *literals* or *goals*. Each goal must be successfully proved in order that the clause succeeds. *Queries* are special rules without head that are always executed.

Figure 8.1 presents a small Prolog code composed by facts, rules and queries. In this example, the predicate *grandfather* has two rules to indicate *what* means to be a grandfather "Someone is grandfather of Y **if** he is father of one of the parents of Y". Both rules explain also *how* to find out if someone is grandfather of another one: "to solve `grandfather(X,Y)`, we need to solve `father(X,Z)` and `father(Z,Y)`, or we need to solve `father(X,Z)` and `mother(Z,Y)`.

```
father(paulo, joao).    ⎫
father(paulo, jose).    ⎪
father(paulo, ana).     ⎬ facts
father(jose, maria).    ⎪
father(ana, lucas).     ⎭

grandfather(X,Y) :-     ⎫
      father(X,Z),      ⎪
      father(Z,Y).      ⎬ rules
grandfather(X,Y) :-     ⎪
      father(X,Z),      ⎪
      mother(Z,Y).      ⎭

:- grandfather(X, maria).  ⎫ query
```

Figure 8.1.   Example of Prolog code

Thus a logic program can have a *declarative* meaning (*"what"*) and a procedural meaning (*"how"*). The declarative interpretation reads the clauses statically, textually and understands the definitions corresponding to the rules and facts. The procedural interpretation tries to understand how the logic program produces a solution.

The procedural interpretation of Horn clauses [65] forms the basis of the operational semantics of logic programming. Indeed, the set of all clauses with the same head predicate symbol **p** and with the same number **n** of arguments (*arity*) can be considered as the definition of the *procedure* **p**, frequently called procedure **p/n**. Each goal in the body of a clause can thus be considered as a procedure call. At a given step of the execution of a logic program, the set of goals that remain to be executed, i.e. the continuation of the computation, is called the current *resolvent*. The execution of a logic program starts by taking the query as initial resolvent and proceeds by transforming the current resolvent into a new one as follows [62]. First select any goal G of the resolvent, and apply a *resolution step*, which consists in finding a clause of the program whose head unifies with G, then replacing G by the body of the unifying clause to produce a new resolvent. The *bindings* of variables resulting from the unification apply to the whole resolvent. This process iterates until either the resolvent is empty, in which case the computation ends with success, or no unifying clause can be found during a resolution step, in which case a *failure* occurs and backtracking takes place. *Backtracking* consists in restoring the previous resolvent and selecting an alternative unifying clause in order to perform an alternative resolution step, i.e. at backtracking Prolog investigates the different alternatives in the order in which they are listed in the program. If all possible alternatives have been tried unsuccessfully and it is not possible to backtrack further, the computation terminates with failure. Figure 8.2 illustrates the execution of the previous example through the *search tree* until the first solution is found. The query is to find the grandfather of maria. The root of the tree represents the query. Each subsequent level of the tree represents the body of the clause that matched to the leftmost goal not yet solved in an immediate lower level of the tree. A special structure called *choice-point* or *or-node* is created for every call to a goal that has more than one candidate clause. This structure stores and controls still available alternatives.

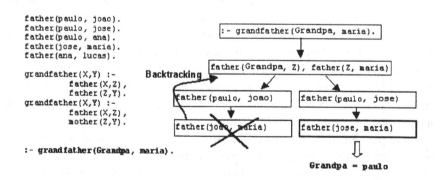

Figure 8.2. Example of Prolog execution

The Prolog sequential computation strategy consists in a depth-first left-to-right search of the program tree. This execution mechanism is responsibility of the execution environment and the programmer does not need to worry about this. Another important concept is the *Prolog semantics* [79] whose meaning is closely related to the execution strategy. Prolog extends pure Horn clause logic with some additional features whose the most relevant are:

- *Meta-logical predicates*: are used to manipulate terms and to query the state of the proof (e.g. instantiation state of the variables).

- *Cut*: represents the only non-logical operator in Prolog which affects the control of the execution. The cut allows to prune parts of the search tree (also called search space).

- *Extra-logical predicates*: are used to perform I/O (e.g. read and write) and to dynamically modify the structure of the program (e.g. adding new clauses to the program using the built-in predicate `assert`).

In this text, *Prolog semantics* refers to the ordering of the exploration of the search tree. Extra-logical predicates whose behaviour depends on the order of execution are called *order-sensitive predicates*.

3.   Parallelism in logic programming

Parallel logic programming [62, 20, 50, 79, 108] is an important research field whose history is closely related to logic programming languages evolution. The main goal in having parallel logic programming environments is to allow programmers to develop faster applications that are suitable to be developed in a declarative language such as optimisation problems, relational databases, inductive logic programming and resource allocation problems among many others. Independently from the specific paradigm considered, in order to execute a program which exploits parallelism, the programming language and/or the runtime environment must supply the means to [79]: (1) *identify parallelism* by recognizing the components of the program execution that will be (potentially) performed by different processors; (2) *start* and *stop* parallel execution; (3) *coordinate* the parallel execution, e.g. specifying and implementing interactions between concurrent components.

These parallel environments could use explicit structures to allow programmers to define parallelism or could explore parallelism automatically without programmers interference. These two types of approaches are called respectively explicit and implicit parallelism.

Explicit parallelism is characterized by the presence of explicit constructs in the programming language, aimed at describing to a certain degree of detail the way in which the parallel computation will take place. The main advantage of this approach is its flexibility which allows to code a wide variety of patterns of execution, giving a considerable freedom in the choice of what should or not run in parallel. However, the management of parallelism is a complex task involving activities as detecting the components of the parallel execution and controlling processes synchronization.

Implicit parallelism allows programmers to write their applications without any concern about the exploitation of parallelism. Parallelism is transparent to the programmer because it is automatically performed by the compiler and/or the runtime system. The main advantage of implicit parallelism is keeping the complexity of parallel software development to special systems, and consequently releasing programmers from additional programming effort. However, the parallelizing system may try to parallelize code which is inherently sequential or to exploit very fine-grained parallelism. These situations normally lead to an inefficient execution.

For Logic Programming, there are both implicit and explicit parallel execution models. Gupta et al [50] point out that there are many proposals for extending a logic programming language with explicit constructs for concurrency. They are largely put into three categories:

1 those that add explicit message passing primitives to Prolog e.g. Delta Prolog [78] and CS-Prolog [44]. Multiple processes are run in parallel that communicate with each other via messages.

2 those that add blackboard primitives to Prolog, e.g. De Bosschere et al [30] and Shared Prolog [21]. These primitives are used by multiple Prolog process running in parallel to communicate with each other via the blackboard.

3 those based on guards, committed choice, and data-flow synchronization, e.g. Parlog [49], GHC [97], and Concurrent Prolog [88].

However, this text will focus only on implicit parallel approaches. In Logic Programming there are some implicit sources of parallelism that facilitate the automatic execution. The two main sources are the following:

- **OR-parallelism**: is the parallel execution of clauses that compose a predicate. The simultaneous development of several resolvents that would be computed successively by backtracking in a sequential execution is called OR-Parallelism since we can consider that the clauses are connected by a logical *or* operator ($\vee$).

- **AND-parallelism**: is the parallel execution of goals that compose a clause. It is called AND-parallelism since goals are connected by a logical *and* ($\wedge$). The main inherent difficulty of exploiting AND-parallelism is to obtain coherent bindings for variables shared by several goals executed in parallel. This has led to the design of execution models where goals which may possibly bind shared variables to conflicting values are either serialised or synchronised. This gives place to two distinct classifications:

- **independent AND-parallelism**: only independent goals which do not share any variable are executed in parallel;

- **dependent AND-parallelism**: concurrent execution of goals sharing variables is possible.

Both kinds are called *control-parallelism* and they will be presented in detail in the next sections. There are other implicit sources of parallelism called *data-parallelism* which exploits parallelism that originates from the data. These are *unification parallelism*, *path parallelism* and *recursion parallelism*. They are very fine-grained, and therefore difficult to achieve good performance in parallel systems and may require specialised architecture in order to yield speedups. Nevertheless, many researchers pursue the objective of exploiting data-parallelism [48, 94, 14].

In the literature we can find several systems that exploit only AND-parallelism [57, 58, 54, 91], only OR-parallelism [4, 76, 16, 72] or a combination of AND- and OR-parallelism. Systems that combine both AND- and OR-parallelism are still under development and research [51, 115, 114, 73, 74, 75, 71], and several problems remain to be solved. For instance, what kind of data structure is more adequate to combine AND-parallelism with OR-parallelism, or what kind of AND-parallelism is more common in application programs, or yet what kind of implementation strategy to use when developing parallel logic programming systems for hybrid parallel architectures such as distributed shared memory systems [80]. Most well-known parallel logic programming systems currently available were implemented for shared-memory architectures, because of its simpler programming model. However some applications need more scalable architectures. Recent work has been done on studying the performance of such systems on modern architectures [83, 92, 17, 84, 59] in order to devise a suitable model that would run efficiently on these modern architectures.

3.1.   OR-Parallelism

OR-parallelism is present in many application areas such as parsing, optimisation problems and databases. OR-parallelism is one of the simplest forms of implicit parallelism, since it can be executed without including any annotation in the code. At the same time, the execution model is close to the sequential one. This helps efficient sequential techniques reuse.

All efficient implementations use the *multi-sequential* model. The multi-sequential approach is characterised by a set of processes (*workers*) that have a full Prolog Engine, i.e. each worker can solve a complete branch of the search tree.

The fundamental problem to solve in an OR-parallel logic programming system is the maintenance of multiple bindings for the same variable in different branches of the search tree [28]. In theory, a new resolvent is generated at each unification of a goal against the head of a clause, by renaming all the variables of the current resolvent. In practice, resolvents are not copied and the same memory locations are used to compute alternative resolvents [62]. As a goal can match to several different clauses and produce different bindings for a variable, a system that allows several clauses to proceed in parallel needs a mechanism to avoid conflicting bindings. In sequential Prolog this problem does not arise because each branch is executed at once and the conditional bindings are annotated in a special stack called *Trail* to be unbound later on backtracking. In parallel, each processor needs to have its own copy of the conditional variable, as processing of each branch can be done separately and simultaneously with other processors. Several solutions have been proposed to cope with this problem. One of the frequently used classifications considers three groups: copying, sharing and recomputation of stacks [62].

1 **copying of stacks**:   In this scheme, each worker maintains a complete copy of the stacks in its workspace. An importer worker copies the stacks of the computation state down to the node providing work. The importer will need to restore the stack bindings to the previous state, i.e. when the imported choice-point was created. This way, they keep their own independent environments sets to produce multiple bindings to a variable. Some systems that use this approach are Muse [4, 5, 61], OPERA [16], PLoSys [76], pclp(FD) [103] and YapOr [81]. The overhead of copying can be reduced through the *incremental copying of stacks* technique. Incremental copying is based on the fact that one idle worker can share a part of the search tree with the exporter. So, both share parts of the stacks. In this technique, the importer backtracks until the last common choice-point before copying the segments of the stacks that are *younger* than the last common choice-point. Incremental copying of stacks can be found in implementations of Muse [4], PLosys [101] and PALS [107].

2 **sharing of stacks**: workers share the parts of the stacks that they have in common. Since several logical variables may have the same identification, which is a location in a shared stack, special data structures are used to store logical variables. Aurora [72] and PEPSys [11, 63]/ECLiPSe [2] are some or-systems that use this technique.

3 **recomputation of stacks**: in this technique each worker computes a pre-determined path of the search tree described by an "oracle" allocated by a specialised process called "controller". To simplify the path description through a bit-string, programs are rewritten to obtain an arity 2. Delphi [24] is the main example of this approach.

Copying and sharing of stacks are the two approaches more successfully employed. Sharing stacks can be done using different techniques as for example hash-windows, variable importation, directory trees, and binding arrays.

The idea of **Hash-windows** [15, 63] is based on the observation that unification of a clause head with its goal arguments, generally, only binds a small number of variables in the ancestor environments. Thus, these variables' bindings can be stored in a small "window" which is only accessible to the process that created the window and its descendant or-processes. Windows are linked for each process and are visible between choice-points (see figure 8.3). The disadvantage of this method is that the access time for each variable is non constant, because when a descendant process does not find the variable on its own window, it needs to scan its ancestor's windows until finding the variable.

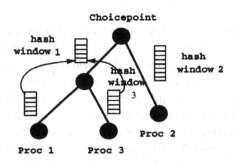

Figure 8.3. Hash-windows technique

Variable importation [69] (see figure 8.4) is based on the idea that the only variables in the ancestor environment which can be affected by unification are those which are unbound. If these variables are "imported" into the local frame for a particular process, the unification can only update variables in the local environment, independently of other OR-parallel processes. The difference between variable importation and hash windows is that in the variable importation method, variables are searched for in environments local to the processor, rather than in special variable tables.

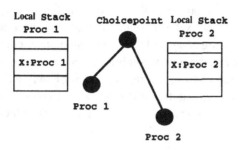

Figure 8.4. Variable importation technique

Directory trees [23] (figure 8.5) are based on the idea that only those environments that contain unbound variables need to be copied to create a separate environment. Ancestor environments which contain no unbound variables can be shared between processes, while copies are made of other environments, with the relevant entries in the directory being changed to point to the new copy.

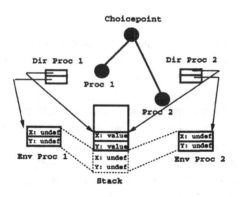

Figure 8.5. Directory trees technique

A **Binding-array** [111] (figure 8.6) is an array private to each worker which contains conditional logical variables. Conditional variables are all variables that were unbound after the creation of a choice-point. Whenever a process needs to trail a variable, it caches its value into the binding array. Every process has the same entry to the same variable. In the Binding Arrays, the operation of taking work is considerably faster since the only activity required is updating the content of the binding arrays. On the other hand, the model has an additional overhead due to the fact that every conditional variable is now stored in binding arrays, and every access to it requires one step of indirection. Additionally, the use of binding arrays requires that, during trailing, both the address of the variable and its current value are saved in the trail stack (which doubles the cost of trailing/untrailing).

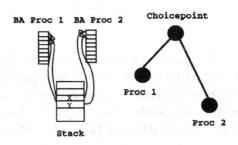

Figure 8.6.   Binding Arrays technique

Each technique has an associated non-constant-time cost that is incurred at task creation time, at variable access time, or at task switching time. All three operations cannot be performed in constant-time [52]. Therefore in order to reduce the overhead costs, systems need to reduce these three costs to as low as possible. Unless the system employs a staticParallel logic programming,scheduling strategy to execute the programs, task switching is always non-constant time, and in this case, it becomes one of the critical parts of the system to be optimised. From the techniques we study here, binding arrays is the only one that keeps both variable access and task creation at a constant time cost.

Another way of maintaining multiple environments for variables is by doing stack-copying. More recently, Gupta et al [53] have proposed the *Stack-splitting* scheme, a modification of the stack-copying technique for environment representation.

The advantage of using stack-sharing is that there is no duplication of amounts of memory used. Copying has the advantage of having exactly a standard sequential execution in each worker with no modification to the WAM memory management, at least as long as this computation is limited to the "private" part of the tree.

If a parallel Prolog system wants to maintain the *Prolog semantics*, the execution of *side-effects* requires special attention: workers should guarantee that the execution of order-sensitive predicates across the branches of the or-tree is performed in the correct order. This typically involves delaying the execution as long as the current branch is not leftmost in the or-tree.

3.2. Independent AND-parallelism (IAP)

The main problem in exploiting independent AND-parallelism is to detect independence between goals. A possible approach consists in performing part of the parallel work detection at compile time to limit the runtime overhead.

Independent AND-parallelism is exploited when the parallel system guarantees that two or more goals in the body of the clause do not share any variables. If the arguments of the goals are found to be independent at runtime then the goals are allowed to proceed independently. A system that implements AND-parallelism generally has the following steps:

- an ordering phase that is needed to detect dependencies among goals;

- a forward execution phase, whose objective is to select goals for AND-parallel execution;

- a backward execution phase that is activated whenever a goal fails.

The ordering phase can be done in several ways. It can use user defined *modes* to indicate dependencies among goals. Arguments of clauses are annotated as being "input" or "output". Another method is to do static data dependency analysis of goals in the clauses. This is done at compile-time and annotates the variables as being independent. This method may lose much parallelism, with the advantage of having minimum runtime overhead. Another method is to build, at runtime, dependency graphs. This method incurs a very high runtime overhead, but can exploit all parallelism available. The final method is to generate code for dependency checking at runtime. This combines static data dependency analysis with runtime dependency checking. Compile-time tools have been built to implement it [77].

The execution model of a system that implements IAP can be done by allowing all independent goals to produce all solutions and then *combine the solutions*, or *recomputing* all goals to the right for every solution produced by goals to the left.

One example of a system that exploits independent AND-parallelism is Hermenegildo's &-Prolog system [58]. In &-Prolog, processing elements called WAM agents cooperate with each other to execute independent goals in parallel. Programs are compiled into graphs called *Extended Conditional Graph Expressions* (ECGE, based on DeGroot's CGEs [33]). The ECGE graphs are independence tests annotated to the Prolog source code at compile-time and checked at runtime as illustrated in table 8.1. The first line, for example, shows a pair of conditions generated at compile-time that will be tested at runtime, i.e., in order to run goals p/2 and q/2 in parallel, the runtime system needs to guarantee that the variable Y is ground (completely instantiated) and that variables X and Z are independent (not shared by goals p/2 and q/2).

Prolog goals	&-Prolog CGEs
p(X,Y), q(Y,Z)	ground(Y), indep(X,Z)
p(Y), q(Y)	ground(Y)
p(Y,Z), q(W)	indep([(Y,W), (Z,W)])
p(Y,Z), q(W)	indep([(W, [Y,Z])])

Table 8.1.   Examples of Prolog mapping to &-Prolog

The generation of the ECGEs represents a complex problem, since many CGEs can be generated for the same clause. The number of variables to be checked for independence contributes to increase the complexity of the computation. In practice even simple runtime checks, such as testing for groundness of a complex nested Prolog term, can be very expensive. Groundness checking can introduce a significant overhead to the system. &-Prolog uses global analysis to infer as much information as possible from the programs [77], and produce reasonable annotated code. The system implements goal recomputation instead of combination of solutions. If cross-product is used, there is no need for recomputation. This increases the quantity of parallelism exploited, but the cross-product algorithm can be a bottleneck. Cross-product can also be the source of an incorrect Prolog execution order, since solutions are produced in any order. If recomputation is used the first solution for the parallel conjunction is found in parallel. Alternative solutions are searched sequentially through backtracking. Some parallelism can be recovered after backtracking has reached the point of restarting solutions for the parallel conjunction. For example, the parallel conjunction a & b & c, will find the first solution a_1, b_1, c_1 in parallel. All alternative solutions for c (for example, c_2, c_3, etc) will be found sequentially through backtracking. When all solutions for c have been found, we backtrack to b to find a new solution for b & c, for example, b_2, c_1. This solution can be found in parallel. Once this new solution is found, all solutions for c (for example, c_2, c_3, etc) will be computed by backtracking. Therefore parallelism is limited by the number of solutions of each goal and by the number of goals in the parallel conjunction.

Another system that also exploits IAP is APEX (And-Parallel EXecution of logic programs) [67]. This model exploits AND-parallelism on shared-memory architectures. The major features of the implementation are (1) dependency analysis between literals of a clause is done dynamically without incurring excessive runtime overhead; and (2) backtracking is done intelligently at the clause level without incurring any extra cost for the determination of the back-track literal. Performance results on a Sequent Balance 21000 show that APEX can achieve linear speedups on dozens of processors.

3.3. Dependent AND-parallelism

Usually, dependent AND-parallelism is exploited in concurrent logic languages. The Parlog system [49] exploits dependent AND-parallelism in a producer-consumer fashion. If two goals share a variable, one is considered to be the producer for the variable and the other one will be the consumer. The goal is selected to be a consumer or a producer through *mode declaration*, i.e. the programmer or a preprocessor annotates the program to indicate the mode (input or output) of the arguments for each predicate. As long as an output argument is assigned one value, the other goal (the consumer) can start execution. If the variable is instantiated to a list or structure, the consumer goal can start to consume the partially instantiated argument. A subset of languages that support dependent AND-parallelism do not support backtracking, such as Parlog. These are called committed-choice languages. Only one clause is allowed to be chosen through the evaluation of *guards*, usually implemented as calls to built-in predicates that make clauses mutually exclusive. Once a guard succeeds, the system commits to the clause containing the guard, and all other clauses are ignored. Other examples of these systems include GHC [97], FGHC [99], Concurrent Prolog [88] and KL1 [64].

Concurrent Prolog (CP) is a logic programming language in that a program is a collection of Horn clauses, and a computation is an attempt to prove a goal – an existentially quantified conjunctive statement – from the clauses of the program. The goal statement describes an input/output relation for which the input is known; a successful (constructive) proof provides a corresponding output. Concurrent Prolog has a different mechanism for controlling the proof construction from other logic programming languages. Prolog uses the order of clauses in the program and the order of goals in a clause to guide a sequential search for a proof, and uses the cut operator to prune undesired portions of the search space. Concurrent Prolog searches a proof in parallel. To control the search, CP embodies two familiar concepts: guarded-command indeterminacy, and dataflow synchronisation. They are implemented using two constructs: the commit operator and the read-only annotation.

KL1 is a system based on the KL1 language, that by its turn is based on GHC (Guarded Horn Clauses). In these languages the control is also added by the user through *guards* that guarantee only one clause to be committed. The KL1 language was designed in the scope of the Fifth Generation Computer Systems Japanese project.

Yang [113] designed a language, P-Prolog, based on mutual exclusion relations where a set of clauses can be "backtrackable". Clauses of the same predicate in this language are syntactically oranised in groups. Clauses in the same group behave in the same way as GHC, but if a fail occurs for a group of clauses, another group can be tried. The P-Prolog language introduced a new concept of programming in committed-choice languages, by allowing a degree of don't know non-determinism to the programs.

Another system that exploits dependent AND-parallelism, but in a different way, is DASWAM [89, 91]. In DASWAM the variable dependency is checked at compile-time and the program is annotated with extended CGEs similar to the ones used in &-Prolog. Goals to the right of a conjunction are allowed to proceed in parallel as long as they do not bind the dependent variable. Only the leftmost goal is allowed to bind the dependent variable in order to keep the operational semantics of Prolog. Hence rightmost goals that try to bind a dependent variable are suspended until become leftmost in the CGE.

3.4.   AND-OR Parallelism

Parallel Logic Programming systems should serve the broadest range of applications to become popular. To achieve this goal, these systems should support both or- and AND-parallelism since most programs exhibit mostly one kind of parallelism. Requirements for a usable parallel WAM engine are [50, 108]: (1) parallelism should be transparent to the programmer; (2) both AND- and OR-parallelism should be supported as a minimum; (3) the performance of the WAM engine on a single processor should be similar to that of a sequential WAM; (4) the parallel WAM should incorporate as many of the known sequential optimisations as possible; and (5) the runtime overhead should be low.

In practice, exploiting one single form of parallelism requires sophisticated system design and exploiting distinct forms of parallelism at the same time is even harder. AND/OR-parallel systems clearly suffer of overhead coming from exploiting either AND-parallelism or OR-parallelism. Unfortunately, this overhead may get worse due to the combination of the two forms of parallelism.

Scheduling becomes, in general, considerably more complex. The presence of both AND- and OR-parallelism requires an additional scheduling step, in order to decide which form of parallelism to exploit next. All additional scheduling requirements can be achieved only through the use of additional data structures which collect the information required by the scheduler.

Since integration of AND-parallelism with OR-parallelism is a very complex task, few systems were successful in doing it:

- Andorra-I [85]: exploits OR-parallelism a la Aurora [72] and determinate AND-parallelism, i.e., goals that match at most one clause in the program are executed firstly and in parallel. The AND-parallel execution is based on JAM Parlog [].

- Penny [73, 74]: a parallel implementation of AKL [56, 60] that also exploits OR-parallelism.

- ParAKL [75]: another parallel implementation of AKL.

- ACE [51]: a system that integrates independent AND-parallelism a la &-Prolog and OR-parallelism a la Muse [4].

- SBA [26]: system that integrates independent AND-parallelism a la &ACE with OR-parallelism a la Aurora [72]. To the best of our knowledge, SBA is the first system to tackle the problem of efficiently representing variable bindings when combining independent AND-parallelism with OR-parallelism. It uses the Paged Binding Array model (PBA) [55].

- BEAM [70, 71]: ambitious system based on the Extended Andorra Model [112] that aims at exploiting dependent and independent AND-parallelism combined with OR-parallelism.

All these systems were designed and developed for shared-memory architectures. As an example of system that integrates AND-parallelism with OR-parallelism designed and developed for distributed machines we can mention the Distributed Andorra Prolog [13].

Systems implemented for centralised memory architectures present two limitations: (1) the hardware is not scalable, (2) algorithms and data structures designed for shared-memory architectures sometimes are not scalable when running on scalable hardwares. Distributed implementations of parallel logic programming systems that combine both AND-parallelism with OR-parallelism require quite complex parallel programming skills.

New software/hardware platforms have been developed that combine the shared memory programming model with a scalable hardware (distributed shared memory systems – DSMs [80]). A new parallel logic programming system that integrates the shared-memory programming model with the distributed-memory programming model, so-called DAOS [18], aims to obtain an efficient system capable of running on a hybrid architecture (for example, a cluster of workstations connected via a fast network that allows a shared-memory programming model). Research carried out about performance of parallel logic programming systems originally written for shared-memory machines, shows that it is possible to achieve significant performance improvements when running these systems on DSM architectures by only making modifications to data layout and making few modifications to the algorithms [83]. These experiments also show that architectural parameters can affect significantly the performance of parallel logic programming systems [92]. These results confirm that parallel runtime models designed to run in DSM platforms can obtain maximum performance from scalable architectures.

These studies indicate that data structures used in scheduling normally lead to false sharing. DAOS innovates by defining a hybrid solution: areas (stacks) which implement the major data-structures are logically shared and areas that are used to implement work management are private. To solve binding conflicts it uses an adaptation of SBA [26].

3.5.   Tools for performance evaluation

The development of efficient parallel logic programming systems is not an easy task. These systems need to tackle some hard problems such as parallel debugging and dynamic distribution of varied sized work among processors. Debugging in parallel environments is difficult due to non-deterministic response time of tasks.

Besides, some applications have very low degree of parallelism, either AND- or OR-parallelism or have very small input data size. Usually programs that present very low degree of parallelism exhibit a sequential programming style, what makes them difficult to parallelise transparently [46]. Programs that present a reasonable amount of parallelism can lead to good performance on parallel systems. However, in practice, there are several issues that contribute to bad performance even when the application has sufficient parallel tasks to supply available processors due normally to poor workload distribution strategies.

One way of dealing with these problems is to provide to parallel system developers and users a visualisation tool for parallel logic programming systems. Several visualisation tools have been developed such as WAMtrace [35], VisAndOr [47], ViMust [96], and VACE [106]. Each one of these tools works for a different parallel logic programming system. VisAll [42] is a more recent tool which attempts at unifying the visualisation paradigms exhibited by the other visualisation tools adding extra features to system optimisation and debugging.

4. Scheduling

Parallel or distributed execution means that different pieces of the program will be divided in several tasks and these tasks will be assigned to available processors. The task scheduler is the entity responsible for deciding what, when and to whom export tasks. All scheduling policies must distribute tasks over the available computational resources improving one or more performance measures. The scheduler is responsible for the efficient utilisation of processors. By efficient utilisation we mean to use only the resources required by the application. A reasonable scheduling strategy will also complete a parallel application in a time close to the earliest possible time. So, scheduling is a fundamental aspect in any parallel system since it is closely related to system performance and thus is an important research field in parallel logic programming.

In parallel logic programming systems, the execution of logic programs generally produces irregular computations that can be mapped into trees. Some applications produce very "bushy" trees, while others produce "thinner" trees. The trees have nodes that are distributed throughout the computing resources to be executed.

In OR-parallel systems the nodes of the execution tree are or-nodes (also called choice-points) that contain alternatives left to be executed. A collection of or-nodes connected in some hierarchical order form an *or-tree*. Finding or-work, i.e., finding new alternatives to execute, is the task performed by an or-scheduler. The *or-scheduler* deals with problems such as which alternative to select next to continue working or which branches need to be pruned.

AND-parallel systems generate the same kind of trees as OR-parallel systems, but with AND-nodes. A collection of AND-nodes connected in some hierarchical order form an *AND-tree*. In AND-parallel systems, the *AND-scheduler* is responsible to select the next goal to be executed.

In systems that exploit both AND-parallelism and OR-parallelism there is a new issue to be tackled which is *what kind* of work to be chosen next: AND- or OR-? There is not much research on this scheduling issue, because the combination of AND- and OR-parallelism is a relatively new issue and there are not many implemented systems that exploit both AND- and OR-parallelism.

Systems that exploit both AND- and OR-parallelism usually generate *AND-OR trees*, although several variations of exploitation of AND-parallelism may lead to different conceptual trees. A *task* to be selected by a scheduler to allocate to a processor is any work done by the system engine between two invocations of the scheduler. In that case, depending on the implementation system, the task sizes may vary. A *task switch* is performed when the scheduler is invoked to find work, and the worker actually switches to another task.

Next we discuss some or-only and AND-only scheduling techniques employed to some well known parallel logic programming systems that exploit either AND- or OR-parallelism, and and/or scheduling techniques proposed for systems that exploit both AND- and OR-parallelism (sections 4.1, 4.2 and 4.3).

4.1. Scheduling OR-parallel work

In a parallel logic programming system an or-scheduler has the function of choosing the best choice-point from which to take an alternative to be explored, in order to utilise resources in a better way. It should not incur too high costs in task switching, and it should cope with a limited number of processors in the system. Another very important issue is to reduce the amount of interaction between the workers. Some schedulers also deal with the problem of "speculativeness". This means that whereas some branches of the search tree are to be executed compulsorily, others (speculative or-branches) might be pruned because of a cut or commit operator. A branch of the or-tree can also be speculative if we can find ways of knowing that it does not lead to a valid solution.

Several schedulers have been proposed to distribute or-work in OR-parallel logic programming systems. In shared memory systems implementations such as Aurora [72] and Muse [4], or-schedulers try to minimize frequency of task switching, and amount of interaction between the workers. In the implementation of other models [110], methods are used that keep the cost of task switching constant by maintaining the cost of task creation or variable access as non constant. Task switching costs are directly related to scheduling. It is known from [22] that it is more efficient to keep the cost of task switching non constant while keeping other costs – variable access and task creation – constant.

Contrary to distributed systems, schedulers for shared memory systems do not need to tackle the processor communication problem. Instead, the main problem to be tackled by an OR-parallel scheduler is to control the frequency of task switching in systems that do not have constant time for this operation. Indirectly, two problems of equal importance arises when we do not control the frequency of task switching: cache corruption and scheduling overheads. These problems can also occur in distributed memory systems where the penalty for not maintaining locality of references is more serious, incurring in memory accesses that can be 10,000 times slower than a shared memory access.

There are several different schedulers for OR-parallel systems, but the main basic strategy to follow, in general, is: *to give preference to mandatory work. If there is no mandatory work select the least speculative work and repeatedly seek for better work.* This is the algorithm that seems to fit all OR-parallel systems nowadays and proved to be the best so far.

There are in the literature several strategies describing different schedulers for different models, and even for the same model. For example, the Aurora system has at least 5 scheduler strategies that compete with each other for parallel efficiency. The strategies vary mainly in the way that the work is chosen in the execution search tree. The schedulers also use different data structures that affect in different ways the performance of the OR-parallel system. Some of them, the most recent ones, deal with speculative work. All these strategies are supposed to be dynamic, i.e., decisions are made based on information collected at runtime.

4.2. Scheduling AND-parallel work

In systems that implement **independent AND-parallelism** the main problem in AND-scheduling is to correctly implement backtracking inside a CGE. If any goal is selected to execute in AND-parallel, and the implementation is WAM-based with contiguous stacks, on backtracking, a worker may not be able to backtrack and continue a previous goal, because there may be another goal being executed after the goal that needs to start again. This is shown in figure 8.7. From goals a & b & c, worker 1 (W1) selects goals a and c, while worker 2 (W2) selects goal b. Assume that worker 1 finishes goal a before worker 2 finishes goal b. In that case, worker 1 starts goal c. Assume that worker 2 finishes execution of goal b. It needs to backtrack and find another solution to a, but a cannot continue, because goal c did not finish yet. This is illustrated as the *trapped goal problem.*

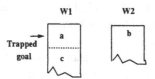

Figure 8.7. The trapped goal problem

The strategy used by the &-Prolog system is random selection of parallel goals from any agent's stack. As shown by the explanation before, this strategy is dangerous if we allow goals inside a ECGE to backtrack. In this case it can cause the problem of "trapped" goals where the topmost goal in an agent's stack backtracks on top of an older goal that did not finish execution yet. In order to solve this problem, there are three approaches. The first one [58] was designed for the &-Prolog system itself, and consists of only selecting a goal for parallel execution if it is the oldest pushed on the goal list. Therefore, WAM agents only execute goals in a certain order that is the sequential Prolog's order of goals. The second approach is to use a more flexible scheduling strategy that instead of restricting selection of goals by age, can select any goal. This is achieved by maintaining a *separator marker* between goals that can be executed in parallel. This scheduling strategy is described by Shen [90] and used in the DASWAM system. A third approach that will probably be used by the PBA implementation is to employ a memory management that allocates memory segments dynamically instead of using contiguous chunks of memory [82].

4.3.   Scheduling AND/OR-parallel work

As new systems that aim at exploiting both AND- and OR-parallelism are developed, the new issue is *what* kind of work to distribute. Some solutions were found that use some kind of global analysis of the program to predict the amount of work to be done in each branch [31, 9]. Other solution is to consider that all Prolog applications when executed in parallel generate a certain amount of work that can be mapped onto a statistical distribution [68]. Assuming that all tasks are to be executed with the same probability distribution, an algorithm is used that obtains the probability to the branches in order to distribute tasks evenly among the processors. This technique although applied to a model that aims to exploit AND- and OR-parallelism, only solves the OR-parallel work distribution.

Another solution is used by Yaoqing Gao et al [45] and consists of generating feedback information to the system by profiling previous interactive system sessions. The solution proposed tries to solve almost all problems by using a combination of static, dynamic, and hybrid approaches to get granularity sizes of Prolog programs, and feed these results into the system together with most-frequent used modes for calls to predicates. The scheduling strategy for OR-parallel tasks is to choose the nodes nearest to the root. The strategy to choose AND-goals is based on goal granularity (given at compile-time). Nothing is reported about how to decide between AND- and or-work, and no results are reported.

Andorra-I [86, 85] is the first and/or parallel logic programming system that has reported results running real applications. Andorra-I is the first reasonably mature implementation of a system that exploits both AND- and OR-parallelism. Andorra-I exploits OR-parallelism as in Aurora by using the binding arrays technique of the SRI model to solve the problem of different conditional bindings per branch of the search tree. It exploits a simple form of AND-parallelism by allowing only *determinate* goals to be executed in parallel. If goals try to bind a common variable with different values the whole branch fails.

Workers in Andorra-I work much the same as in the Aurora system, but instead of having a single worker to explore a branch, several workers cooperate with each other in a *team* in order to explore AND-parallelism if this exists in the branch. The first version of Andorra-I used a fixed configuration of workers into teams, and once the user started the execution with a given configuration, this would never change. If the application had a tree that did not suit the chosen configuration, Andorra-I would lose parallelism. Andorra-I received a reconfigurer [38] whose objective is to rearrange workers into teams according to the parallelism existing in the application. This strategy allows workers to be redeployed from one team to another to look for work. It also decides **what** kind of work to choose – AND- or OR- – in order to achieve reasonable load balancing and overall better performance. The other new issue in parallel logic programming systems, either AND-parallel systems or OR-parallel systems, that is tackled by the reconfigurer is the efficient utilisation of resources by the parallel system.

4.4.   Compile-time analysis

Regarding compile-time analysis tools for parallelisation support, we will concentrate on granularity and complexity analysis. Consider as a discussion case the GRANLOG model [9, 10] that extends and generalises previous models proposed in the literature [98, 32, 31]. The GRANLOG tool composed of several modules accepts a Prolog program as input and generates an annotated Prolog program containing information about complexity of goals and clauses. The information produced by GRANLOG is very useful to guide scheduling decisions. For example, based on the number and grain size of alternatives in a choice-point, the scheduler can select better pieces of work to assign to processors. Global information about amounts of AND-parallelism and OR-parallelism can be inferred from the grain sizes provided by the GRANLOG tool, in order that we can choose between sources of AND-parallel work or sources of OR-parallel work [36, 37, 39]. In the context of the APPELO project [46] three research works were carried out utilising GRANLOG information: (1) the centralised scheduler used by the PLoSys system was modified in order to support GRANLOG information [41, 102], (2) the Andorra-I reconfigurer [38] and the Andorra-I engine were modified to support GRANLOG information [40], (3) the DSLP scheme (Distributed Scheduling for Logic Programming) was designed and developed using information provided by GRANLOG [27]. Results are very encouraging and show that logic programming can be an alternative to efficient software development producing very good performance at a very low programming cost. Static analysis can produce useful information to guide scheduling decisions, and its runtime model can be more efficient than systems developed for specific parallel platforms (shared-memory only or distributed-memory only).

4.5.   Discussion

Systems that exploit OR-parallelism implemented for centralised memory platforms are quite mature and their scheduling techniques are improving with each new algorithm and implementation. Researchers have vastly studied or-scheduler systems and it is becoming clear what is required of an OR-parallel scheduler to make an OR-parallel system efficient. General conclusions are that low communication between workers is important, in order to be able to perform well with both speculative applications and non-speculative applications. We also need to have an efficient way of keeping the left-to-right order of the branches in the tree and have an efficient algorithm for searching for deepest leftmost work. It is also clear that an OR-parallel system can benefit from schedulers that are capable of disposing some processors if the application does not need them. The Bristol-scheduler [12] and the Dharma-scheduler [93], used in Aurora, and the Muse-scheduler [6], used in the Muse system, have been successful in doing their jobs. It remains to investigate the effectiveness of these techniques for modern architectures such as distributed-shared memory architectures and softwares or clusters of SMP workstations. Some work is being done in this direction with the evaluation of several scheduling strategies for OR-parallel systems [19, 27, 105, 104].

Other or-scheduling methods such as the one proposed by Lin [68] were implemented. Lin proposed to remove structural imbalance from logic programs through compile-time analysis. The efficiency of such systems is arguable since the scheduler needs to be centralised in order to keep track of the number of idle processors and the number of tasks not yet taken. Besides, these methods assume that the scheduling is done at once for all processors. In that case, processors remain blocked while tasks are being distributed by the scheduler.

AND-schedulers were not so studied as or-schedulers. Common sense is to maintain a distributed pool of AND-goals, and in systems that exploit IAP to prevent the trapped goal problem.

Tick [98] introduced a compile-time technique to guide scheduling decisions at runtime. His method produced very simple information about granularity of clauses, by considering each call to a goal, including recursive calls, as having cost 1. Tick studied his method for the FGHC system and did not get very good results. His conclusion was that the small granularity of AND-parallel tasks in FGHC caused too much overheads that could not be hidden by good scheduling decisions. The same method was implemented by Wai-Keong [43] to schedule OR-parallelism and produced very good results.

Compile-time granularity analysis can be used successfully if the compiler can detect regular pieces of computation in the source code. The information produced by the compiler could then help the schedulers, both AND- and or-schedulers. If the compiler cannot detect anything about the program computation pattern, some strategy that does not rely on compile-time analysis data, needs to be used anyway. Some solutions assume that the amount of work in the applications follows a probabilistic distribution, but this seems a very unpractical approach since the shape of Prolog application search trees varies widely. Other solutions propose the idea of profiling parallel executions and can be the source for future research.

With respect to scheduling both AND-parallelism and OR-parallelism, Andorra-I [114] is an and/or parallel logic programming system that tackles this problem. So far the results have been quite encouraging. Andorra-I is capable of running a wide range of applications that contain both AND- and OR-parallelism or either kind alone and achieve good speedups without the need for manually adjusting the number of workers to exploit AND-parallelism and OR-parallelism [38].

5. Conclusion

This text summarised the main issues in exploiting parallelism from logic programs. We discussed the main problems related to exploit AND-parallelism, to exploit OR-parallelism, and the main problems related to combine the exploitation of both kinds of parallelism. We also discussed about schedulers used in some parallel logic programming systems and compile-time analysis techniques that can help better exploitation of parallelism and better scheduling decisions.

The main issues in exploiting parallelism from logic programming that were discussed in this text are:

- *What kind of parallelism to exploit?* Regarding the programming paradigm, parallelism can be exploited implicitly or explicitly.

- *What kind of architecture to choose?* The system can run in any architecture as long as it runs *efficiently* which means to suit the system model to the architecture chosen.

Implicit parallel logic programming has the advantage of allowing program reuse and transparency at user level. However to exploit parallelism implicitly is not a trivial task and several techniques and approaches have been developed in the last years as discussed in this document. Nowadays, some important research fields in parallel logic programming are: (1) efficient ways to combine AND- and OR-parallelism; (2) efficient scheduling and execution in scalable systems; (3) compile-time (static) analysis to help run-time decisions; and (4) extensions of logic programming to prune the search space (such as tabling and constraints). Systems should support both or- and AND-parallelism to serve the broadest range of applications since most programs exhibit mostly one kind of parallelism. In practice, exploiting one single form requires sophisticated system design and exploiting distinct forms of parallelism is even harder.

In order to exploit parallelism implicitly, it is necessary to detect sources of parallelism in the applications. Since tasks in logic programming are created dynamically, we need a good runtime support to identify sources of parallelism. Compile-time granularity analysis can help to improve performance in some cases.

Results of few parallel logic programming systems show that it is possible to achieve good performance on parallel architectures with a minimum effort from the end user.

Besides the extensive bibliography we provide with this text we would indicate several interesting pointers to the Logic Programming World. One very interesting and comprehensive site is located at http://www.comlab.ox.ac.uk/archive/logic-prog.html. For a comprehensive bibliography site goto http://liinwww.ira.uka.de/bibliography/LogicProgramming/. The ALP (Association for Logic Programming) site can be found at http://www.cwi.nl/projects/alp/index.html. There you have access to the ALP Newsletter. There is also the Compulog-Americas initiative that contains information about logic programming groups and research in the Americas: http://www.cs.nmsu.edu/~compulog. The Journal of Logic Programming is the most important publication in the area. Several important conferences are: ICLP (International Conference on Logic Programming), PPDP (International Conference on Principles and Practice of Declarative Programming, old PLILP – Programming Languages and Implementation of Logic Programming), PAP (Practical Applications of Prolog), PACLP (Practical Applications of Constraint Logic Programming), and PADL (Practical Aspects of Declarative Languages). Pontelli keeps in his home page an interesting chapter on "Adventures in Parallel Logic Programming" (http://www.cs.nmsu.edu/~epontell/advent.html). Enjoy your trip!

References

[1] Special issue: Applications of logic programming. *The Journal of Logic Programming*, 26(2), February 1996.

[2] A. Aggoun et al. *ECLiPSe 3.5 User Manual*. ECRC, December 1995.

[3] H. Aït-Kaci. *Warren's Abstract Machine — A Tutorial Reconstruction*. MIT Press, 1991.

[4] K. A. M. Ali and R. Karlsson. The Muse Or-parallel Prolog Model and its Performance. In *Proceedings of the North American Conference on Logic Programming*, pages 757–776. MIT Press, October 1990.

[5] K. A. M. Ali and R. Karlsson. Performance of Muse on the BBN Butterfly TC2000. In *LNCS 605, PARLE'92 Parallel Architectures and Languages Europe*, pages 603–616. Springer-Verlag, June 1992.

[6] K. A. M. Ali and R. Karlsson. Scheduling speculative work in muse and perfomance results. Internal Report, SICS, 1993.

[7] J. Andersson, S. Andersson, K. Boortz, M. Carlsson, H. Nilsson, T. Sjoland, and J Widén. SICStus Prolog User's Manual. Technical report, Swedish Institute of Computer Science, November 1997. SICS Technical Report.

[8] K. R. Apt. *From Logic Programming to Prolog*. Prentice Hall, 1997.

[9] J. L. V. Barbosa. Granlog: Um modelo para análise automática de granulosidade na programação em lógica. Dissertação de mestrado, Universidade Federal do Rio Grande do Sul, 1996.

[10] J. L. V. Barbosa, P. K. Vargas, C. F. R. Geyer, and I. C. Dutra. Granlog: an integrated model to granularity analysis in logic programming. In *Workshop on Parallelism and Implementation Technology for (Constraint) Logic Programming Languages*, July 2000. In conjunction with the Intl. Conference on Computational Logic, 2000, London, England.

[11] U. Baron, J. C. Kergommeaux, M. Hailperin, M. Ratcliffe, P. Robert, J.-C. Syre, and H. Westphal. The Parallel ECRC Prolog System PEPSys: an Overview and Evaluation Results. In *International Conference on Fifth Generation Computer Systems 1988*, pages 841–850. ICOT, Tokyo, Japan, November 1988.

[12] A. Beaumont. *Scheduling in Or-Parallel Prolog Systems*. PhD thesis, University of Bristol, Department of Computer Science, 1994.

[13] V. Benjumea and J. M. Troya. A Static Implementation of the Basic Andorra Model for Distributed Memory Systems. In *Post-JICSLP96 Workshop on Parallelism and Implementation of Logic Programming Systems*, 1996. Bonn, Germany.

[14] J. Bevemyr, T. Lindgren, and H. Millroth. Reform Prolog: The Language and its Implementation. In *Proceedings of the Tenth International Conference on Logic Programming*, pages 283–298. MIT Press, June 1993.

[15] P. Borgwardt. Parallel prolog using stack segments on shared memory multiprocessors. In *International Symposium on Logic Programming*, pages 2–12, Silver Spring, MD, February 1984. Atlantic City, IEEE Computer Society.

[16] J. Briat, M. Favre, C. F. R. Geyer, and J. C. Kergommeaux. *Implementations of Distributed Prolog*, chapter OPERA: Or-parallel Prolog System on Supernode, pages 45–63. John Wiley & Sons Ltda, 1992.

[17] Calegario, V. M. and Dutra, I. C. Performance Evaluation of Or-Parallel Logic Programming Systems on Distributed Shared Memory Architectures. In *Proceedings of the EUROPAR'99*, pages 1484–1491, Aug-Sep 1999.

[18] L. F. P. Castro, V. Santos Costa, C. F. R. Geyer, F. Silva, P. K. Vargas, and M. E. Correia. DAOS – Scalable And-Or Parallelism. In *Proceedings of the EUROPAR '99*, pages 899–908, 1999.

[19] A. P. B. Centeno and C. F. R. Geyer. PENELOPE: Um modelo de Escalonador Hierárquico para o Sistema PLoSys. In *X Simpósio Brasileiro de Arquitetura de Computadores e Processamento de Alto Desempenho*, 1998. Búzios, RJ.

[20] P. Ciancarini. Parallel Programming with Logic Languages: A Survey. *Computer Languages*, 17(4), October 1992.

[21] P. Ciancarini. Blackboard Programming in Shared Prolog. In *Languages and Compiler for Parallel Computing*. MIT Press, June 1993.

[22] A. Ciepielewski, K. A. M. Ali, and P. Biswas. Execution Models and Parallel Implementations of Prolog. unpublished, June 1993.

[23] A. Ciepielewski and S. Haridi. A Formal Model for Or-Parallel Execution of Logic Programs. In *IFIP 83 Conference*, pages 299–305. North Holland, 1983.

[24] W. F. Clocksin. Principles of the DelPhi parallel inference machine. *Computer Journal*, 30(5):386–392, 1987.

[25] J. Cohen. A View of the Origins and Development of Prolog. *Communications of the ACM*, 31(1):26–36, 1988.

[26] M. E. Correia, F. M. A. Silva, and V. Santos Costa. The SBA: Exploiting orthogonality in OR-AND Parallel Systems. In *ILPS'97*, Outubro 1997. Also published as Technical Report DCC-97-3, DCC - FC & LIACC, Universidade do Porto, April, 1997.

[27] C. A. Costa and C. F. R. Geyer. Uma Proposta de Escalonamento Distribuído para Exploração de Paralelismo na Programação em Lógica. In *X Simpósio Brasileiro de Arquitetura de Computadores e Processamento de Alto Desempenho*, 1998. Búzios, RJ.

[28] J. A. Crammond. A Comparative Study of Unification Algorithms for OR-Parallel Execution of Logic Languages. In D. DeGroot, editor, *Int. Conf. on Parallel Processing*, pages 131–138, St. Charles, Ill., August 1985. IEEE.

[29] L. Damas, V. Santos Costa, R. Reis, and R. Azevedo. *YAP User's Guide and Reference Manual*, 1989.

[30] K. De Bosschere and P. Tarau. Blackboard-based Extensions in Prolog. *Software — Practice and Experience*, 26(1):49–69, January 1996.

[31] S. K. Debray, P. L. García, M. V. Hermenegildo, and N. Lin. Estimating the computacional cost of logic programs. In *International Static Analysis Symposium*, volume 864 of *Lecture Notes in Computer Science*, Namur, Belgium, September 1994. Springer-Verlag.

[32] S. K. Debray and N. Lin. Cost analysis of logic programs. *ACM Transactions on Programming Languages and Systems*, 15(5):826–875, November 1993.

[33] Doug DeGroot. A Technique for Compiling Execution Graph Expressions for Restricted And-Parallelism in Logic Programs. In C. Polychronopoulos, editor, *Lecture Notes in Computer Science 297, Supercomputing*, pages 1074–1093, 1987.

[34] P. Deransart, A. Ed-Dbali, L. Cervoni, and A. A. Ed-Ball. *Prolog, The Standard: Reference Manual*. Springer Verlag, 1996.

[35] T. Disz and E. Lusk. A Graphical Tool for Observing the Behavior of Parallel Logic Programs. In *Proceedings of the 1987 International Logic Programming Symposium*, pages 46–53, 1987.

[36] I. C. Dutra. A Flexible Scheduler for Andorra-I. In A Beaumont and G. Gupta, editors, *Lecture Notes in Computer Science 569, Parallel Execution of Logic Programs*, pages 70–82. Springer-Verlag, June 1991.

[37] I. C. Dutra. Strategies for Scheduling And- and Or-Work in Parallel Logic Programming Systems. In *Proceedings of the 1994 International Logic Programming Symposium*, pages 289–304. MIT Press, 1994.

[38] I. C. Dutra. *Distributing And- and Or-Work in the Andorra-I Parallel Logic Programming System*. PhD thesis, University of Bristol, Department of Computer Science, February 1995. PhD thesis.

[39] I. C. Dutra. Distributing And-Work and Or-Work in Parallel Logic Programming Systems. In *Proceedings of the 29th Hawaii International Conference on System Sciences*, pages 646–655. IEEE, 1996.

[40] I. C. Dutra, V. Santos Costa, J. L. V. BARBOSA, and C. F. R. GEYER. Using Compile-Time Granularity Information to Support Dynamic Work Distribution in Parallel Logic Programming Systems. In *X Simpósio Brasileiro de Arquitetura de Computadores, SBAC-PAD*, October 1999.

[41] D. N Ferrari, P. K. Vargas, J. L. V. Barbosa, and C. F. R. Geyer. Modelo de Integração Plosys-Granlog: Aplicação da Análise de Granulosidade na Exploração do Paralelismo OU. In *XXV Conferência Latino Americana de Informática/V Encontro Chileno de Computação da Sociedade Chilena de Computação (CLEI - PANEL'99)*, September 1999. Assuncion, Paraguay.

[42] N. Fonseca, I. C. Dutra, and V. Santos Costa. VisAll: A Universal Tool to Visualise Parallel Execution of Logic Programs. In *Joint International Conference and Symposium on Logic Programming*, pages 100–114, 1998. Manchester, Inglaterra.

[43] Wai-Keong Foong. Or-Parallel Prolog with Heuristic Task Distribution. In *Lecture Notes in Artificial Intelligence 592, Logic Programming Russian Conference*, pages 193–200, 1991.

[44] I. Futó. Prolog with Communicating Processes: From T-Prolog to CSR-Prolog. In *Proceedings of the Tenth International Conference on Logic Programming*, pages 3–17, June 1993.

[45] Y. Gao et al. Intelligent Scheduling And- and Or- Parallelism in the Parallel Logic Programming System RAP/LOP-PIM. In *Proceedings of the 1991 International Conference on Parallel Processing*, August 1991.

[46] C. F. R. Geyer, J. L. V. Barbosa, I. C. Dutra, V. Santos Costa, and G. Gupta. The APPELO Project – Parallel Environment for Logic Programming. In *Protem-CC III*, pages 30–55. CNPq, 1998. Available at http://www.inf.ufrgs.br/procpar/opera/APPELO/.

[47] L. Gómez, M. Carro, and M. Hermenegildo. Some Paradigms for Visualising Parallel Execution of Logic Programs. In *Proceedings of the Tenth International Conference on Logic Programming*, pages 184–202, 1993.

[48] A. González and J. Tubella. The Multipath Parallel Execution Model for Prolog. In *Proceedings of PASCO'94 (Parallel Symbolic Computation Conference)*, pages 426–432. World Scientific Publishing, 1994.

[49] S. Gregory. *Parallel Logic Programming in PARLOG*. Addison–Wesley, 1987.

[50] G. Gupta, K. A. M. Ali, M. Carlsson, and M. V. Hermenegildo. Parallel logic programming: a survey. available from ftp://ftp.cs.nmsu.edu/pub/papers/lp/survey.ps.Z, 1997.

[51] G. Gupta, M. V. Hermenegildo, E. Pontelli, and V. Santos Costa. ACE: And/Or-parallel Copying-based Execution of Logic Programs. In *Proceedings of the Eleventh International Conference on Logic Programming*, Italy, June 1994.

[52] G. Gupta and B. Jayaraman. On Criteria for Or-parallel Execution of Logic Programs. In *Proceedings of the 1990 North American Conference on Logic Programming*, pages 737–756. MIT Press, October 1990.

[53] G. Gupta and E. Pontelli. Stack-Splitting: a Simple Technique for Implementing Or-Parallelism and And-parallelism on Distributed Machines. In *Proceedings of the Sixteenth International Conference on Logic Programming*, pages 290–304. MIT Press, 1999.

[54] G. Gupta, E. Pontelli, and M. V. Hermenegildo. &ACE: A High Performance Parallel Prolog System. In *Proceedings of the First International Symposium on Parallel Symbolic Computation, PASCO'94*, 1994.

[55] G. Gupta and V. Santos Costa. And-Or Parallelism in Full Prolog with Paged Binding Arrays. In *LNCS 605, PARLE'92 Parallel Architectures and Languages Europe*, pages 617–632. Springer-Verlag, June 1992.

[56] S. Haridi and S. Jansson. Kernel Andorra Prolog and its Computational Model. In D.H.D. Warren and P. Szeredi, editors, *Proceedings of the*

Seventh International Conference on Logic Programming, pages 31–46. MIT Press, 1990.

[57] M. V. Hermenegildo. An Abstract Machine for Restricted And-Parallel Execution of Logic Programs. In Ehud Shapiro, editor, *Proceedings of the Third International Conference on Logic Programming*, pages 25–39. Springer-Verlag, 1986.

[58] M. V. Hermenegildo and K. J. Greene. The &-Prolog System: Exploiting Independent And-Parallelism. *New Generation Computing*, 9(3,4):233–256, 1991.

[59] Z. Huang, C. Sun, A. Sattar, and W-J. Lei. Parallel Logic Programming on Distributed Shared Memory System. In *Proceedings of the IEEE International Conference on Intelligent Processing Systems*, October 1997.

[60] S. Jansson and S. Haridi. Programming Paradigms of the Andorra Kernel Language. In *Proceedings of the 1991 International Logic Programming Symposium*, pages 167–186. MIT Press, October 1991.

[61] R. Karlsson. *A High Performance OR-parallel Prolog System*. Swedish Institute for Computer Science, SICS Dissertation Series 07, The Royal Institute of Technology, 1992.

[62] J. C. Kergommeaux and P Codognet. Parallel logic programming systems. *Computing Surveys*, 26(3):295–336, September 1994.

[63] J. C. Kergommeaux and P. Robert. An Abstract Machine to Implement Or-And Parallel Prolog Efficiently. *The Journal of Logic Programming*, 8(3), May 1990.

[64] Y. Kimura and T. Chikayama. An Abstract KL1 Machine and its Instruction Set. In *International Symposium on Logic Programming*, pages 468–477. San Francisco, IEEE Computer Society, August 1987.

[65] R. A. Kowalski. Predicate Logic as a Programming Language. In *Proceedings IFIPS*, pages 569–574, 1974.

[66] R. A. Kowalski. Algorithm = logic + control. *Communications ACM*, 22(7), July 1979.

[67] Y-J. Lin and V. Kumar. An Execution Model for Exploiting And-Parallelism in Logic Programs. *New Generation Computing*, 5(4):393–425, 1988.

[68] Z. Lin. Self-Organising Task Scheduling for Parallel Execution of Logic Programs. In *Proceedings of the 1992 International Conference on Fifth Generation Computer Systems*, pages 859–868. ICOT, 1992.

[69] G. Lindstrom. Or-Parallelism on Applicative Architectures. In *Proceedings of the 1985 International Logic Programming Symposium*, pages 159–170, July 1984.

[70] R. Lopes and V. Santos Costa. *Parallelism and Implementation of Logic and Constraint Logic Programming*, chapter The BEAM: Towards a first EAM Implementation. Nova Science, Inc., 1999.

[71] R. Lopes, F. Silva, V. Santos Costa, and S. Abreu. The RAINBOW: Towards a Parallel BEAM. In *Workshop on Parallelism and Implementation Technology for (Constraint) Logic Programming Languages*, July 2000. In conjunction with the Intl. Conference on Computational Logic, 2000, London, England.

[72] E. Lusk et al. The Aurora or-parallel Prolog system. *New Generation Computing*, 7(2,3):243–271, 1990.

[73] J. Montelius. *Penny, A Parallel Implementation of AKL*. PhD thesis, Swedish Institute for Computer Science, SICS, Sweden, May 1997.

[74] J. Montelius and P. Magnusson. Using SIMICS to evaluate the Penny system. In Jan Małuszyński, editor, *Proceedings of the International Symposium on Logic Programming (ILPS-97)*, pages 133–148, Cambridge, October 13–16 1997. MIT Press.

[75] R. Moolenaar and B. Demoen. A parallel implementation for AKL. In *Proceedings of the Programming Language Implementation and Logic Programming: PLILP '93, Tallin, Estonia*, pages 246–261, 1993.

[76] E. Morel, J. Briat, J. Chassin de Kergommeaux, and C. Geyer. Side-Effects in PloSys OR-parallel Prolog on Distributed Memory Machines. In *JICSLP'96 Post-Conference Workshop on Parallelism and Implementation Technology for (Constraint) Logic Programming Languages, Bonn, Germany*, September 1996.

[77] K. Muthukumar and M. V. Hermenegildo. The CDG, UDG, and MEL Methods for Automatic Compile-time Parallelization of Logic Programs for Independent And-parallelism. In *Proceedings of the Seventh International Conference on Logic Programming*, pages 221–237. MIT Press, June 1990.

[78] L. M. Pereira, L. Monteiro, J. Cunha, and J.N. Aparício. Delta Prolog: a distributed backtracking extension with events. In Ehud Shapiro, editor, *Third International Conference on Logic Programming, LNCS 225*, pages 69–83. Springer-Verlag, 1986.

[79] E. Pontelli. Adventures in parallel logic programming, 1996. Available at http://www.cs.nmsu.edu/~epontell/adventure/.

[80] J. Protic, M. Tomasevic, and V. Milutinovic. Distributed Shared Memory: Concepts and Systems. *IEEE Parallel and Distributed Technolog*, 4(2), 1996.

[81] R. Rocha, F. Silva, and V. Santos Costa. YapOr: an Or-Parallel Prolog System based on Environment Copying . In *9th Portuguese Conference on Artificial Intelligence – EPIA 99*, pages 178–192, September 1999.

[82] V. Santos Costa. COWL: Copy-On-Write for Logic Programs. In *Proceedings of the IPPS/SPDP99*, pages 720–727. IEEE Computer Press, May 1999.

[83] V. Santos Costa and R. Bianchini. Optimising Parallel Logic Programming Systems for Scalable Machines. In *Proceedings of the EUROPAR'98*, pages 831–841, Sep 1998.

[84] V. Santos Costa, R. Bianchini, and I. C. Dutra. Parallel Logic Programming Systems on Scalable Architectures. *Journal of Parallel and Distributed Computing*, 60(7):835–852, 2000.

[85] V. Santos Costa, D. H. D. Warren, and R. Yang. Andorra-I: A Parallel Prolog System that Transparently Exploits both And- and Or-Parallelism. In *Third ACM SIGPLAN Symposium on Principles & Practice of Parallel Programming*, pages 83–93. ACM press, April 1991. SIGPLAN Notices vol 26(7), July 1991.

[86] V. Santos Costa, D. H. D. Warren, and R. Yang. The andorra-i preprocessor: Suporting full prolog on the basic andorra model. In *1991 International Conference on Logic Programming*. MIT Press, June 1991.

[87] M. Schneider. Links on objects and components/prolog, 1999. Available at http://www.cetus-links.org/oo_prolog.html.

[88] E. SHAPIRO, editor. *Concurrent Prolog–Collected Papers*. MIT Press, 1987.

[89] K. Shen. Exploiting And-parallelism in Prolog: the Dynamic Dependent And-parallel Scheme (DDAS). In *Proceedings of the 1992 Joint International Conference and Symposium on Logic Programming*, 1992.

[90] K. Shen. Memory Management and Goal Scheduling for Parallel Execution Revisited. Internal Report, Bristol University, 1993.

[91] K. Shen. Overview of DASWAM: Exploitation of Dependent And-parallelism. *The Journal of Logic Programming*, 29(1–3), 1996.

[92] M. G. Silva, I. C. Dutra, R. Bianchini, and V. Santos Costa. The Influence of Computer Architectural Parameters on Parallel Logic Programming Systems. In Gopal Gupta, editor, *Workshop on Practical Aspects of Declarative Languages (PADL99), LNCS 1551*, pages 122–136, January 1999.

[93] R. Y. Sindaha. *Branch-Level Scheduling in Aurora: The Dharma Scheduler*. PhD thesis, University of Bristol, Department of Computer Science, 1995.

[94] D. A. Smith. MultiLog: Data Or-Parallel Logic Programming. In *Proceedings of the Tenth International Conference on Logic Programming*, pages 314–331. MIT Press, 1993.

[95] L. Sterling and E. Shapiro. *The Art of Prolog*. MIT Press, 1994.

[96] J. Sunberg and C. Svensson. MUSE TRACE: A Graphic Tracer for Or-Parallel Prolog. Technical report, SICS, 1990.

[97] J. Tanaka, K. Ueda, T. Miyazaki, A. Takeuchi, Y. Matsumoto, and K. Furukawa. Guarded Horn Clauses and Experiences with Parallel Programming. In *1986 Proceedings Fall Joint Computer Conference*, pages 948–954. IEEE Computer Society Press, November 1986.

[98] E. Tick. Compile Time Granularity Analysis for Parallel Logic Programming Systems. *New Generation Computing*, 7(2,3):325–337, 1990.

[99] K. Ueda and M. Morita. A New Implementation Technique for Flat GHC. In *Proceedings of the Seventh International Conference on Logic Programming*, pages 3–17. MIT Press, June 1990.

[100] P. Van Roy. *Can Logic Programming Execute as Fast as Imperative Programming?* PhD thesis, University of California at Berkeley, November 1990.

[101] P. K. Vargas. Exploração de Paralelismo OU em uma Linguagem em Lógica com Restrições. Dissertação de mestrado, CPGCC - UFRGS, August 1998.

[102] P. K. Vargas, J. L. V. Barbosa, D. N Ferrari, J. C. Kergomeaux, and C. F. R. Geyer. Distributed or scheduling with granularity information. In *12th Symposium on Computer Architecture and High Performance Computing*, October 2000. to appear.

[103] P. K. Vargas and C. F. R. Geyer. Introduzindo o paralelismo ou na programação em lógica com restrições. *IX Simpósio Brasileiro de Arquitetura de Computadores e processamento de Alto Desempenho*, 1997. Campos do Jordão, SP.

[104] P. K. Vargas and C. F. R. Geyer. Uma política de escalonamento distribuída no modelo pclp(fd) de exploração de paralelismo ou. *XXIV Conferência Latino Americana de Informática/V Encontro Chileno de Computação da Sociedade Chilena de Computação (CLEI - PANEL'98)*, 1998. Quito, Ecuador.

[105] P. K. Vargas, C. F. R. Geyer, and I. C. Dutra. TAMAGOSHI - Plataforma para Avaliação de Escalonamento de Tarefas em Programação em Lógica Paralela. In *X Simpósio Brasileiro de Arquitetura de Computadores e Processamento de Alto Desempenho*, 1998. Búzios, RJ.

[106] R. Vaupel, E. Pontelli, and G. Gupta. Visualization of And/Or-Parallel Execution of Logic Programs. In *Proceedings of the Fourteenth International Conference on Logic Programming*, pages 271–285, 1997.

[107] K. Villaverde, H. Guo, E. Pontelli, and G. Gupta. Incremental stack-splitting. In *Workshop on Parallelism and Implementation Technology for (Constraint) Logic Programming Languages*, July 2000. In conjunction with the Intl. Conference on Computational Logic, 2000, London, England.

[108] M. P. von Lahuizen. Parallel Logic Programming Techniques. Technical report, Delft University of Technology, PDS-1998-003, February 1998. Available at http://pds.twi.tudelft.nl/reports/.

[109] D. H. D. Warren. An Abstract Prolog Instruction Set. Technical Note 309, SRI International, 1983.

[110] D. H. D. Warren. Or-Parallel Execution Models of Prolog. In *TAP-SOFT'87, The 1987 International Joint Conference on Theory and Practice of Software Development, Pisa, Italy*, pages 243–259. Springer-Verlag, March 1987.

[111] D. H. D. Warren. The SRI model for or-parallel execution of Prolog— abstract design and implementation issues. In *Proceedings of the 1987 Symposium on Logic Programming*, pages 92–102, 1987.

[112] D. H. D. Warren. The Extended Andorra Model with Implicit Control. Presented at ICLP'90 Workshop on Parallel Logic Programming, Eilat, Israel, June 1990.

[113] R. Yang. *P-Prolog a Parallel Logic Programming Language*, volume 9. World Scientific, Singapore, 1987. Series in Computer Science.

[114] R. Yang, A. Beaumont, I. Dutra, V. Santos Costa, and D. H. D. Warren. Performance of the Compiler-Based Andorra-I System. In *Proceedings of the Tenth International Conference on Logic Programming*, pages 150–166. MIT Press, June 1993.

[115] R. Yang, V. Santos Costa, and D. H. D. Warren. The Andorra-I Engine: A parallel implementation of the Basic Andorra model. In *Proceedings of the Eighth International Conference on Logic Programming*, pages 825–839. MIT Press, 1991.

Chapter 9

PARALLEL ASYNCHRONOUS TEAM ALGORITHMS

Benjamín Barán

*National University of Asuncion**

bbaran@cnc.una.py

Abstract   Solution of today large complex problems may need the combination of several
different methods, algorithms and techniques in a distributed computing system
with heterogeneous processors, usually interconnected through a communication
network. In this context, Team Algorithm is presented as a general technique
used to combine different methods and algorithms in a distributed computing
system composed of different workstations, personal computers and/or parallel
computers. Moreover, experimental results have proved that in many real world
problems, Team Algorithms can benefit from the use of asynchronous implementations, speeding up the whole process with an important synergy effect, in a new
appealing technique known as Parallel Asynchronous Team Algorithms.

The main idea behind Parallel Asynchronous Team Algorithms is very simple:
to partition a large complex problem in small sub-problems that can be solved in
different processors of a distributed system with well known sequential methods,
combining the partial results of each sub-problem in such a way that a good global
solution is finally found. Because each processor works at its own speed and
eventually, with its own algorithms, an asynchronous implementation eliminates
idle synchronization times speeding up the whole process. Team Algorithms
have been successfully applied in the resolution of many engineering problems
in which a synergetic effect may take place between different processors.

Keywords: Distributed system, asynchronous implementation, parallelism, network, algorithm combination.

1.   Introduction

Parallel asynchronous implementations of iterative algorithms are nowadays proving themselves as good choices for high performance computation

---

*National Computing Center, National University of Asuncion, POB:1439, San Lorenzo, Paraguay.

*R. Corrêa et al. (eds.), Models for Parallel and Distributed Computation. Theory, Algorithmic Techniques
and Applications, 243–277.*

in distributed-memory environments. Among their most attractive features are their ease of implementation, facility for computational load balancing, shorter convergence times and so on (see [4, 9]).

A new context for asynchronism arose with the introduction of the so-called *Team Algorithms*Team Algorithms (TAs), which are hybrids of different algorithms. Such combinations of algorithms were first proposed in the context of power system applications [14] for sequential computers. Later on, they were proposed in the context of parallel asynchronous computing, where they were first called team algorithms, in which such methods have a natural implementation [30]. A formal study of this idea, with a convergence analysis, has already been published by Baran et al. in [4].

TA is a technique which combines distinct algorithms interacting in the solution of the same global problem. Team Algorithms can be naturally implemented in parallel, assigning different sub-problems to each processor of an asynchronous distributed system like a computer network. The afore mentioned parallel asynchronous combination of different algorithms is known as A-Team (Asynchronous Team) [1, 2, 4]. Using this concept, some works [1, 7, 8, 26] proposed a synergetic combination of traditional numerical methods with Genetic Algorithms (GA), forming an A-Team capable of finding very good solutions, outperforming methods in use nowadays.

The objective of this chapter is to present Team Algorithm technique and its advantages in the resolution of various engineering problems.

Section 2 introduces the concept of Team Algorithms and a possible classification. It also provides a framework for TA convergence analysis. Section 3 presents a case, the load flow of electrical networks, as an example engineering problem for which the use of TA is an appealing choice. Computational results of a TA implementation for the load flow problem are discussed in Section 4. Section 5 briefly presents two engineering examples of A-Teams combining Genetic Algorithms with numerical methods, with excellent experimental results. Conclusions are presented in Section 6.

2.   Framework for convergence analysis of Team Algorithms

In this section a system of equations given by:

$$\phi(x) = 0, \qquad x \in \Re^n, \qquad \Phi : D \to \Re^n, \qquad (9.1)$$

where D is a closed subset of $\Re^n$, is considered.

The main idea is to use a system of p processors in such a way that each processor solves only a part of the whole system and communicates its partial result to the other processors to finally solve the 'global' problem (9.1). To formalize this, the following notation and assumptions are needed.

Let a Cartesian decomposition of $\Re^n$ be given by:

$$\Re^n = \Re^{n_1} \times \cdots \times \Re^{n_m}, \qquad n_1 + \cdots + n_m = n. \qquad (9.2)$$

Let $\mathbf{m} := \{1, 2, \ldots, m\}$ and $D \subset \Re^n$ be a domain such that

$$D = D_1 \times \cdots \times D_m, \qquad D_i \subset \Re^{n_i}, \qquad \forall i \in \mathbf{m}. \qquad (9.3)$$

A vector $x \in D$ is conformingly partitioned as:

$$x = \left[x_1^T, x_2^T, \ldots, x_m^T\right]^T, \qquad x_i \in D_i, \qquad \forall i \in \mathbf{m}, \qquad (9.4)$$

and $\Phi(x)$ as:

$$\Phi(x) = \left[\Phi_1^T(x), \Phi_2^T(x), \ldots, \Phi_m^T(x)\right]^T, \qquad \Phi_i : D \to \Re^{n_i}. \qquad (9.5)$$

Equation (9.1) may then be rewritten as:

$$\forall i \in \mathbf{m}, \qquad \Phi_i(x) = 0. \qquad (9.6)$$

Each *subproblem*Subproblem $\Phi_i(x) = 0$ is solved by an iterative algorithm represented below by the maps G_i, each of which updates x_i, i.e.

$$\forall i \in \mathbf{m}, \qquad x_i \leftarrow G_i(x). \qquad (9.7)$$

The iterative algorithm is chosen such that the fixed point $x^* = G(x^*)$ of the map

$$\forall i \in \mathbf{m}, G(x) = \left[G_1(x)^T, \ldots, G_m(x)^T\right]^T, \qquad G_i(x) : D \to D_i, \qquad (9.8)$$

is a solution of (9.1), i.e. $\Phi(x^*) = 0$.

The mathematical expression for $G_i(x)$ depends on the specific algorithm chosen to solve the corresponding subproblem. In general, no one algorithm can solve all subproblems equally well and, if convergence can be guaranteed, then it is clearly better to choose, for each subproblem, the algorithm that is best suited for it. When two or more different algorithms are chosen, this hybrid of algorithms is called a *Team Algorithm*Team Algorithms. As a first example, the case where each subproblem is solved in a different processor $(p = m)$ of a distributed (asynchronous) computing environment, known as a *Block (Asynchronous) Team Algorithm* is considered below.

2.1.   Block Team Algorithm

Each processor i used in the implementation of a block Team Algorithm attempts to solve its 'local' subproblem (9.6) using a map G_i that updates x_i, transmitting the updated value to the other processors; thus, a synchronous version of a block TA can be written as:

$$\forall i \in \mathbf{m}, \qquad x_i(k+1) = G_i\Big(x(k)\Big), \qquad k = 0, 1, 2, \ldots \qquad (9.9)$$

The main interest of this paper is in a block asynchronous implementation that implies the possibility of using the variables x_i in a desynchronized manner when applying an algorithm represented by map (9.8) (see [20]). To be more specific, it will be assumed that, at the time of iteration k, processor i (which updates x_i) receives information from another processor j with a time-varying delay of $k - d_j^i(k)$ units. The assumption will be made that the delays are uniformly bounded in time by a positive integer d. This is also referred to as partial asynchronism [9] and stated formally as a restriction on the range of the positive integer-valued functions $d_j^i(\cdot)$ in Assumption 9.1 below.

Assumption 9.1 (Uniform bound on delays)*Uniform bound on delays*
$\exists d \in \mathbb{N}, \quad \forall k \in \mathbb{N}, \quad \forall i, j \in \mathbf{m}, \quad$ s.t. $\quad d_j^i(k) \in \{k, k-1, \ldots, k-d\}.$

The version of the vector x, available in processor i at iteration k, is denoted as $x^i(k)$ and given by:

$$\forall i \in \mathbf{m}, \qquad x^i(k) := \Big[x_1(d_1^i(k))^T, \ldots, x_m(d_m^i(k))^T\Big]^T. \qquad (9.10)$$

Using this notation, the general bounded-delay asynchronous block-iterative algorithm based on the update laws (9.7) may be written as:

$$\forall i \in \mathbf{m}, \qquad x_i(k+1) = G_i(x^i(k)). \qquad (9.11)$$

Equation (9.11) represents the implementation of an *asynchronous* block algorithm in which each processor i tries to solve its 'local' subproblem (9.6) by updating x_i, i.e. applying map G_i and using the most recently received values of x_j for all $j \neq i$. After obtaining the new value of x_i, processor i communicates its results to those processors that need it, without blocking or interruptions of the latter, and initiates a new iteration. The process continues until a given stopping criterion is met.

Throughout this paper the following assumption will be also made:

Assumption 9.2 (Uniqueness of fixed point)*Uniqueness of fixed point   There exists a closed set $D \subset \Re^n$, such that $G(D) \subset D$, where the map G is defined by equations (9.7) and (9.8), which contains exactly one fixed point x^* of the map G.*

In the notation introduced above, $x^* = \left[x_1^{*T}, x_2^{*T}, \ldots, x_m^{*T} \right]^T$, $x_i^* \in D_i$, for all $i \in \mathbf{m}$, $x_i^* = G_i(x^*)$, and, since it is being assumed that G represents an iterative method to solve (9.1), then $\Phi(x^*) = 0$.

The concept of *block-Lipschitz continuity*Block-Lipschitz continuity that is equivalent to Lipschitz continuity, but often more convenient in the present context, is now defined:

Definition 9.1 (Block-Lipschitz continuity)   *A function $G(x)$ satisfying (9.8) is called* block-Lipschitz continuous *in a given norm with respect to a Cartesian decomposition $D_1 \times \cdots \times D_m$ defined in (9.3) if $\forall x, y \in D$ satisfying (9.4) the following inequality holds:*

$$\|G(x) - G(y)\| \leq \sum_{i=1}^{m} L_i \|x_i - y_i\|. \tag{9.12}$$

The constants L_i are called block-Lipschitz constants.

The lemma below follows immediately.

Lemma 9.2 *A function $G(x)$ is block-Lipschitz continuous iff it is Lipschitz continuous.*   □

It will be assumed throughout this paper that a single norm $\| \cdot \|$ is used in all spaces $\Re^{n_i}$, $i = 1, \ldots, m$ and on $\Re^n$ as well. The following assumption will also be made:

Assumption 9.3 (Block-Lipschitz continuity):   *Each map $G_i(x)$ of equation (9.11) is block-Lipschitz continuous, i.e.*

$$\forall i \in \mathbf{m}, \qquad \forall x, y \in D, \qquad \|G_i(x) - G_i(y)\| \leq \sum_{j=1}^{m} l_{ij} \|x_j - y_j\|. \tag{9.13}$$

An *error vector* $e = [e_1^T, \cdots, e_m^T]^T \in \Re^n$ is defined as:

$$\forall i, j \in \mathbf{m}, \qquad \left\{ \begin{array}{l} e_i(k+1) := x_i(k+1) - x_i^*, \\ e_j(d_j^i(k)) := x_j(d_j^i(k)) - x_j^*. \end{array} \right. \qquad (9.14)$$

In addition, a *reduced error vector* is defined as $z = [z_1, \cdots, z_m]^T \in \Re^m$

$$\forall i, j \in \mathbf{m}, \qquad \left\{ \begin{array}{l} z_i(k+1) := \|e_i(k+1)\|, \\ z_i(d_j^i(k)) := \|e_i(d_j^i(k))\|. \end{array} \right. \qquad (9.15)$$

Then, as in [4], the following result is obtained:

Lemma 9.3 *Under assumptions 9.1 and 9.2, given a non-negative matrix $H :=$ $(h_{ij}) \geq 0$ and a non-negative time-varying vector $z \geq 0$, satisfying the difference inequality*

$$\forall i \in \mathbf{m}, \qquad z_i(k+1) \leq \sum_{j=1}^{m} h_{ij} z_j(d_j^i(k)), \qquad (9.16)$$

then $\rho(H) < 1$ is a sufficient condition for $z(k)$ to tend to zero exponentially as $k \to \infty$, i.e.

$$\lim_{k \to \infty} z(k) = 0 \qquad if \qquad \rho(H) < 1,$$

where $\rho(H)$ denotes the spectral radius of matrix H.

Definition 9.4 (Comparison Matrix) *Comparison Matrix   The matrix H associated with the inequality (9.16) is referred to as a Comparison Matrix.*

The following theorem is an immediate consequence of Lemma 9.3 and will be used to obtain convergence conditions for several different methods treated in this paper.

Theorem 9.5 (Sufficient condition for convergence)*Sufficient condition for convergence   Under assumptions 9.1 to 9.3, the asynchronous block-iterative algorithm given by*

$$\forall i \in \mathbf{m}, \qquad x_i(k+1) = G_i(x^i(k)),$$

converges to the unique fixed point x^* in D if

$$\rho(H) < 1,$$

where $H := (h_{ij}) \in \Re^{m \times m}$, with $h_{ij} = l_{ij} \geq 0$, for all i, j, since the $l_{ij}s$ are the block-Lipschitz constants of the $G_i s$. $\qquad \square$

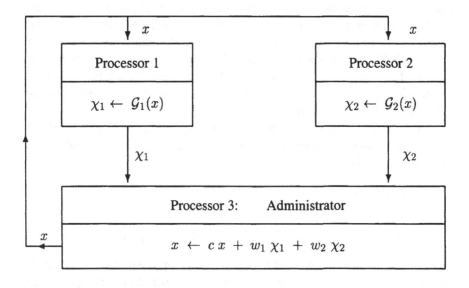

Figure 9.1. Fully Overlapped Team Algorithm.

Remark 9.1 For a non-negative matrix H, there are several conditions equivalent to $\rho(H) < 1$, some of which may be easier to check (see Lemma 1 of [20]).

Remark 9.2 This theorem gives a very general convergence condition for asynchronous fixed-point iterations and can be considered to be a more general formulation of the earlier result of Bhaya et al. [10]. It will be shown below that, in the present form, this theorem is easily applicable to all the variants of team algorithms that are in this paper, and this, indeed, is the major point of interest in the result.

2.2. Fully Overlapped Team Algorithm

Several problems may be solved by more than one algorithm, each with its own advantages. The use of several algorithms in parallel, in an asynchronous environment, with the objective of combining the best properties of each one, was first suggested by Talukdar et al. [30], who proposed to solve the algebraic system of equations (9.1), using two (or more) iterative methods which solve the whole problem in different processors, combining the partial results in a process (or processor) called the *Administrator*, as represented in Figure 9.1.

Keeping in mind an implementation on a parallel computer with a shared-memory architecture available at the time, [30] used a so-called *blackboard* model [22] and presented several numerical examples showing that a combination of the Newton method with the Steepest Descent method, i.e. a Team Algorithm, is able to solve problems in a given domain where the Newton method fails because the Jacobian is singular, and the Steepest Descent method alone needs a very large number of iterations to converge. It will now be shown how the framework introduced above can be used to formalize the methods proposed by [30] in a straightforward manner.

An asynchronous implementation of the TA represented in Figure 9.1 that is implementable in a distributed memory computing system, can be mathematically represented as:

$$
\begin{aligned}
\chi_1(k+1) &= G_1\left(x^1(k)\right) \\
\chi_2(k+1) &= G_2\left(x^2(k)\right) \\
x(k+1) &= c\,x(k) + w_1\,\chi_1^3(k) + w_2\,\chi_2^3(k),
\end{aligned}
\tag{9.17}
$$

where, using the notation defined in (9.10), $x^i(k)$ represents the most recently received value of x in processor i, at iteration k, while $\chi_i^3(k)$ $(i = 1, 2)$ represents the most recently received value of χ_i, at iteration k, in processor 3 (the Administrator).

The TA given by (9.17) works as follows: each processor i $(i = 1, 2)$ attempts to solve the given problem using an iterative algorithm represented by the map G_i, and utilizes the most recently received value of x received from the Administrator, to which it transmits its updated value χ_i. At the same time, the Administrator updates x combining the values of χ_i it has last received, using properly chosen positive 'weights' $(w_1, w_2 \geq 0)$. The constant c is chosen to ensure that, for the solution x^*, the following holds:

$$
x = \chi_1 = \chi_2 = x^* \in \Re^n,
$$

which, in turn, implies that:

$$
c = 1 - w_1 - w_2.
\tag{9.18}
$$

To derive a sufficient condition for the convergence of the TA represented by (9.17), assume that each map G_i has a Lipschitz constant L_i in the given domain D, and define an *Expanded State Vector* Expanded State Vector $W \in \Re^{3 \times n}$ as:

$$
W = \begin{bmatrix} W_1 \\ W_2 \\ W_3 \end{bmatrix} := \begin{bmatrix} \chi_1 \\ \chi_2 \\ x \end{bmatrix}, \quad \text{where} \quad \begin{cases} W_1 &= \chi_1 \in \Re^n, \\ W_2 &= \chi_2 \in \Re^n, \\ W_3 &= x \in \Re^n. \end{cases}
\tag{9.19}
$$

Using the above notation, (9.17) may be rewritten as:

$$
\begin{bmatrix} W_1(k+1) \\ W_2(k+1) \\ W_3(k+1) \end{bmatrix} = \begin{bmatrix} G_1\left(W_3^1(k)\right) \\ G_2\left(W_3^2(k)\right) \\ c\,W_3(k) + w_1\,W_1^3(k) + w_2\,W_2^3(k) \end{bmatrix}, \tag{9.20}
$$

which may be considered as a block TA, of the form (9.11) with the Comparison Matrix H given by:

$$H = \begin{bmatrix} 0 & 0 & L_1 \\ 0 & 0 & L_2 \\ w_1 & w_2 & |c| \end{bmatrix}. \tag{9.21}$$

The following corollary of Theorem 9.5 may now be stated:

Corollary 9.6 *Under assumptions 9.1 and 9.2 (properly reformulated to consider W instead of x), the (Fully Overlapped) Team Algorithm given by (9.17), (or equivalently by (9.20)), converges to the unique solution x^* in D, if $\rho(H) < 1$, where H is given by (9.21).* □

The TA given by (9.17) is called *Fully Overlapped* because both processors evaluate updates of the whole vector x, and these updates are then combined ('overlapped') in some pre-specified way to yield a 'global' update that is then sent to both processors, this being the task of the so-called administrator.

In order to appreciate the advantage of solving the same problem in parallel using a Fully Overlapped TA, consider a problem that may be solved using a fast algorithm G_1 (such as the Newton method) with $L_1 \geq 1$, but which is not able to ensure convergence in the whole region D (because, for example, it has a singular Jacobian in a subset of D), or alternatively may be solved using a very slow algorithm represented by G_2 with $L_2 < 1$ (such as a gradient method) which converges in the whole domain D.

Using a Fully Overlapped TA, it is possible to assure convergence using Corollary 9.6, by choosing the following weights:

$$w_1 = \epsilon; \qquad w_2 = 1 - \epsilon; \qquad \text{where} \quad 0 < \epsilon < \frac{1 - L_2}{L_1 - L_2}. \tag{9.22}$$

Such a combination allows advantage to be taken of the velocity of G_2 without the risk of using an algorithm that is not able to ensure convergence in the whole region D. In fact, several examples presented in [30] show that the Fully Overlapped TA combining Newton and Steepest Descent methods can solve, in a few iterations, problems that the Newton method alone (G_1) fails to solve (because $L_1 > 1$), while the Steepest Descent method (G_2 with $L_2 < 1$) needs hundreds of iterations to get close to the solution. The above analysis provides a consistent explanation of these experimental results.

To conclude the discussion on fully overlapped team algorithms, an *A-team* (asynchronous team), described in [26], may be considered as a special case of a fully overlapped team algorithm with variable weights (where $w_i(k) \in \{0, 1\}$). An A-team works in the following way: each processor tries to solve equation (9.1) using a different algorithm, occasionally writing its partial result

x in a global memory. A *shell* monitors the improvement of the estimation to the solution. If the algorithm does not improve beyond a given threshold, the shell reads a new value of the vector x from the global memory and continues from this new point (see [31, 32]).

2.3. Generalized Team Algorithm

The use of asynchronous block TAs opens up the possibility of solving large scale problems efficiently using distributed-memory parallel computers. In addition, the use of fully overlapped TAs makes possible the solution of problems that no one of the individual algorithms is able to solve independently. Usually, however, the fully overlapped TA does not lead to an appreciable speedup, leading to the idea of so-called partially overlapped TAs, to be described below.

Block, fully and partially overlapped TAs can be conveniently classified as special cases of a *generalized TA* that will be discussed in this section. The basic idea is as before: a generalized TA is a block TA that admits all possible combinations of overlapping, i.e. one or more processors may calculate the value of one or more variables using different algorithms, and in case there is overlapping in this sense, then an *administrator*, that calculates a weighted sum of the overlapped variables for use in subsequent calculations, must be specified.

In order to bookkeep information on which processors update which subvectors x_i (associated to subproblem $\Phi_i = 0$), an *assignment matrix*Assignment matrix $\Psi = (\psi_{ij})$ associated to the corresponding Generalized Team Algorithm is also introduced.

Definition 9.7 (Generalized team algorithm)   *A generalized team algorithm consists of four items:*

(i) *a specification of p different algorithms represented by the maps*
   $G_1, \ldots, G_p$;

(ii) *a partition of the problem $\Phi(x) = 0$ into m subproblems $\Phi_i(x) = 0$,*
   $i = 1, \ldots, m$;

(iii) *an $m \times p$ matrix $\Psi = (\psi_{ij})$, called an* assignment matrix *associated to the generalized TA, where $\psi_{ij} = 1$ (respectively 0) if the subvector x_i is (resp. is not) updated by the map G_j in processor j, and the rows of Ψ are ordered such that the first r are associated to subvectors that are updated by only one processor;*

(iv) *a specification of an* administrator, *i.e. indication of how and in which processor(s) the overlapped variables are combined for use in subsequent calculations.*

A simple example illustrates the above definition:

Example 9.8 A TA combines two different algorithms ($p = 2$) that are represented by their corresponding maps G_1 and G_2. This TA is used to solve a problem partitioned in three blocks ($m = 3$), in such a way that x_1 is updated only by G_1, x_2 is updated only by G_2, and x_3 is updated by both G_1 and G_2.

The following is the assignment matrix associated to this TA:

$$\Psi = \begin{bmatrix} 1 & 0 \\ 0 & 1 \\ 1 & 1 \end{bmatrix}. \tag{9.23}$$

Column 1 indicates that G_1 updates x_1 and x_3, while row 3 indicates that x_3 is updated by both G_1 and G_2. Note that $r = 2$ because only x_1 and x_2 are updated in separate processors (x_1 by G_1 and x_2 by G_2).

Remark 9.3 The fully overlapped Team Algorithm proposed by Talukdar et al. [30], and discussed in subsection 2.2, is a special case of a Generalized TA, where: $r = 0$, $m = 1$, $p = 2$, and $\Psi = \begin{bmatrix} 1 & 1 \end{bmatrix}$. The block TA discussed in subsection 2.1, is a special case, of a Generalized TA, where the assignment matrix is an identity matrix and there is no overlapping; i.e. $\Psi = I$, and $p = m = r$. Given the dimension n of the system (1), the *expanded dimension*Expanded dimension $\bar{n}$, (defined in [18] in order to account for the number of equations together with their multiplicity, in the case of overlapping), satisfies the relation: $\bar{n} = \sum_{i=1}^{m} \sum_{j=1}^{p} \psi_{ij} n_i \geq n = \sum_{i=1}^{m} n_i$.

Assuming overlapping in a partially asynchronous environment (Assumption 9.1), a subvector x_i, may be updated by more than one processor. Thus there may exist several different *versions* of x_i depending on the evaluating processor (or map). In this context, the following notation will be adopted:

χ_{ij} ... denotes the version of subvector x_i calculated using map G_j;

x_i ... denotes the subvector calculated by the administrator from the χ_{ij}s.

As an example, a generalized TA with associated assignment matrix (9.23) is illustrated in Figure 9.2. Note that the administrator calculates x_3 using a weighted sum of different versions of subvector 3, and may be represented as:

$$x_3(k+1) = c\,x_3(k) + w_1\,\chi_{31}(d_1^3(k)) + w_2\,\chi_{32}(d_2^3(k)), \tag{9.24}$$

where $c = 1 - w_1 - w_2$. It should be noted that $x_3, \chi_{31}, \chi_{32} \in \Re^{n_3}$, because they are different versions of the same subvector.

The Generalized TA illustrated in Figure 9.2, referred to henceforth as Version A, dedicates a processor to the exclusive task of administrator (evaluation of a weighted sum of vectors). In most situations, the other processors will

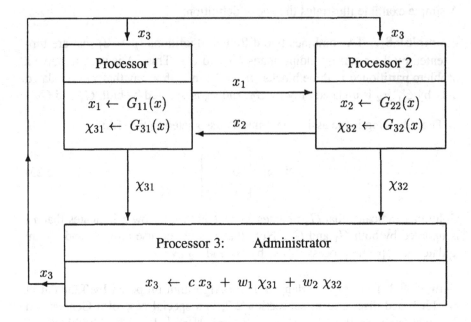

Figure 9.2. Version A of a Generalized Team Algorithm.

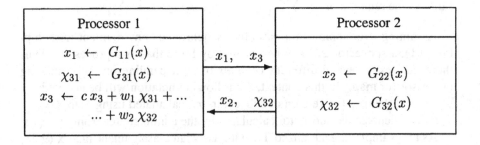

Figure 9.3. Version B of a Generalized Team Algorithm.

have the more complex tasks of solving subproblems, so that such an allocation will lead to unbalancing of processor loads. One possible remedy is to include the administrator's task in the task list of one or more of the processors. One implementation of this idea, denominated version B, is illustrated in Figure 9.3.

In versions A and B, the administrator's task is performed in only one processor; however its work may be done by several processors (as in an A-team). For

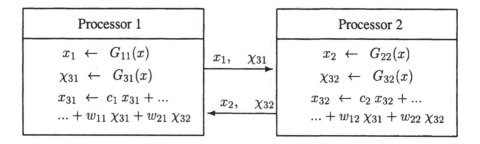

Figure 9.4. Version C of a Generalized Team Algorithm.

example, Figure 9.4 illustrates version C of a Generalized TA, with a distributed (and replicated) Administrator.

Remark 9.4 The overlapping decomposition technique of [18]Overlapping decomposition technique can be interpreted as a special case of a TA (version C) with the following weights defining the administrator: $w_{11} = w_{22} = 1$, and $w_{12} = w_{21} = 0$. This technique is well suited to solve large-scale systems with some tightly coupled blocks as Example 2, in the next section, will illustrate.

More details on these and other versions of Generalized TAs can be found in [2]. Of special interest are the versions with the so-called *implicit* Administrators, in which the weighted sum of the different versions of a subvector is carried out as part of the algorithm to be applied and not as a special process (or another equation) as illustrated in Figure 9.5. When this is done, the dimension of the comparison matrix H is reduced, eliminating a line whose elements sum to unity or more (this is propitious for the satisfaction of the convergence condition $\rho(H) < 1$).

In this paper, Version A is chosen to illustrate the procedure used to derive a sufficient condition for the convergence of a generalized TA using Theorem 9.5. Similar detailed convergence analyses of the other versions can be found in [2].

In the context of block TAs analyzed above, G_j represented the algorithm executed in processor j, updating x_j. However, when a Generalized TA is considered, a given processor j may update any subvector x_i of x; and this operation is represented by a map G_{ij}. Thus processor j executes an algorithm represented by a map G_j given by:

$$\forall i \in \mathbf{m}, \quad \forall j \in \mathbf{p}, \quad G_j(x) = \left[G_{1j}(x)^T, \dots, G_{mj}(x)^T \right]^T, \quad G_{ij}(x) : D \to D_i \quad (9.25)$$

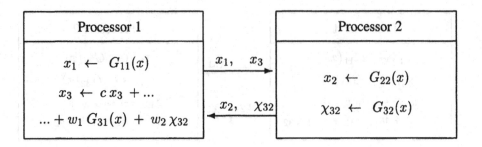

Figure 9.5.   Version E of a Generalized Team Algorithm.

With the above notation, the asynchronous iterations of the Generalized TA illustrated in Figure 9.2 are represented by the following equations:

$$\begin{bmatrix} x_1(k+1) \\ \chi_{31}(k+1) \end{bmatrix} = \begin{bmatrix} G_{11}(x^1(k)) \\ G_{31}(x^1(k)) \end{bmatrix}$$

$$\begin{bmatrix} x_2(k+1) \\ \chi_{32}(k+1) \end{bmatrix} = \begin{bmatrix} G_{22}(x^2(k)) \\ G_{32}(x^2(k)) \end{bmatrix} \tag{9.26}$$

$$[x_3(k+1)] = [c\,x_3(k) + w_1\,\chi_{31}(d_1^3(k)) + w_2\,\chi_{32}(d_2^3(k))],$$

where $c = 1 - w_1 - w_2$. To analyze the convergence of this TA, it suffices to define an *expanded state vector* $W(k)$ as:

$$W(k) = \begin{bmatrix} W_1(k) \\ W_2(k) \\ W_3(k) \\ W_4(k) \\ W_5(k) \end{bmatrix} := \begin{bmatrix} x_1(k) \\ x_2(k) \\ x_3(k) \\ \chi_{31}(k) \\ \chi_{32}(k) \end{bmatrix}, \text{where} \begin{cases} W_1 = x_1 \in \Re^{n_1}, \\ W_2 = x_2 \in \Re^{n_2}, \\ W_3 = x_3 \in \Re^{n_3}, \\ W_4 = \chi_{31} \in \Re^{n_3}, \\ W_5 = \chi_{32} \in \Re^{n_3}. \end{cases} \tag{9.27}$$

Recalling that $x \in D = D_1 \times D_2 \times D_3$; it follows that the expanded state vector W belongs to a *expanded domain* D_w, where $D_w = D_1 \times D_2 \times D_3 \times D_3 \times D_3$; since χ_{31}, χ_{32} and x_3 are different versions of the same subvector. Equation (9.26) may thus be considered to describe a block-iterative algorithm in the variable W. Thus Theorem 9.5 gives a sufficient condition for its convergence, provided the assumptions listed below hold.

(i) **Uniform bound on delays (Assumption 1)**

(ii) **Uniqueness of fixed point in D (Assumption 2).** *Given the system of equations (9.1) defined in a closed set $D \subset \Re^n$, there exist p maps G_i $(i \in \mathbf{p})$*

satisfying equation (9.25) such that

$$\forall i \in \mathbf{m}, \qquad \forall j \in \mathbf{p}, \qquad G_{ij}(D) \subset D_i,$$

that is, for all $j \in \mathbf{p}$, $G_j(D) \subset D$, and such that D contains exactly one fixed point

$$x^* = \left[x_1^{*T}, \ldots, x_m^{*T}\right]^T, \qquad x_i^* \in D_i, \qquad \forall i \in \mathbf{m},$$

for every map G_j, which is a solution of (9.1).

(iii) *Each map $G_{ij}(x)$ is* block-Lipschitz continuous:

$$\forall i \in \mathbf{m}, \quad \forall j \in \mathbf{p}, \quad \forall x, y \in D, \quad \|G_{ij}(x) - G_{ij}(y)\| \leq \sum_{k=1}^{m} l_{ijk} \|x_k - y_k\|.$$

$$(9.28)$$

(iv) *The domain D considered in the assumptions above is convex and the weights chosen by the administrator are such that every subvector calculated by the Administrator remains in D.*

Now consider version A (with centralized administrator in a dedicated processor, as illustrated in Figure 9.2) of a generalized TA that has an expanded state vector W_a given by:

$$W_a = \left[W_1^T, W_2^T, \ldots, W_m^T\right]^T, \qquad (9.29)$$

$$\text{with} \qquad W_i = \begin{cases} [x_i] & \in \Re^{n_i} & \text{for} \quad 1 \leq i \leq r, \\ \begin{bmatrix} x_i \\ \psi_{i1}\chi_{i1} \\ \vdots \\ \psi_{ip}\chi_{ip} \end{bmatrix} & \in \Re^{(p+1)n_i} & \text{for} \quad r < i \leq m, \end{cases}$$

where χ_{ij} is multiplied explicitly by $\psi_{ij} \in \{0, 1\}$ to emphasize that some components of W_a are zeros (indicating thereby that some variables are deleted when analyzing a specific problem).

The corresponding asynchronous TA (that updates the expanded state vector W_a) may be represented by:

$$\begin{array}{rcl} [\, x_i(k+1)\,] & = & [\, G_{ii}(x^i(k))\,] \end{array}$$
for $1 \le i \le r$,

$$\begin{bmatrix} x_i(k+1) \\ \psi_{i1}\chi_{i1}(k+1) \\ \vdots \\ \psi_{ip}\chi_{ip}(k+1) \end{bmatrix} = \begin{bmatrix} cx_i(k) + \Sigma_{j=1}^p w_j \psi_{ij}\chi_{ij}(d_j^i(k)) \\ G_{i1}(\, x^1(k)\,) \\ \vdots \\ G_{ip}(\, x^p(k)\,) \end{bmatrix}$$
for $r < i \le m$,

$$(9.30)$$

where $c + \Sigma_{j=1}^p w_j = 1$.

Given the representation of the TA, Version A, it is possible to derive the corresponding nonnegative comparison matrix

$$H_a = \left(H_{ij} \right), \qquad i,j = 1,\ldots,m \qquad (9.31)$$

where the submatrices H_{ij} that represent the dependency between W_i and W_j are given by:

$$H_{ij} = \begin{cases} l_{iij} & \in \Re & \text{for } 1 \le i \le r \\ & & \text{and } 1 \le j \le r \\[2mm] \begin{bmatrix} l_{iij} \; 0 \cdots 0 \end{bmatrix} & \in \Re^{1\times(p+1)} & \text{for } 1 \le i \le r \\ & & \text{and } r < j \le m \\[2mm] \begin{bmatrix} 0 \\ l_{i1j} \\ \vdots \\ l_{ipj} \end{bmatrix} & \in \Re^{p+1} & \text{for } r < i \le m \\ & & \text{and } 1 \le j \le r \\[2mm] \begin{bmatrix} |c| & \psi_{i1}w_1 & \cdots & \psi_{ip}w_p \\ l_{i1i} & 0 & \cdots & 0 \\ \vdots & \vdots & \ddots & \vdots \\ l_{ipi} & 0 & \cdots & 0 \end{bmatrix} & \in \Re^{(p+1)\times(p+1)} & \text{for } r < i \le m \\ & & \text{and } j = i \\[2mm] \begin{bmatrix} 0 & 0 & \cdots & 0 \\ l_{i1j} & 0 & \cdots & 0 \\ \vdots & \vdots & \ddots & \vdots \\ l_{ipj} & 0 & \cdots & 0 \end{bmatrix} & \in \Re^{(p+1)\times(p+1)} & \text{for } r < i,j \le m \\ & & \text{and } j \neq i. \end{cases}$$

For example, the comparison matrix for the TA of Figure 9.2 is given by:

$$H_a = \begin{bmatrix} l_{111} & l_{112} & l_{113} & 0 & 0 \\ l_{221} & l_{222} & l_{223} & 0 & 0 \\ 0 & 0 & |c| & w_1 & w_2 \\ l_{311} & l_{312} & l_{313} & 0 & 0 \\ l_{321} & l_{322} & l_{323} & 0 & 0 \end{bmatrix}. \qquad (9.32)$$

Considering a Generalized TA as a block TA, in the expanded state vector W, the following corollary can be derived from Theorem 9.5:

Corollary 9.9 *Version A of a Generalized Team Algorithm (with an administrator in a dedicated processor) given by (9.30), converges to the unique solution x^* in D, under the given assumptions, if $\rho(H_a) < 1$, where the comparison matrix H_a is given by (9.31).* □

To conclude this discussion it is emphasized that the technique illustrated above may also be applied to derive sufficient convergence conditions in an asynchronous environment for any version of a Generalized Team Algorithm.

3. The load flow problem

The Load Flow problem of electrical power networks can be formulated as a so-called almost linear equation [10]:

$$Yx = I(x), \tag{9.33}$$

where matrix Y is the *admittance matrix*Admittance matrix of the network given by:

$$Y = (y_{ki}) \in \mathbf{C}^{n \times n}, \qquad \text{with} \quad y_{ki} = G_{ki} + jB_{ki} \in \mathbf{C}; \tag{9.34}$$

$x \in \mathbf{C}^n$ is the voltage vector and $I(x)$ is the current vector in $\mathbf{C}^n$. Note also the following terminology: y_{ki} for $k \neq i$ is the *admittance* of the *link* between *node k* and *node i* of the network; its reciprocal, the impedance, is denoted $z_{ki} = R_{ki} + jX_{ki}$ and the ratio of the *resistance R_{ki}* to the *reactance X_{ki}* is referred to as the *R/X ratio* of the link. A complete survey on the problem formulation and classical sequential methods of solution are given in [28].

Remark 9.5 *The voltage x_k of a node k may be given in polar coordinates by its magnitude V_k and its phase θ_k, i.e., $\forall k \in \mathbf{n}$, $x_k = V_k e^{j\theta_k} \in \mathbf{C}$.*

The static operating state of the system is specified by the constraints on the power and/or voltage at the network buses (or nodes). A bus k may be of one of the three types, namely:

PQ bus: when the injected power $S_k = P_k + jQ_k$ is specified.
PV bus: when the active power P_k and the voltage magnitude V_k are specified.
slack bus: when the complex voltage x_k itself is specified.

The slack bus is a fictitious concept created to compensate power losses and used as a phase reference. The number n_{PV} of PV buses is in general very small, compared to the total number of buses n.

Since the current I_k of a node k is not given, it may be calculated as:

$$\forall k \in \mathbf{n}, \quad I_k(x_k) = \left(\frac{S_k}{x_k}\right)^* = \frac{P_k - jQ_k}{V_k\, e^{-j\theta_k}}, \tag{9.35}$$

where $*$ denotes the complex conjugate.

A quasi-Newton method called the *Fast Decoupled method*Fast Decoupled method (or FD) [29] is by far the most used in practical applications. Under some standard hypotheses, the method assumes that the magnitude and the phase of the voltage vector are decoupled, so that the following system of equations is solved at each iteration:

$$\Delta P/V = B' \Delta\theta \tag{9.36}$$

$$\Delta Q/V = B'' \Delta V, \tag{9.37}$$

where B' and B'' are matrices derived from the admittance matrix Y.

Although the FD method works well in a sequential environment on most problems, it is not easy to parallelize and has convergence problems in special cases where the system has transmission lines (links) with a high resistance/reactance (R/X) ratio or when neighboring nodes have highly different phases (see discussion in [36]).

A block-Jacobi version of the *Y-Matrix method*Y-Matrix method (or BJ) described in [28] can, on the other hand, be easily parallelized, with reasonable performance when no PV bus is considered [10]. However, when PV buses are present, the BJ method may not converge. In fact, two examples are presented in this section where the method does not converge. A discussion on several variations of the method that try to ensure convergence even when PV buses are present may be found in [28].

As examples of the potential of TAs, this section presents two example problems that cannot be conveniently solved using either the fast decoupled, or the block-Jacobi methods working alone. However, it is shown that these problems can be easily solved using a *Team Algorithm* that combines the advantages of the BJ and FD methods. The idea, based on [14], is to use the fast decoupled method in blocks which contain PV buses (which, in good partitions, are assigned to as few blocks as possible), while the block-Jacobi method is used in the other blocks, where no PV buses are present, where high R/X ratios or highly different neighboring node phases are encountered. Thus, the hope is to exploit all the good features of the fast decoupled method, while avoiding its known convergence problems in certain specific situations.

Example Problem 1: An electrical system of 616 buses based on real data and consisting of 8 different interconnected subsystems with the following characteristics:

(i) all PV buses belong to subsystem 1;

(ii) subsystems 2 to 8 have several links with a high R/X ratio and neighboring buses with highly different phases.

This problem was solved using the following TA:

SUBSYSTEM 1: Fast Decoupled method ($G_1 = G_{FD1}$);
SUBSYSTEMS 2 to 8: Block-Jacobi method ($G_i = G_{BJi}$, $i = 2, \ldots, 8$).

In case the 8 subsystems are not strongly interconnected, the following asynchronous block TA can be implemented, choosing, for each subsystem, the algorithm that works best for it:
PROCESSOR 1:

$$x_1(k+1) = G_{FD1}\Big(x^1(k) \Big),$$

PROCESSOR i, $\quad i = 2, \ldots, 8$:

$$x_i(k+1) = G_{BJi}\Big(x^i(k) \Big).$$

The experimental results presented in the next section, show that the Team Algorithm proposed solves the given problem while the FD and BJ methods do not.

However, harder problems that cannot be easily solved by a block TA sometimes arise. In such cases the option of using a Generalized Team Algorithm with properly chosen partial overlapping may be exercised, as illustrated in following example.

Example Problem 2: The electrical system of 616 buses, also based on real data, is sketched in Figure 9.6. It consists of 5 different interconnected subsystems with the following characteristics:

(i) all PV buses belong to subsystem 1;

(ii) subsystems 2 to 4 have several links with a high R/X ratio and neighboring buses have highly different phases;

(iii) subsystem 5 is the smallest in size and is strongly connected with subsystems 1 and 4.

The option of using one processor for each block is not very attractive here, because it implies cutting several *strong* links (generating large Lipschitz constants), which, in general, does not facilitate convergence. One solution, illustrated in Figure 9.6, is to use partial overlapping to avoid cutting strong links;

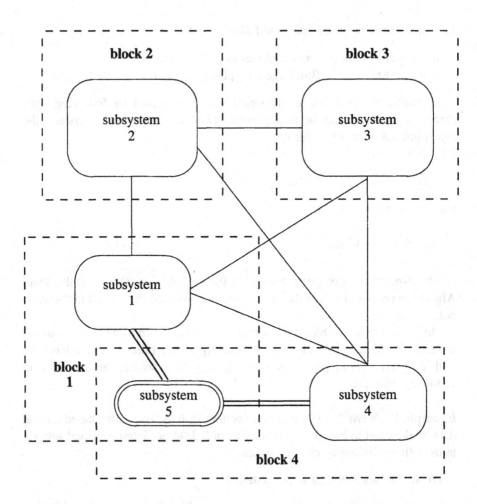

Figure 9.6. Electrical system of Example Problem 2 consisting of 5 subsystems (double lines indicate a strong interconnection). Note the use of partial overlapping in the TA, since subsystem 5 is solved by both processors 1 and 4 working on blocks **1** and **4**, respectively.

in other words, a Generalized TA is used with the following assignment matrix:

$$\Psi = \begin{bmatrix} 1 & 0 & 0 & 0 \\ 0 & 1 & 0 & 0 \\ 0 & 0 & 1 & 0 \\ 0 & 0 & 0 & 1 \\ 1 & 0 & 0 & 1 \end{bmatrix}. \tag{9.38}$$

Note that $p = 4$ processors were used to solve the $m = 5$ subsystems, because subvector x_5 is updated by processors 1 and 4 which also update their corresponding subvectors. To choose the algorithm for each processor the reasoning of Example Problem 1 was repeated. The resulting asynchronous TA was implemented using several different versions.

The most efficient version is the one in which the administrator is implicit (see version E, Figure 9.5), i.e., its calculations are performed in a processor that also calculates the updates of a given partition of the vector x:

PROCESSOR 1:

$$\begin{bmatrix} x_1 \\ x_5 \end{bmatrix} \leftarrow G_{FD1}(\hat{x}^1) = \begin{bmatrix} G_{FD11}(\hat{x}^1) \\ G_{FD51}(\hat{x}^1) \end{bmatrix}, \text{ where } \hat{x}^1 = [x_1^T, \ldots, x_4^T, (w_1 x_5 + w_2 \chi_{54})^T]^T.$$

PROCESSOR i : $\quad i = 2, 3$

$$x_i \leftarrow G_{BJi}(x^i), \quad \text{where} \quad x^i = [x_1^T, \ldots, x_5^T]^T.$$

PROCESSOR 4 :

$$\begin{bmatrix} x_4 \\ \chi_{54} \end{bmatrix} \leftarrow G_{BJ4}(x^4) = \begin{bmatrix} G_{BJ44}(x^4) \\ G_{BJ54}(x^4) \end{bmatrix}, \quad \text{where} \quad x^4 = [x_1^T, \ldots, x_5^T]^T.$$

Experimental results presented in the following section show that, for Example Problem 2, the FD and BJ algorithms as well as several block TAs fail, while the Generalized TA above not only solves the problem correctly, but does so with a considerable speedup.

4.   Computational results

This section presents the experimental results obtained for implementations of the algorithms discussed above on an eight node Intel iPSC/860 (hyper)cube. In order to organize the presentation of these results, a few *Figures of Merit* are defined below. The standard definition of *speedup of an algorithm* [27] is:

Definition 9.10 Speedup, *Speedup* S_p := $\dfrac{\textit{best sequential time of any algorithm}}{\textit{parallel time of the algorithm considered}}$

In the context of this paper, the best sequential time to solve the Load Flow problem considered was found to be that of a sequential TA.

For many applications, it is not only the speedup that is of interest, but also the efficiency, η, defined in [27] as:

Definition 9.11 Efficiency,*Efficiency* $\eta \quad := \quad 100 \; S_p/p \quad$ *(in percent)*, *where* p *represents the number of processors utilized.*

To measure the advantage of using asynchronous implementations instead of classical synchronous ones, the concept of *Speedup due to Asynchronism* is introduced:

Definition 9.12 Speedup due to asynchronism, $S_a := \dfrac{parallel \; synchronous \; time}{parallel \; asynchronous \; time}$

As a figure of merit convenient for measuring the advantage of using a Team Algorithm, the *Relative Speedup of a TA* is defined.

Definition 9.13 Relative speedup of a TA,

$$S_r := \frac{best \; sequential \; time \; in \; the \; set \; of \; TA \; members \; working \; alone}{parallel \; time \; of \; the \; TA \; considered}$$

The corresponding efficiency measure η_r is defined as:

Definition 9.14 Relative efficiency of a TA, $\eta_r \quad := \quad 100 \; S_r/p \; (in \; percent)$, *where* p *represents the number of processors utilized.*

Finally, a figure of merit that measures the gain in using a TA with respect of any of the member algorithms, in a sequential environment, is defined.

Definition 9.15 Sequential speedup of a TA,

$$S_s := \frac{best \; sequential \; time \; in \; the \; set \; of \; TA \; members \; working \; alone}{time \; using \; a \; sequential \; TA}$$

Experimental results of several implementations running on an Intel iPSC/860 eight node (Hyper) cube are presented in the following subsections. The programs were coded in Fortran iPSC/860, an extension of Fortran 77 for the parallel computer used. This allowed the re-use, with small modifications, of standard routines already developed for solving (sequentially) the Load Flow Problem.

Table 9.1. Best experimental times.

METHOD	p	iter	time [s.]	S_r	η_r (%)	S_p	η (%)	S_a
Example Problem 1 $\varepsilon = 0.001$ and $x(0) = $ 'flat start'								
Matrix Y (Jacobi)	1	∞	∞	-	-	-	-	-
Fast Decoupled	1	131	9.918	**1**	**100**	0.12	12	-
sequential TA	1	33	1.165	8.52	852	**1**	**100**	-
synchronous TA	2	30	1.017	9.8	488	1.15	57	1
	4	30	0.808	12.3	307	1.4	36	1
	8	31	0.498	19.9	249	2.34	29	1
asynchronous TA	2	65	0.648	15.3	766	1.8	90	1.56
	4	43	0.484	20.5	512	2.4	60	1.67
	8	32	0.382	25.9	324	3.1	38	1.30

Table 9.2. Partitions for which the best times were obtained.

METHOD	p	PARTITION
Example Problem 1 $\varepsilon = 0.001$ and $x(0) = $ 'flat start'		
Matrix Y (block-Jacobi)	1	616
Fast Decoupled	1	616
sequential TA	1	77; 539
synchronous TA	2	77; 539
	4	154; $\cdots$; 154
	8	77; $\cdots$; 77
asynchronous TA	2	77; 539
	4	77; 154; 154; 231
	8	77; $\cdots$; 77

4.1. Numerical results using a Block TA

This subsection presents experimental results obtained in the solution of Example Problem 1, using BJ and FD methods, as well as several different implementations of block team algorithms (without overlapping) using 1, 2, 4 and 8 processors.

A typical example using the well known 'flat start' condition ($x_k = 1$ for all k) as a starting point with a tolerance $\varepsilon = 0.001$, is shown in Table 9.1. Similar results were obtained with other tolerances and initial values.

The main trends observed are discussed below:

- the best execution time is not always obtained for a well-balanced partition; a small block-Lipschitz constant is usually more important than good load balancing;

- a block TA is clearly better than any of the member algorithms working in isolation. Moreover, under some conditions, only the Team Algorithm was able to solve the Example Problem 1;

- an asynchronous TA is significantly better than its synchronous counterpart, as illustrated in Table 9.1.;

- the *best sequential time*Best sequential time was obtained using a sequential TA, while the *best parallel time*Best parallel time was obtained by an asynchronous implementation of this TA. The following relation holds:

$$S_r = S_s \times S_p \times S_a.$$

Thus the large values of S_r (relative speedup of a TA) obtained in the asynchronous TA implementations (see Table 9.1), sometimes referred to as a synergetic effect of the team, are a consequence of the combined effect of three factors: the use of a Team Algorithm (S_s), the use of parallelism (S_p), and the use of asynchronism (S_a);

- S_p increases with the number p of processors $(p \leq 8)$, while η decreases;

- the use of an asynchronous TA results in an increase in S_p when a smaller tolerance ε is imposed or an initial point $x(0)$ closer to the solution is used. In fact, for a *bad* initial condition, the asynchronous version may not converge even when the sequential TA converges.

4.2.   Numerical results using a TA with Partial Overlapping

In this subsection, the solution of Example Problem 2 is considered, using the algorithms utilized for example problem 1, as well as several versions of generalized TAs. Typical results are shown in Table 4.

The main conclusions concerning Example Problem 1 are also valid for Example Problem 2; in addition, it was observed that:

- asynchronous implementations have the highest variance: table 4 shows the range (minimum and maximum) of time (and number of iterations) for successive runs of the same problem, with the same data;

- in general, methods without adequately chosen overlapping do not converge (even the block TA does not converge). An exception to this is the asynchronous block TA that occasionally converges depending on the

Table 9.3. Best sequential time.

METHOD	'overlap'	p	iter	time [sec]	S_p	η (%)
Example Problem 2						
$\varepsilon = 0.001$ and $x(0)$ = 'flat start'						
Matrix Y (BJ)			-		-	-
Fast Decoupled	no	1	-	does not converge	-	-
sequential TA			-		-	-
sequential TA	yes	1	53	2.33379 to 2.33396	**1**	**100**
synchronous	no	2	-	does not	-	-
TA		4	-	converge	-	-
asynchronous	no	2	-	occasional	-	-
TA		4	-	convergence	-	-
synchronous	yes	2	64	1.82564 to 1.82789	1.3	64
TA		4	84	1.32385 to 1.32602	1.8	44
	version A		138 to 184	1.73598 to 2.30847	1.3	45
asynchronous	version D	2	146 to 189	1.57115 to 2.03278	1.5	74
TA	version E		172 to 194	1.76724 to 2.10294	1.3	66
with	version A		62 to 93	1.11098 to 1.64973	2.1	42
overlapping	version D	4	83 to 410	0.97077 to 1.26550	2.4	60
	version E		68 to 79	0.79059 to 0.89672	3	74

asynchronism permitted for each specific execution. Therefore, the use of overlapping seems to be the only way to ensure convergence to the solution of Example Problem 2;

- considering the different versions of the asynchronous TA implemented, the best performances were obtained using version E, followed by version D (duplicated implicit administrator);

As a final observation, Table 5 illustrates typical times used in communication and synchronization by a synchronous implementation, displaying a clear tendency towards increasing penalty with the number of processors. Thus, it may be concluded that asynchronous implementations become more important for parallel computers with large numbers of processors.

5.   A-Teams with Genetic Algorithms

A-Teams (Asynchronous Team), have been successfully used to solve complex engineering problems [1, 5, 6, 7, 8]]. This section briefly presents two cases where the combination of Genetic Algorithms (GA) with numerical methods have achieved excellent experimental results: hydroelectric generation optimization [7] and design of reliable computer networks [8]. These real world

Table 9.4.   Typical synchronization plus communication times.

Example Problem 1			
p	total time [sec]	synchronization + communication time	Percentage
2	1.92	0.07	4 %
	2.01	0.10	5 %
	2.10	0.13	6 %
4	0.81	0.07	9 %
	0.90	0.11	12 %
	0.99	0.17	17 %
8	2.56	2.13	83 %
	2.57	2.17	84 %
	2.60	2.20	85 %

problems have been solved using a standard network of personal computers, available in most organizations.

5.1.   Hydroelectric generation optimization

Paraguay exports electricity thanks to the generation of several large scale hydroelectric plants as Itaipu, Yacyreta and others. The optimization of the generation process is very complex, due to the large number of generation units with different physical characteristics, operational and maintenance restrictions and so forth. The above optimization problem may be stated as minimizing the turbine flow (Q) in a hydroelectric power station, given the demanded power (P_D) and the number of generators (N_T). The solution is not trivial because each turbine has its own operative restrictions characterized by the way in which the turbine flow (Q) turns into useful electrical power (P_i) and the height (h) of the water in the reservoir. The power generated by each one of the turbines (P_i) and the flows (Q_i) required in each one of them are related by known functions (f_i), which are experimentally obtained by "index-tests".

$$P_i = f_i(Q_i, h) \tag{9.39}$$

There are several optimization methods to solve the problem [15, 35], each one presents advantages and disadvantages depending on the particular character-istics of each problem. However, in some cases none of the available methods is able to solve the problem. A strategy to overcome this inconvenience is to use A-Teams which propose a synergetic combination of traditional numerical methods (NM) with Parallel Genetic Algorithms (PGA).

In this optimization example the A-Team applied is a combination of a well-known sequential numeric method, the Gradient Method [23], and a distributed version of a Genetic Algorithm [17], similar to the one proposed in [24]. More specifically, the proposed A-Team works as follows [7]:

- The GA will calculate the Pi power to be generated by the first k generating units, where $\forall i \in \{1, ..., k\} : k < N_T$;

- The NM will receive as parameter the remaining required power: $P_D - \sum P_i$, to determine the P_i powers, where $\forall i \in \{k + 1, ..., N_T\}$;

- Known all the P_i Powers, the efficiency η is calculated.

This solution enabled, on the one hand, that all analyzed points satisfy the problem's restriction, and, on the other hand, that NM is used in a reduced search space with a smaller computational complexity, when an adequate k value is used.

It should be noticed that the performance of the selection operator is limited due to the fact that, according to the definition in [4], the extreme values of η are too close to each other. For this reason the objective function is scaled using a linear scaling function. Finally, the parallel asynchronous implementation uses a $Master - Slave$ model [21].

Experimental results

Experimental results were obtained with a platform of 5 personal computers (Pentium 75 MHz, 8 MB RAM) on an Ethernet network (10 Mbps). The global population, with 200 individuals, was split into 4 sub populations of 50 individuals. Each sub-population was assigned to each one of 4 PC's, remaining the fifth computer as the manager (Master). 20 runs were conducted for both, the combination of the sequential GA with the numerical method (here called combined GA) and the A-Team. Results were averaged.

For example, Table 9.5 presents experimental results of a hydroelectric dam with 9 turbines (each one of them generating from 300 MW up to 710 MW of electrical power), $P_D = 1500MW$ and $h = 116.5$ m. Regarding the quality of the solution, (i.e. efficiency) Table 9.5 indicates that the average results of both the $GA + NM$ and the A-Team are better than those obtained by the NM. In addition, the speedup for the A-Team is considerably greater than the sequential implementations. This situation may be also appreciated in Figure 9.7, which shows the advantages of using a computer network to solve large complex problems. In fact, when the dimension N_T of the search space increases, the problem is more complex and the A-Team widely outperforms the other methods.

Table 9.5.   Time vs. Number of turbines for $PD = 1500MW$.

Algorithm	Variable	$N_T=3$	$N_T=4$	$N_T=5$	$N_T=6$	$N_T=7$	$N_T=8$	$N_T=9$	AVERAGE
NM	η_{opt}	85.964	85.964	85.964	85.964	85.964	85.964	-	85.964
	t_{total} (sec.)	0.14	2.04	15.05	136.59	3688.51	136576.3	Too Big	23.403.097
GA+NM	η_{max}	85.818	85.991	85.991	85.991	85.991	85.991	85.991	85.962
	η_{prom}	85.718	85.991	85.991	85.991	85.908	85.987	85.983	85.931
	t_{prom} (sec.)	13.61	38.256	128.256	566.71	2984.1	9018.02	9058.38	2.124.825
A-Team	η_{max}	85.991	85.991	85.991	85.991	85.991	85.991	85.991	85.991
	η_{prom}	85.989	85.989	85.988	85.989	85.988	85.990	85.984	85.989
	t_{prom} (sec.)	8.663	16.502	75.982	171.46	223.488	1313.89	3500.18	301.664

NM: numerical method (without GA) t_{prom}: time average

η_{prom}: efficiency average η_{max}: maximum efficiency

GA+NM: sequential combination of GA and NM t_{max}: total time to obtain η_{max}

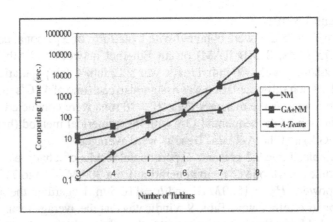

Figure 9.7.   Time vs. Number of Turbines for $P_D = 1500MW$.

Upon considering the computation time to obtain similar solutions, the NM presents better performance when the dimension of the problem (number of turbines in service) is small. However, when the dimension NT of the search space increases, the problem is more complex and the A-Team outperforms widely the other considered methods (Table 9.5). Indeed, Figure 9.7 shows that for $N_T \geq 6$ the A-Team is faster than the NM for equivalent solutions and this advantage increases quickly with the size of the problem.

5.2. Topological optimization of reliable networks

Network planning is concerned with the design of sufficiently reliable networks capable of delivering high capacity and speed, at a reasonable cost [11]. Due to the lack of good network designing tools, engineers have applied experience and intuition to design networks in the fields of telecommunications, electricity distribution, gas pipeline and computer networks. However, for very large-scale networks being designed nowadays, the traditional trial and error approach is not applicable and computational automatic tools are required.

In this section, a reliable network design problem is defined as an all-terminal network reliability (also known as uniform or overall reliability). In this approach, every pair of nodes needs a communication path to each other [12, 19]; that is, the network forms at least a spanning tree.

Thus, the primary design problem is to choose enough links to interconnect a given set of nodes with a minimal cost, given a desired minimum network reliability. This minimization design problem is NP-hard [16] with as a further complication because the calculation of all-terminal reliability is also NP-hard. Although several papers have been published [8] on this problem or similar ones, no known method (e.g.: branch and bound, simulated annealing and Tabu search) is efficient enough to deal with real large networks. Due to the above mentioned difficulties, several Genetic Algorithms (GA) approaches have been proposed for different network design problems [8].

Considering the complexity of the problem and the existence of different known methods, an A-Team seems to be a good candidate for designing reliable networks. The proposed A-Team combines parallel genetics algorithms (GA), with different reliability calculation approaches using a network of personal computers. The network to be optimized is modeled by a probabilistic undirected graph $G = (N, L, p)$, in which N represents the set of nodes, L a given set of possible links, and p the reliability of each link. It is assumed that there is only one bi-directional link between each pair of nodes; that is, there is no redundancy between nodes. The optimization problem may be stated as follows:

$$Z = \sum_{i-1}^{N-1} \sum_{j=i+1}^{N} c_{ij} x_{ij} \qquad \text{subject to:} \quad R_x \geq R_0 \qquad (9.40)$$

where x_{ij} is a decision variable $\{0, 1,$ ci$\}$j is the cost of a link (i, j), R_x is the network reliability, and R_0 is the minimum reliability requirement.

Proposed A-Team

The algorithm consists of two kinds of processes, a Coordinator and the PGAs (Parallel Genetic Algorithms). There is only one coordinator, which is responsible of creating the PGA processes, collect their results and take note of the global statistics. The PGAs do the real work. Once the coordinator initializes all the processes, each PGA computes its solutions, broadcasts its partial results to the others, and receives what its peers have sent to it.

Each PGA has its own population, randomly generated by each independent process. The population size is maintained constant in each PGA process, even though it may receive any number of candidates from the peer processes. For that purpose, a genetic selection operator is used after asynchronously receiving any number of candidates, to choose between the own population and the incoming candidates. As a result, the interaction between processes is performed, interchanging good candidates. Every g generations, or when the PGA find a new best solution, a PGA broadcast his best solution to its peers. If g is too large, there will be little interaction, but if it is too small, there may be large overhead with premature convergence.

A PGA process has five main tasks: (1) to generate the initial population (a specialized method is used to ensure random and feasible network candidates); (2) to receive candidates from the peers, and to apply selection to maintain a constant population size; (3) to perform the regular GA operations (crossover and mutation); (4) to broadcast its best solution (in that way, a good candidate may be exported to other processes several times. If the network is the global best, all PGAs will adopt it until a new best network appears). Each PGA transmits its best solution every g generations or when a new best appears; and (5) to inform the coordinator when a given stopping criteria is satisfied.

Experimental results

Experimental results were obtained using a 10 Mbps Ethernet network, with three personal computers with different configurations. Programs were written in C and parallel implementations were written using PVM (Parallel Virtual Machine). The work load (population) was distributed among processors according to their relative performance to obtain a reasonable balance.

Figure 9.8 shows a typical running of a GA and of the implemented A-Team, plotting the temporary best solution as a function of time. It can be seen that the A-Team converges much faster than the GA. In fact, the A-Team satisfies a stopping criterion (by population homogenization) more than 400 s before the GA finds a similar quality solution. It can be noted that the A-Team not only gets good solutions, but also a more predictable running time when compared to sequential GA.

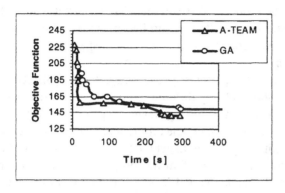

Figure 9.8. Typical running showing a fast convergence of an A-Team.

6. Conclusions

An increasing interest in methods that combine different algorithms to solve several real world problems has been observed in the last few years, specially since distributed systems have become widely available.

The solution of algebraic systems of equations combining different methods has been considered at an experimental level in several earlier papers such as [14, 30, 26]. The main idea of a team algorithm is to solve a large problem efficiently by choosing the best algorithm for each subproblem, specifically allowing the choice of different methods for each subproblem. Each algorithm is thus used where it best serves the purpose, without the difficulties of using it where it may have convergence problems. Examples have been presented to show that an appropriate combination of algorithms can efficiently solve problems that no one algorithm alone solves satisfactorily.

For some problems it is convenient to use more than one algorithm to solve the same subproblem, making it possible to take advantage of a fast algorithm that does not possess guaranteed convergence properties if it were to be used as an isolated member of a team. Moreover, the artifice of duplicating (or overlapping) some equations may help to get better convergence; for example, when there exist blocks strongly coupled to their neighbors.

The ease of implementing asynchronous team algorithms and their effectiveness encourage their use in distributed memory systems with several processors. The solution of large problems in such distributed systems becomes attractive, since the high penalty of synchronizing processors is not incurred.

Asynchronous Team Algorithms (A-Teams) represent a technique to exploit available computer power of completely asynchronous distributed systems, as computer networks. For example, the A-Team proposed in section 5.1 presents average results that are clearly better than the ones obtained by a sequential implementations. The proposed A-Team reduces considerably the calculation time required to optimize the efficiency in hydroelectrical energy generation. Moreover, this improvement grows up as the search space (number of turbines in service) increases.

Furthermore, section 5.2 presents a case where the complexity of the design and the variety of completely different algorithms, each one with its own strength, make it an ideal field for testing Asynchronous Team Algorithms. The idea behind this problem was to combine different algorithms, as the two implemented in the reliability calculation, to get the best of each one. In this context, a parallel specialized Genetic Algorithm was proposed for the design problem, while an upper bound calculation and Monte Carlo simulation were used for network reliability estimation. By combining all those algorithms in a network of available personal computers, good experimental results were reported [8], with considerable speedup that scales very well with the size of the problem.

Considering that today most modern organizations have access to a good number of computers interconnected through a communication network, A-Team techniques give an ideal approach to solve very large complex problems such as the design of reliable networks [8], or optimization of the resources used for electrical energy generation [6, 7], between several other real world applications [1].

Finally, it should be emphasized that the systematic study of team algorithms in asynchronous environments has been initiated relatively recently. The potential is promising, but more research needs to be done in areas as: the use of intelligent administrators; application of team algorithms for other engineering problems characterized by subproblems with highly different properties; and combinations of different artificial intelligence techniques with traditional algorithms.

References

[1] A-Teams Project Home Page, Carnegie Mellon UNiversity. URL: http://www.cs.cmu.edu/afs/cs/project/edrc-22/project/ateams/WWW/.

[2] B. Barán, *A study of parallel asynchronous team algorithms*, PhD thesis, Federal Univ. of Rio de Janeiro, COPPE, CS Dept., Rio de Janeiro, Brasil, 1993 (*in Portuguese*).

[3] B. Barán, E. Kaszkurewicz and D.M. Falcão, *Team Algorithms in Distributed Load Flow Computations*, IEE Proceedings on Generation, Transmission and Distribution. Vol. 142, N° 6, November 1995, pp. 583–588.

[4] B. Barán, E. Kaszkurewicz and A. Bhaya, *Parallel Asynchronous Team Algorithms: Convergence and Performance Analysis*, IEEE Transactions on Parallel and Distributed Systems, Vol.7, Number 7, July 1996, pp. 677–688.

[5] B. Barán and E. Chaparro, *Algoritmos Asíncronos combinados en un Ambiente Heterogéneo de Red*, XXIII Conferencia Latinoamericana de Informática, Valparaiso, Chile, 1997 (*in Spanish*).

[6] B. Barán, E. Chaparro and N. Cáceres, *A-Teams en la Optimización del Caudal Turbinado de una Represa Hidroeléctrica*, Conferencia Iberoamericana de Inteligencia Artificial IBERAMIA-98, Lisbon, Portugal, 1998 (*in Spanish*).

[7] B. Barán, E. Chaparro and N. Cáceres, *Hydroelectric optimization using A-Teams*, International Conference in Electrical Power System Operation and Management (EPSOM-98), Zurich, Switzerland, 1998.

[8] B. Barán and F. Laufer, *Topological Optimization of Reliable Networks using A-Teams*, Symposium: Architecture, Tools and Algorithm for Networks, Parallel and Distributed Systems, Internacional Conference Systemics, Cybernetics and Informatics SCI'99, Orlando - Florida, United States of America, 1999.

[9] D. P. Bertsekas and J. N. Tsitsiklis, *Parallel and distributed computation – numerical methods*, Prentice Hall, Englewood Cliffs, New Jersey, 1989.

[10] A. Bhaya, E. Kaszkurewicz, and F. C. Mota, *Asynchronous block-iterative methods for almost-linear equations*, Lin. Algebra Appl., 1991, pp. 487–508.

[11] C. J. Coulbourn, *Reliability Issues in Telecommunication Network Planning*, University of Vermont. URL: http://www.emba.uvm.edu/colbourn.

[12] C. J. Coulbourn, *The Combinatorics of Network Reliability*, Oxford Univ. Press, 1987.

[13] J. Calvet and A. Titli, *Overlapping vs. partitioning in block-iteration methods: application in large-scale system*, Automatica 25, 1989, pp. 137–145.

[14] Y. P. Dusonchet, S. N. Talukdar, and H. E. Sinnot, *Load flows using a combination of point Jacobi and Newton's methods*, IEEE Trans. Power Apparatus and Systems, PAS-90, 1971, pp. 941–949.

[15] D.M. Falcão, A. L. B. Bomfim, C. R. R. Dornellas, and G. N. Taranto, Genetics Algorithms in Power Systems Optimization, V SEPOE, Recife - Brazil, 1996.

[16] M.R. Garey and D. S. Johnson, *Computers and Intractability: A Guide to the Theory of NP-Completeness*, V SEPOE, Recife - Brazil, 1996.

[17] D.E. Goldberg, *Genetic Algorithm in Search, Optimization and Machine Learning*, San Francisco, Freeman, 1979.

[18] M. Ikeda and D. D. Šiljak, *Overlapping decomposition, expansions and contractions of dynamic systems*, Large Scale Systems, Vol. 1, 1980, pp. 29–38.

[19] R. H. Jan, *Design of reliable networks*, Comput. Oper. Res. Vol 20, 1993, pp. 29–38.

[20] E. Kaszkurewicz, A. Bhaya, and D. D. Šiljak, *On* the convergence of parallel asynchronous block-iterative computations, Lin. Algebra Appl., 1990, pp. 139–160.

[21] M. Mejía and E. Cantú, *DGENESIS: Software para la Ejecución de Algoritmos Genéticos Distribuidos*, XX Conferencia Latinoamericana de Informática , Mexico City, Mexico, 1994 (*in Spanish*).

[22] N. H. Penny, *Blackboard systems: the blackboard model of problem solving and the evolution of blackboard architectures*, The AI Magazine, 1986.

[23] W. H. Press, B. P. Flannery, S. A. Teukolsky, and W. T. Vetterling, *Numerical Recipes in C: The Art of Scientific Computing*, Cambridge University Press, 1988.

[24] P. S. Souza and S. N. Talukdar, *Genetics Algorithms in Asynchronous Teams*, ICGA - 91, San Diego - California, 1991, pp. 392–397.

[25] M. E. Sezer and D. D. Šiljak, *Nested epsilon decompositions of linear systems: weakly coupled and overlapping blocks*, SIAM J. Matrix Anal. Appl., 12 (1991), pp. 521–533.

[26] P. S. Souza and S. N. Talukdar, *Genetic algorithms in asynchronous teams*, in Proc. Fourth Intl. Conf. on Genetic Algorithms, California, 1991, pp. 392–397.

[27] H. S. Stone, *High Performance Computer Architectures*, Addison-Wesley, 1971.

[28] B. Stott, *Review of load-flow calculation methods*, Proc. IEEE, 1974, pp. 916-927.

[29] B. Stott and O. Alsaç, *Fast decoupled load flow*, IEEE Trans. Power Apparatus and Systems, PAS-93, 1974, pp. 859–869.

[30] S. N. Talukdar, S. S. Pyo, and R. Mehrotra, *Designing algorithms and assignments for distributed processing*, Tech. Rep. EPRI EL-3317 RP1764-3, Electric Power Research Institute, 1983.

[31] S. N. Talukdar, V. C. Ramesh, and J. C. Nixon, *A distributed system of control specialists for real-time operations*, in Proc. Third Symp. Expert System Appl. to Power Systems, Japan, 1991.

[32] S. N. Talukdar, V. C. Ramesh, R. Quadrel, and R. Christie, *Multiagent organizations for real-time operations*, in Proc. of the IEEE, 1992, pp. 765-778.

[33] M. H. M. Vale, D. M. Falcão, and E. Kaszkurewicz, *Electrical power network decomposition for parallel computations*, in Proc. Int'l. Conf. Circuits and Systems (ISCAS), IEEE, San Diego, California, 1992.

[34] A. Vanelli, *Solution techniques for 0-1 indefinite quadratic programming problems with applications to decomposition*, PhD thesis, University of Waterloo, 1983.

[35] A. J. Wood and B. F. Wollenberg, *Power Generation, Operation and Control*, John and Sons, 1983.

[36] F. F. Wu, *Theoretical study of the convergence of the fast decoupled load-flow*, IEEE Trans. Power Apparatus and Systems, PAS-96, 1977, pp. 268-275.

Chapter 10

PARALLEL NUMERICAL METHODS FOR DIFFERENTIAL EQUATIONS

Carlos A. de Moura*

Universidade Federal Fluminense[†]

demoura@lncc.br

Abstract   Some of the methods designed for the numerical solution of differential equations and which present an efficient implementation within a parallel environment are briefly surveyed. Included among these are: the Domain Decomposition and Multigrid hyper-algorithms, the Piecewise Parabolic method, the spectral (frequency) approach, some strategies for finite-element, and finite-difference higher accuracy techniques.

Keywords: Parallel algorithms, numerical methods, differential equations.

1.   Introduction

1.1.   Initial remarks

Let us start mentioning that, despite all the years gone by since [1] was published, it still remains as a very important piece of literature on the parallel numerical treatment of differential equations. Many of nowadays seminal ideas on the subject were discussed there, so that it definitely is not out-dated, except for its list of references, quite complete at publishing time but still a rich source today.

We also ought to refer to [2], which was indeed a breakthrough for Scientific Computing, having been a big thrust to pave the way for the flourishing parallel computing has taken over throughout this ending decade. With respect to this change associated to the 90's, let us quote [3]:

---

* Av. Rui Barbosa 170/2201 Rio de Janeiro 22250-020 RJ, Brazil.
[†] Instituto de Computação - UFF, Niterói, Brazil.

R. Corrêa et al. (eds.), Models for Parallel and Distributed Computation. Theory, Algorithmic Techniques and Applications, 279–313.

"During the past two years dramatic changes took place in the development and use of parallel computers. Parallel Computing has entered main stream computing, and parallel computers have become part of the product lines of virtually all major computer companies."

Maybe more important than this look at the hardware trend is the software state-of-the-art: as reported at ICIAM'99 [1], portions of LAPACK – the parallel scientific package for linear algebra – is part of the software included in a sensible portion of the computers sold currently.

The dominant property differential equations exhibit and that is often exploited while drawing numerical algorithms for their solution is the "local" character of the constraints they impose. The finite element methodFinite element, finite differenceFinite difference schemes, as well as the domain decomposition technique are strongly based on such a property. Compare the techniques lying on these approaches to the ones oriented towards the algorithms themselves, or rather, some of their implementation. That is the case for reordering techniques, such as coloring (nodes renumbering), or nested dissection (changing the operations order). In this context it is worth recalling J. Rice's vision [4], which ascertains that Linear Algebra is only a *tool*, claiming that the deep ideas for parallelism in PDEPDE's must be sought at the *models* themselves.

Let us close by quoting Keyes' words [5]:

"It is not sufficient to be a computer scientist to do parallel computation of PDE's. PDEPDE problems must be mathematically understood before an appropriate algorithm can be selected or even an appropriately parallel computer designed. The parallel computational destiny of PDE's is not in the language, environment, or architectures of the day, but in the physics of the problems they are modeling. The latter is unchanging, and all others are in the process of converging to it."

1.2.   Final remarks

These notes are an outgrowth and, at the same time, a shrinkage of [6], an attempt having been made to improve and update the chapters brought from that version to this short course. A 3-hour lecture coupled to a not-more-than-30-page constraint has forced us to be very selective. But please be aware that this does not mean, by no means, that any excluded material has a rating lower as compared to those included. Just put the blame on the constraints ... More details can be read either at [6, 7] or at the literature we have managed to refer to.

---

[1] J.Dongarra, Invited Lecture

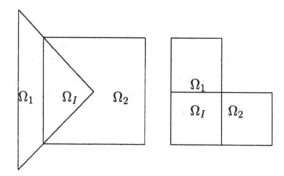

Figure 10.1. Decomposed domains.

It is a pleasure to thank the invitation and the patience from the organizing committee of the School, as well as the support from CIMPA.[2]

Once again, as in [6], "it is impossible to find words" to thank Sandra, my wife, always my love sustainer.

2. Domain decomposition – DD

For the numerical solution of PDEPDE's the "divide–to–conquer"Divide-and-conquer paradigm is often – maybe *always* should be said – employed, with different attires:

- finite element methodsFinite element

- Fourier analysis (discrete and continuous)

- operator splitting

    - alternate directions

    - method of lines or, in general, semi-discretizations

    - Galerkin methods or, in general, simple function spaces projections

    - frequency splitting or, more recently, wavelets expansions

- the most clearly posed division strategy, the domain decomposition – DD – methods.

---

[2]Except for a couple of updatings and some deleted mistypings, this chapter nearly contains the notes presented during September/October 1999 CIMPA School in Natal.

2.1.   General facts

Although regarded today as an important *"bona fide"* parallelization tool, the DD technique was first devised more than a century ago by Hermann Schwarz [8] (1843-1921), with no concurrent computing in mind. His aim was rather to replace problems defined on not simple-shaped domains by others more amenable to the treatment then available, which required always simple geometries, cf. Fig. 10.1. The algorithm he proposed may be described as follows. Let

$$Lu = f \qquad (10.1)$$

be a given PDEPDE on the domain Ω, with boundary condition

$$u = \phi \quad \text{on} \ \ \partial\Omega. \qquad (10.2)$$

Then consider the partition of Ω into two subdomains (cf. Fig. 10.1)

$$\Omega_1 \cup \Omega_2 = \Omega, \quad \Omega_I = \Omega_1 \cap \Omega_2.$$

Set

$$\Gamma_i = \partial\Omega_i \cap \overset{\circ}{\Omega}$$

and:

- **attribute** an initialization value $\tilde{u}_1$ for $u|_{\Gamma_1}$

- until convergence **do**

    - on Ω_1 **solve**
    $$Lu_1 = f$$
    $$u_1 = \phi \ \text{on} \ \Gamma \cap \partial\Omega_1$$
    $$u_1 = \tilde{u}_1 \ \ \text{on} \ \ \Gamma_1$$

    - **get** from the solution u_1
    $$\tilde{u}_2 = u_1|_{\Gamma_2}$$

    - on Ω_2 **solve**
    $$Lu_2 = f$$
    $$u_2 = \phi \ \text{on} \ \Gamma \cap \partial\Omega_2$$
    $$u_2 = \tilde{u}_2 \ \ \text{on} \ \ \Gamma_2$$

    - **put**
    $$\tilde{u}_1 = u_2|_{\Gamma_1}$$

- **end do**

This scheme may be thought of as made by fractional steps, see Section 2.3. Observe that the values of u on $\Omega_I = \Omega_1 \cap \Omega_2$ give the coupling between the iteration processes on the subdomains. It is thus quite reasonable to expect that the more generous the overlapping Ω_I is, the faster the convergence will be, as more information is being transmitted from a process (or a domain) to the other. This indeed happens, as it can be proved with the maximum principle. On the other hand, the huger Ω_I is, the costlier are the problems to be solved on the subdomains, because many and many variables will belong simultaneously to different problems. A decision must be made to balance this cost effect against a better convergence rate.

The main advantages of the DD technique are:

- Complex geometries may be exchanged by simpler ones, where such techniques as Fast PoissonPoisson solvers or FFTFFT may be used.

- Different numerical schemes may be quite simply employed in different subdomains, according to the different behavior of the modeled phenomena, like in meteorology, cf. [9, 10], or air pollution, cf. [11], this latter dealing with a hardly tractable problem if out of a parallel environment.

- Multi-scale solutions may be conveniently (*i.e.*, adaptively) treated.

- Different kinds of equations or different types of parameter behavior laws may be used. For example:

    - Heat transfer of compressible flow over a heated plate. Far away from the plate the thermal conductivity parameter may be left out, thus switching from a full advection-diffusion equation to a simpler advection one.

    - External flow around an airfoil. Far form the airfoil, instead of using the full N-S (Navier-Stokes) equations, the kinematic viscous terms become dimensionally irrelevant and consequently the Euler equations may be used as a dependable model.

    - Subsonic-supersonic transition for potential flows [12].

    - Molecular-continuous flow in the upper atmosphere. As the flow passes from the molecular state to its continuous state, we can interchange Boltzmann and N-S equations.

For many applications of DD in Mechanics, an important source of references is [13]. The author has introduced the so-called Balanced DD technique, see also [14, 15], and [16].

DD technique is a rather general strategy, being feasible its coupling: to different solvers – approximate or exact, iterative or direct ones – in different sub-

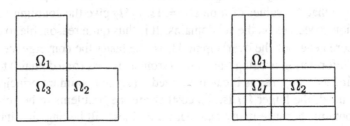

Figure 10.2.   Substructuring = non-overlapping   Overlapping

domains; to varied discretization techniques (finite elementsFinite element [17], finite differencesFinite difference [10], collocation methods, even multigrids). Besides, it supports different schemes of definition for the partitions themselves (stripes, cells; overlapping or non-overlapping domains), see Fig. 10.2.

How to make the choice between overlapping and non-overlapping domains? The conclusions by Chan & Resasco [18] should be mentioned. They introduced the non-overlapping DD, remarking that the savings in the system size could outweigh the slowness of convergence. This type of decomposition is also known in the engineering *milieu* as sub structuring, see Fig. 10.2, or segmentation technique, cf. [19]. In '89 Bjørstadt and Widlund [20] showed that we can think of their method just as an accelerated form of the earlier associated overlapping pattern.

Further, DD is not only employed for direct problems (to get the state solution) but also for inverse problems (to deal with parameter identification) [21], control problems [22], and also recent applications in non-traditional areas, like image restoration, cf. [23], and integral equations [24]. It may as well be used in 1–d settings, as with the multi-shooting treatment of BVP for systems of ODE's [25][3].

A point often under-evaluated is the gain in sequential computational complexity that DD can yield. When the work for solving a problem grows more than linearly with respect to its size, splitting it up into two problems of half the initial size will produce a faster method provided the sub-solutions can be combined efficiently to generate the global solution.

---

[3] As regards to parallel solutions to IVP for ODE's, a nice set of papers are quoted in [26].

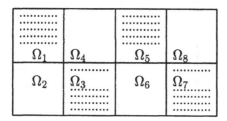

Figure 10.3. Coloring.

2.2. Parallelization

It is clear that the above described way of constructing the iterations is sequential, not (explicitly) parallel. One way to have it in parallel is to decouple the solving with the scheme

$$Lu_1^{k+1} = f \quad on \quad \Omega_1$$

$$u_1^{k+1} = \Theta\tilde{u}_1^k + (1 - \Theta)\tilde{u}_2^k \quad on \quad \Gamma_1$$

and

$$Lu_2^{k+1} = f \quad on \quad \Omega_2$$

$$u_2^{k+1} = \Theta\tilde{u}_1^k + (1 - \Theta)\tilde{u}_2^k \quad on \quad \Gamma_2.$$

An alternate way is to look at the finest DD possible – that is, when the subdomains are grid nodes – and use one of the parallelization techniques devised in this case, *v. g.* coloring. All domains with the same color may be treated at the same time (parallel-wise), provided the number of processors allows it. Of course, as more colors are used, less parallelization will be available, see Fig. 10.3.

It is proven – see [27] and references therein – that:

1 For 2nd order elliptic problems

$$Lu = -\sum_{i,j=1}^{2} \frac{\partial}{\partial x_i}\left(a_{ij}\frac{\partial u}{\partial x_j}\right)u + a_0 u$$

$$u = g_D \quad on \quad \Gamma_D$$

$$\frac{\partial u}{\partial n} = g_N \quad on \quad \Gamma_N,$$

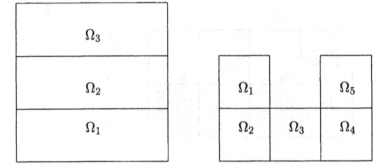

Figure 10.4. Stripes and cells.

with coefficients a_{ij}, a_0 bounded on Ω, we have convergence of the DD scheme, with a red-black implementation and coupling with convergent finite elementFinite element or collocation methods. The convergence rate is independent of the mesh-size parameter, provided internal vertices are shared by no more than two subdomains (no *cross-points* exist)[4]. This condition practically allows only stripes, no cells, in the decomposition, cf. Fig. 10.4.

2 For Navier-Stokes equations, as long as the constant $\beta > 0$ in Babuska-Brezzi-Ladyzenskaia *inf-sup* condition is uniformly bounded from below for all approximating spaces V_h^ℓ corresponding to all subdomains Ω_ℓ, the two-level DD iterative procedure coupled to (convergent) fini te element or collocation methods converges at a rate that independs of the mesh size h.

The tear-and-interconnect technique for finite elements[5]Finite element has its counterpart within the general framework of DD. Here, just like there, the formulation proposed in [29] leads to a saddle point problem, the interface coupling being related to the Lagrange multipliers.

We should observe that the DD paradigm goes in the opposite direction as the clusters of heterogeneous processing nodes do: these are oriented towards making the nodes to do the same computations on all data (they get restricted to those calculations they perform the best), while for DD the aim is to have the nodes to handle all the computations for a fixed subset of the data. This keeps the moving of data at a quite low level.

---

[4]Observe that when defining a domain partition geometry one is tempted to stick to the triangularization constraints FEM imposes, but it should be emphasized that DD does not require such restrictions. In [28] the splitting of the original domain is carried out with a partition of unity.

[5]see Section 4

2.3.   Additive × multiplicative Schwarz methods

Consider the simplest prototype for (1), namely

$$-\Delta u = f \quad on \quad \Omega, \quad u|_{\partial\Omega} = 0 \tag{10.3}$$

and write the fractional steps form of Schwarz algorithm [30, 31] as

$$-\Delta u^{2n+1} = f \quad on \quad \Omega_1, \quad u^{2n+1}|_{\Gamma_1} = u^{2n}|_{\Gamma_1},$$

$$-\Delta u^{2n+2} = f \quad on \quad \Omega_2, \quad u^{2n+2}|_{\Gamma_2} = u^{2n+1}|_{\Gamma_2}.$$

Introduce the usual notation

$$a(u; v) = \int \Delta u \cdot \Delta v,$$

as well as

$$V_1 = H_0^1(\Omega_1), \qquad V_2 = H_0^1(\Omega_2)$$

and denote by P_i the projections defined through

$$a(P_i v; w) = a(v; w), \quad \forall w \in V_i.$$

Extend the functions on V_i to $H_0^1(\Omega)$ by requiring them to vanish outside of Ω_i and then get the following relations for the errors:

$$a(u^{2n+1} - u; v_1) = 0, \quad \forall v_1 \in V_1$$

$$u^{2n+1} - u^{2n} \in V_1$$

$$a(u^{2n} - u; v_2) = 0, \quad \forall v_2 \in V_2$$

$$u^{2n} - u^{2n-1} \in V_2.$$

It follows that

$$u - u^{2n} = (I - P_2)(u - u^{2n-1})$$

$$u - u^{2n+1} = (I - P_1)(u - u^{2n})$$

$$\Rightarrow u - u^{2n+1} = (I - P_1)(I - P_2)(u - u^{2n-1}).$$

This means that the error propagation operator for this multiplicative Schwarz method is

$$(I - P_2)(I - P_1)$$

and consequently the algorithm can be viewed as an iterative method for solving

$$(P_1 + P_2 - P_2 P_1)u_h = g_h$$

for some conveniently defined g_h.

This operator, being a polynomial of degree 2, is not fit to parallel computing, as it requires two sequential steps:

$$u^{2n} = (I - P_2)u^{2n-1}$$

$$u^{2n+1} = (I - P_1)u^{2n}.$$

This becomes even worst if more than two subspaces (subdomains) are used, even though the corresponding polynomial will not have maximal degree (the product of projections associated to non overlapping regions vanishes).

The essential idea from Widlund *et al.* [32, 33, 34] to deal both theoretically as algorithmically with this point was to replace (3) by

$$Pu_h = \sum P_i u_h = g_h, \qquad (10.4)$$

where P_i are the projections and g_h is chosen so as that (4) has the same solutions as (3). This is the so-called additive Schwarz method. The derivation of error estimates becomes simpler.

2.4.   A question by Chan; words from Keyes; final remarks

How does in practice DD work? We have a double loop of iterations, since in each subdomain an iterative solver must be used, normally a PCG – preconditioned conjugate gradient – method. The efficiency of this technique rests thus upon the efficiency of the preconditioner used, the cost being high if the number of iterations can not be kept at a minimum. Consequently a natural question to be raised is in [35]:

> What would be better, parallelizing the PCG in the whole domain, or applying the PCG to the parallelized decomposed domains?

In other words, rephrasing [4] as quoted in the Introduction:

> Transform the PDEPDE problem in a Linear Algebra one and then parallelize it, or rather parallelize the PDE problem and use whatever LA tools are available?

In a paper by Chan and Goovaerts [36], it is shown with numerical experiments that the DD approach is much more efficient, particularly when discontinuous coefficients are present.

A bottleneck is inherently associated to DD, as synchronization is implicitly required. An asynchronous approach has been proposed in [37].

3.   The multigrid method

The multigrid method was invented about two decades ago, cf. [38, 39]. Its discovery had a surprise look, as the main idea behind its strategy goes in the opposite direction as regards as what all discretization algorithms do: multigrids need to fly from finer to coarser meshes, while standard numerical schemes always dive into finer and finer ones.

3.1.   Introductory facts

Consider a scalar elliptic boundary value problem which upon discretization on a uniform grid with mesh size h leads to the $N \times N$ system of linear algebraic equations

$$L^h u^h = f^h$$

where N is proportional to h^2. It is better to think of L^h, u^h and f^h as grid operators and functions. We shall denote by U the exact solution, U^h standing for its restriction to the grid, so that we will be employing the reverse of the standard convention. Usually the system to be solved is quite large so that we must look to efficient methods, and this may not be the case for Gauss-SeidelGauss-Seidel, for example, which would require a computational burden proportional to $N^2 = 1/h^4$. Even SORSOR will not fare much better: it will ask for an order $N^{3/2}$ work load.

It is seen that with multigrid methods we can reach truncation error accurate solutions with O(n) operations, $n =$ number of unknowns, and the goal is to have it implemented in parallel so as to reach solutions in a time span of order O($\log n$), supposedly the minimum time required to attain the solution for an algebraic linear system.

The important observation that gave rise to multigrid methods was that re-laxation schemes have an *error smoothing property* which may be heuristically explained as follows [40].

Take Gauss-SeidelGauss-Seidel method applied to a 5-point discretization of a 2-dimensional PoissonPoisson equation:

$$u_{i,j}^h = \frac{1}{4}\left(u_{i-1,j}^h + u_{i+1,j}^h + u_{i,j-1}^h + u_{i,j+1}^h - h^2 f_{i,j}\right)$$

where lower indices indicate space discretization defined by the grid with mesh size h and on the right hand size the most recent updated must be chosen. The error

$$v^h = U^h - u^h$$

(recall that U is the exact solution) satisfies

$$v_{i,j}^h = \frac{1}{4}\left(v_{i-1,j}^h + v_{i+1,j}^h + v_{i,j-1}^h + v_{i,j+1}^h\right),$$

which is actually an averaging process, a smoother. And it works out more and more as a mollifier thanks to the regularizing property of elliptic equations: the solution U is two level smoother – in terms of Sobolev space norms – than f, and we expect that the discretization operator would carry along this property, so that the error will be at least as smooth as U (or even more). Coarser grids may then be used to efficiently produce good corrections. This smoothing property is related to the grid mesh size: the finer the grid, smoother is the error, while coarser ones will allow the discrepancies to show up much clearly.

3.2.   A simple example – the main idea

Suppose we are solving the linear system obtained from a discretization of an elliptic equation and have already chosen an iterative method (as well as its implementation). Then the only speeding factor at hand will be the initial guess, and one possibility is to consider a coarser grid, solve the associated lighter system and bring back the obtained values for the original grid (by interpolation, say) to act as the initial guess.

For example, in a rectangular $2N \times 2N$ grid we could skip half of the grid points and get restricted to a $N \times N$ discretization. (This is called the standard coarsening.) In fact, to get the initial values for this new system we could again repeat recursively the procedure, and keep doing that until a small enough system would be reached and then solved (perhaps even exactly). This whole procedure is called *nested iteration* and it is expected that the additional cost to solve the systems associated to the coarser grids will be justified by having quicker convergence for the actually sought one.

To fully describe the multigrid method let us take a sequence of grids denoted by $G_1, G_2, ..., G_q$, where G_1 is the coarsest one and G_q is the finest one we will consider [41, 42]. The system on the p grid will be denoted by

$$L_p u_p = f_p. \tag{10.5}$$

Suppose that with the above procedure we were led to the approximation u_q on G_q. At this point we recall the iterative improvement scheme associated to direct methods. After directly getting a solution x_0 to

$$Ax = b, \tag{10.6}$$

if A is badly conditioned we may have a large residual

$$r_0 = b - Ax_0,$$

which is associated to the error

$$\epsilon_0 = x - x_0.$$

Of course ϵ_0 is not known (computable) exactly, otherwise the exact solution x would be known, but

$$b = Ax = A(x_0 + \epsilon_0) = b - r_0 + A\epsilon_0 \Rightarrow A\epsilon_0 = r_0.$$

Therefore we can try to solve

$$A\epsilon_1 = r_0 := b - Ax_0 \tag{10.7}$$

and put

$$x_1 = x_0 + \tilde{\epsilon}_1,$$

where $\tilde{\epsilon}_1$ stands for the computed solution. Then we would have

$$Ax_1 = b,$$

were the solution $\tilde{\epsilon}_1$ exact. We continue with this procedure and it normally gives a 2-digit improvement at each iteration.

Now, in the multigrid method we also consider the residual on grid G_q,

$$r_q = f - L_q u_q$$

and then have it restricted to G_{q-1} (a procedure called *injection*), denoting the result by $\tilde{r}_{q-1}$. Just like in (6) we approximately solve the system

$$L_{q-1}e = \tilde{r}_{q-1}, \tag{10.8}$$

denoting by e_{q-1} this approximate solution. It would give a correction to u_q as long as we have it *extended* to G_q, just like previously shown for iterative refinements. But now we'd better repeat this process to the coarser grids. Since (8) was not exactly solved, we form

$$r_{q-1} = \tilde{r}_{q-1} - L_{q-1}e_{q-1},$$

restrict r_{q-1} to G_{q-2}, which generates $\tilde{r}_{q-2}$, approximately solve

$$L_{q-2}e = \tilde{r}_{q-2}$$

to get e_{q-2}, and go on until e_1 is gotten on G_1. Now it is time to move up. Extend e_1 to G_2, with the operator p_2, then form

$$\hat{e}_2 = e_2 + p_2 e_1.$$

This value is then used as initial value for iterating on the previous equation

$$L_2 e = \tilde{r}_2.$$

This gives $\tilde{e}_2$. Have its prolongation to G_3, put $\hat{e}_3 = p_3 \tilde{e}_2 + e_3$ and then solve

$$L_3 e = \tilde{r}_3,$$

having $\hat{e}_3$ as a new initialization. Continue going all the way up to G_{q-1}, extend $\tilde{e}_{q-1}$ to G_q and add it to u_q to obtain the multigrid approximate solution to

$$L_q u_q = f.$$

This whole procedure is called a V-cycle, see Fig. 10.5. It may be also modified by the following strategy: instead of going all the way up, make a detour while in the middle of the road, going down, then go up again, and so on: this is known as the W-cycle.

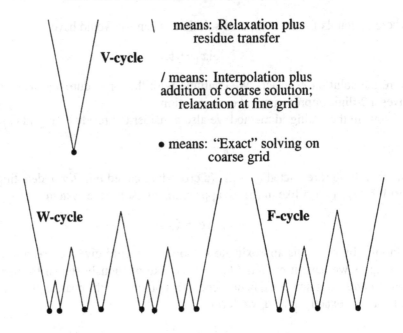

Figure 10.5.   V, W and F - cycles.

It should be emphasized that during the V-cycle, in both senses, the equation for the residue is the one to be used, and by no means (5). It is for the residue equation that the smoothing property holds (since it is seen that the error $V - v$ fulfills it), and it is this property that allows all the approximations to be maintained and further being corrected. Were (5) being used we would just be redoing the nested iteration procedure, nothing being added or corrected to the already obtained approximations. (Recall that this is just as in the iterative improvement, where the residue ought to be computed with A and b, and absolutely not with the transformed $\tilde{A}$ and $\tilde{b}$, of the triangular equivalent system $\tilde{A}x = \tilde{b}$ which is the one actually solved.)

3.3.   Parallel environment – a super convergent algorithm

The main steps in a multigrid algorithm may be listed as follows [43]:

1 smoothing (starting the solving)

2 computing the residue

3 projecting onto a coarser grid

4 interpolating from a coarser grid

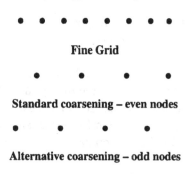

Figure 10.6. One-dimensional grids.

The first two steps may be thought as *intra-grid*, while the other two are *inter-grid* [44].

Observe that a clear potential for parallelism exists in the last two steps because the projection and interpolation operators (restriction and prolongation, in the previous notation) are local, so that we can perform them naturally in parallel. For the systems solving (phases 1. and 2.) we can use any one of the parallel approaches for general systems.

The main concern: the coarser get the grids, the less communications should be allowed (needed), so that they would not outweigh the computing load. It is also clear that the coarser the grid, less tasks exist, so that larger is the probability of having unbalanced loads. It was exactly this worry that guided McBryan and Fredericksen [45, 46] to propose an algorithm that would allow even for massively parallel machines an equilibrated work load. Their basic idea may be explained more simply by considering a 1-dimensional grid [47]. Standard multigrid techniques work with a series of coarser grids, each typically obtained by eliminating every other point of the finer (previous) grid, see Fig.6.

The proposed algorithm, called PSMG – Parallel Super convergent Multigrid – rests on the observation that for each fine grid there correspond two natural coarser grids – the one defined by the even nodes and that one related to the odd nodes. Assume that periodic boundary conditions hold, so that either set may be used to give rise to the coarser grid solution, and supposedly the corresponding solutions will be quite close. Traditionally the choice rests upon the even nodes.

It is quite natural to pose the question: why not trying to combine both solutions, so that we may be led to another solution that is better than each previous one separately? This is heuristically justified, as the odd and even nodes may have different *visions* of the solutions, just like the 3-d slides are built with slightly different pictures in order to give the viewer the 3-dimensional impression. And it is indeed possible to find a combination of the two coarser solutions that is significantly better than each one separately. It then follows that such a global scheme will converge faster than the standard one. For example, if the arithmetic average is taken, the corresponding algorithm will be dealing with a convex combination of the other two, and thus will converge at least as fast as they do.

The important consequence in terms of load balancing is that all processors that used to stay active while dealing with the finer grid will not become idle: both coarse-grid problems will be solved simultaneously and with the same instruction sets. It pays the price to switch from standard multigrid to PSMG as long as the number of grid points is comparable to or less than the number of processors.

This idea may be carried over to higher dimensions. Just do with each dimension as explained above for the one-dimensional problem. The authors have compared their algorithm with standard multigrids and concluded that it is a super convergent technique, yielding at least 4 times faster convergence, but even reaching in higher dimensions 40 times faster rates and over. Of course this is a method that normally makes sense to be employed only in a parallel environment.

3.4.   Scattered remarks

We must mention the study of adaptive multigrids carried on distributed memory computers. One of the main concerns is that the assignment of tasks is done at run time, so that a risk of communication overload and load unbalance exists [48]. For meshes that carry a complicated structure, load balancing may not be easy to achieve, and in this context [49] introduces an automatic load balancing strategy reported as feasible.

Studies have been carried to compare the performance obtained when the detour to a too coarse grid is avoided, more accuracy being required then from the iteration procedure at finer meshes, the so-called cascading multigrids, cf. [50] as well as [51].

The package Madpack [52] was developed for a quite general framework (abstract multigrids) and it uses a mixed language based on C and Fortran, trying to get away from the difficulties inherent to either language (this includes also C++). It is strongly based on experience from the author's researches and users of the code previous versions.

The Multigrid network[6] edits roughly each month an electronic newsletter with useful up-to-date information on the subject, as well as to the public files and code repository on multigrids available through *ftp*[7]. The following piece of bibliography is also worth being read: [53, 54, 55].

---

[6]Subscription through mgnet-requests@cs.yale.edu
[7]Anonymous ftp site: casper.cs.yale.edu

3.5.   Comparisons

Observe that both the MG and the DD techniques may be thought of as *hyper-algorithms*, as they must be combined to other numerical schemes, such as FEM, FD, spectral methods, cf. [56], or boundary elements, cf. [57]. How are their performance compared, this is a question many numerical experiments were conducted so as to present an answer, cf. [58, 59]. But it should be kept in mind that these *hyper-algorithms* can also be combined themselves, see for example [5].

We should remark here that real life problems can not usually be dealt with in a linearly ordered fashion, as within many mathematical domains. It is worth presenting the conclusions drawn from [60], where an assessment of semi-automatic parallelization[8], data parallel processing[9] and message passing programming[10] was made. He concludes: better have all options at hand, discarding none!

4.   The finite element method

Finite element We will discuss two approaches to the finite element method in a concurrent computing environment, namely the Element-by-element method and the tearing-and-interconnecting algorithm, described respectively in [61] and [62].

4.1.   Element-by-element method

In fact this technique was not developed for parallel processors, it is older than them: its aim is to be able to deal with large problems even with a PC.

Consider a linear boundary value problem

$$\text{find } u \left| \begin{array}{ll} Lu = f & \text{in} \quad \Omega \\ Bu = g & \text{on} \quad \partial\Omega \end{array} \right.$$

which is stated in the weak (variational) form as

$$\text{find } u \in H \left| \begin{array}{c} a(u; v) = b(v) \\ \forall v \in \tilde{H} \end{array} \right.$$

and approximated (rather, discretized) as

$$\text{find } u_h \in H^h \subset H \left| \begin{array}{c} a(u_h; v_h) = b(v_h) \\ \forall v_h \in \tilde{H}^h \subset \tilde{H} \end{array} \right.$$

---

[8]PTRAN, Parafrase-2, etc
[9]HPF, Vienna Fortran, and so on
[10]PVM, MPI, Linda, et al.

with H^h, $\tilde{H}^h$ piecewise-polynomial function spaces on a finite element discretization of Ω.

First the element matrix and vector contributions A_e, b_e are computed, then accumulated (assembled) to form the global system

$$Au = b.$$

The independence of each element contributions let the computation be made parallel-wise.

Before assembling the right-hand-side b (the left-hand-side would better not be assembled), boundary conditions are applied. They are usually enforced by modifying the global system but we should do it at the element level. It corresponds to modifying the element matrix A_e as well as the right-hand side contribution b_e for each element affected by the boundary conditions. Natural conditions are done parallel-wise while essential conditions must be done globally. The reason for taking this choice is that the natural boundary conditions are already incorporated to the system, on the right-hand side, while essential conditions appear as constraints on the nodes, or on the coefficients, or even on the basis functions. Given a node we need to figure out which elements it belongs to, and then perform the necessary mesh changes. But this does not have to inhibit parallelism.

In short: A_e, b_e can be computed independently and concurrently for $e = 1, 2, ..., E$. Observe that this strategy is completely independent of domain partitioning, mesh topology or element ordering. The elements may be assigned to the processors in any order. In general we will have much more elements than processors, so that a natural load balancing will occur, unless there are elements of different types and the more computation expensive ones get clustered in some processors.

The inclusion of boundary conditions may also carry an influence upon load balancing. This would show up whenever a single processor would be in charge of all – or a large amount of – the boundary elements, that is, the elements adjacent to the boundary. But this will be indeed of minor weight, as boundary conditions assembling corresponds to around 1% of the computing load, while element formation rises up to 30%. A reordering of the elements will fix it up, though, in both cases.

Assume now that we are dealing with a non-self-adjoint elliptic partial differential equation. The corresponding matrix is non-symmetric and we must use a preconditioned biconjugate gradient method, which we summarize as:

Given an arbitrary initial guess u^0, compute, for $n = 1, 2, ...$, the iterates

$$
\begin{aligned}
u^{n+1} &= u^n + \beta_n p^n \\
p^0 &= r^0 \\
p^n &= r^n + \alpha_n p_{n-1}, \quad n \geq 1 \\
\bar{p}^0 &= r^0 \\
\bar{p}^n &= \bar{r}^n + \alpha_n \bar{p}_{n-1}, \quad n \geq 1
\end{aligned}
$$

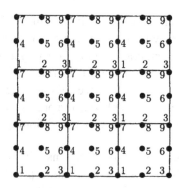

Figure 10.7. Structured Grid Numbering.

where

$$\beta_n = \frac{(r^n)^T \bar{r}^n}{(\bar{p}_n)^T Q^{-1} A p^n}$$

$$\alpha_n = \frac{(r^n)^T \bar{r}^n}{(r_{n-1})^T \bar{r}^{n-1}}$$

and

$$
\begin{aligned}
r^0 &= b - Au^0 \\
r^{n+1} &= r^n - \beta_n Q^{-1} A p^n, \quad n \geq 1 \\
\bar{r}^0 &= r^0 \\
\bar{r}^{n+1} &= \bar{r}^n - \beta_n (Q^{-1}A)^T \bar{p}_n, \quad n \geq 1
\end{aligned}
$$

Q = preconditioning matrix for A.

The costly matrix-vector products can be done in parallel thanks to the element-by-element strategy. If we follow down the tasks it amounts to, in order to form v_e, A_e, and then $A_e v_e = w_e$, it is seen that all this may indeed be done concurrently.

In the most general case, parallelization could be difficult if the local to global numbering creates storage conflicts and memory overwrites. But if the grid is built in a quite structured fashion, it would be feasible. Take as an example the biquadratic elements in Fig. 10.7.

Of course all first nodes in each element may be treated in parallel, as they correspond to different global nodes, the same for the second, and so on. But observe that rather than taking 9 loops – *par-loops*, we should say – it is possible to use clusters and require only 4 *par-loops*. Indeed, take the nodes { 1,2,4,5 }, then { 3,6 }, then { 7,8 }, and finish with { 9 }. No memory conflict will show up.

Besides renumbering the nodes, fictitious constraint equations, nodes and/or degrees of freedom can be added in order to have more work done in parallel, if needed. And also a coloring method could be employed (this is not the "red-black" ordering). Here we color the nodes so that whenever a global node belongs to different elements, in each of these elements a different color is assigned to it. Then elements with the same color may be scattered through different processors, without requiring any message passing.

The choice of the preconditioning Q is important, both for convergence and parallel implementation. A good performance attained with Jacobi preconditioner is reported in [61]. It is well known that this is an easy choice for a parallel job (JacobiJacobi method is the prototype parallel iterative scheme and this is a property inherited by its preconditioning matrix).

4.1.1   Applications.

Applications of this method to different areas may be illustrated by [63], where optimization problems are treated, and by [64], which has successfully dealt with a model for hydrodynamic lubrication where the cavitation phenomenon may show up, thus giving rise to a free boundary problem.

4.2.   Tear-and-interconnect method

A very performing parallel strategy for the solution of DE[11] with the finite element method – FEM – is the *tear-and-interconnect*, developed by Farhat and Roux. Besides the technique in the previous subsection, FEM are dealt with roughly through an approach to non-overlapping domain decomposition method, also known as substructuring which can be described as:

> To solve $Lu = f$ on a domain $\Omega \subset \mathbf{R}^n$ consider $\Omega = \cup\Omega_i$ where $\Omega_i \cap \Omega_j = \Gamma_{ij}$ has dimension $< n$. The prescribed equation is then solved on each Ω_i, the interfaces Γ_i giving rise to an additional problem, often solved iteratively.

Here the main idea is to consider all subdomains *really split-up*, replacing the interface constraint by the introduction of Lagrange multipliers. The mesh partition already defined for the FEM gives then rise to "floating" domains (those that remain without any defined boundary conditions) and these lead to local singularities. They are associated to rigid body motions.

A more complete introduction to this strategy was presented in [6].

4.2.1   Mesh decomposition.

This method is a hybrid one, as part of the system solution is done directly (subdomains), part iteratively (interfaces).

When partitioning the domain it is important to have in mind that if the bandwidth of each subdomain has the same order as the unpartitioned system, the algorithm performs more operations and thus it may be slower, even if its iterative part converges very quickly, see Fig. 10.8. The convergence is affected by:

- number of interface nodes

---

[11]herein static ones will be considered, the remarks previously made on time advancing being fit to be applied in connection to this method

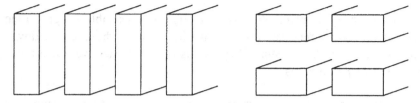

Figure 10.8. Stripwise and boxwise partitioning of a parallelepiped.

- connectivity: new information is exchanged only from the subdomains that have a plane (edge) in common. We mentioned above the positive points of this, but on the other hand it may affect negatively the convergence, as the needed information is passed along indirectly. Thus, for the interface, bandwidth should be large.

The following heuristical laws are proposed, based on the authors testing cases:

- the bandwidth for each subdomain must be a small fraction of that for the whole domain

- the number of interface points should be kept small

- domains should interconnect with as many others as possible

- "ugly" domains (a dimension much bigger than another) should be avoided

- floating subdomains must also be avoided

- the more the subdomains are balanced, the most balanced will be any implementation

Of course it is hard to meet all of these conditions, in fact some of them push towards opposite directions than another, *v.g.* second and third ones. Nevertheless some good results are presented to support the authors' claims about the algorithm performance.

5. Evolutionary equations

5.1. Parallelism in time scale?

It became a "*cliché*" to say that parallel programs rest mainly upon a single technique: "divide to conquer". Such a strategy was always present – even before any thought on parallel programming – on the finite element method, on the domain decomposition technique, on substructuring, and even on finite differencesFinite difference. But – except for this latter – all of them deal with space variables, the time variable having to bear a "curse": modeled phenomena are deterministic and thus sequential algorithms are the only way to mimic time flow. But in many problems this brings really an undesirable weight to the numerical solution: time span must be long, time steps ought to remain tiny.

A lot of effort has been spent to bypass this bottleneck and we discuss now some of the proposed roads.

5.1.1   Frequency domain – spectral method.   This is a quite well performing technique, very often employed for weather forecasting [65, 66, 67] or for geophysical problems [68, 69, 70, 71, 72, 73, 74, 75, 76, 77], both heavy number-crunching environments. It deals with the Fourier transform, we should say either discrete FT, or FFTFFT, or chopped transforms, many different approaches having been developed for dealing with it. It must be observed that similarly, there are other techniques which were considered as superseded in a not so far-away past, then brought back to attention by the parallel processing environment. That is the case for the separation of variables, which has been recently looked up, cf. [78], for example.

We shall exemplify with the linear wave equation, but some more general non-linear models may be dealt with, different linearizations being used.

Fourier transform with respect to the time variable[12] takes the equation

$$u_{tt} - cu_{xx} = f$$

into

$$-\tau^2 \hat{u}(x, \tau) - c\hat{u}_{xx}(x, \tau) = \hat{f}.$$

Thus we are left with a family of ordinary differential equations with respect to the space variable and – most important of all for the parallel environment – they are uncoupled, being possible to solve them on different processors, no communication needed.

5.1.2   Binary bisection method.   Consider now a linear ODE with variable coefficients

$$y'' + ay' + by = 0, \quad y(0) = y_0, \quad y'(0) = y_1$$

and take the simple difference scheme

$$\frac{y^{n+1} - 2y^n + y^{n-1}}{(\Delta t)^2} + a^n \frac{y^{n+1} - y^{n-1}}{2\Delta t} + b^n y^n = 0$$

which amounts to

$$y^{n+1} = A^n y^n + B^n y^{n-1}.$$

---

[12]one of the space variables may also be taken, and this is the case in meteorology models

This can be rewritten as

$$\begin{bmatrix} y^{n+1} \\ y^n \end{bmatrix} = \begin{bmatrix} A^n & B^n \\ 1 & 0 \end{bmatrix} \begin{bmatrix} y^n \\ y^{n+1} \end{bmatrix} \tag{10.9}$$

or, with an obvious notation,

$$\tilde{y}^{n+1} = \tilde{A}^n \tilde{y}^n, \tag{10.10}$$

so that

$$\tilde{y}^{n+1} = \tilde{A}^{n+1}\tilde{y}^n = \tilde{A}^{n+1}\tilde{A}^n\tilde{y}^{n-1} = ... = \tilde{A}^{n+1}\tilde{A}^n\tilde{A}^{n-1}...\tilde{A}^2\tilde{y}^1.$$

Depending on the communication overhead, it may be advantageous to perform the computation through the "bisection" pattern, each processor performing the multiplication of two matrices $\tilde{A}^{n+1}\tilde{A}^n$, $\tilde{A}^{n-1}\tilde{A}^{n-2}$, and so on. This idea was due to Evans [79] back in '88 and recently extended [80, 81] to arbitrary explicit finite difference schemesFinite difference that generate solutions to initial-boundary value problems assigned to linear partial differential equations, as long as the boundary constraints are linear.

This "bisection" technique – which features a *fan-in* pattern – may be applied to different problems and algorithms. We feel it is not sufficiently exploited in numerical environments, so that we would like to emphasize its features, sometimes in unexpected applications. With this aim let us discuss the solution of a tri-diagonal system

$$Ax = \beta.$$

Bringing A to triangular form requires $O(n)$ operations. If $A = LU$, their rows satisfy

$$(0, ..., 0, c_i, a_i, b_i, 0, ...0)$$

$$(0, ..., 0, l_i, 1, 0, ...0)$$

$$(0, ..., 0, u_i, b_i, 0, ...0)$$

and

$$u_i = a_i - a_i b_{i-1}/u_{i-1}.$$

In this formula the dependence of the i-th level on the $(i-1)$-th one inhibits vectorization and parallelization. A truly parallel approach to circumvent this deadlock was proposed by Stone [82] already in '73. Define

$$q_0 = 1, \ q_1 = a_1,$$

$$q_i = a_i q_{i-1} - c_i b_{i-1} q_{i-2}, \ i = 2, ..., n$$

so that

$$u_i = q_i/q_{i-1}.$$

Again we deal with the recurrence equation as in (10):

$$\begin{bmatrix} q_i \\ q_{i-1} \end{bmatrix} = \begin{bmatrix} a_i & -c_i b_{i-1} \\ 1 & 0 \end{bmatrix} \begin{bmatrix} q_{i-1} \\ q_{i-2} \end{bmatrix}$$

or

$$Q_i = G_i Q_{i-1} = \left(\prod_{j=2}^{i} G_j \right) Q_1, \quad i = 2, 3, \dots$$

Observe that all partial results $u_j, j = 1, 2, \dots$, are needed, so that the algorithm can be thought of as one to generate the products $p_{ij} = \prod_{l=1}^{j} d_l$, with $i = 1$ and $j = 2, \dots, N$. A non-sequential scheme to calculate p_{1j} is described below for $N = 8$.

First step:

$$[d_1 \; d_2 \; \dots \; d_8] \; * \; diag[1 \; d_1 \; d_2 \; \dots \; d_7] = [p_{11} \; p_{12} \; p_{23} \; p_{34} \; p_{56} \; p_{67} \; p_{78}]^t$$

Second step:

$$\Pi := [p_{11} \; p_{12} \; p_{23} \; p_{34} \; p_{45} \; p_{56} \; p_{67} \; p_{78}] * diag[1 \; 1 \; p_{11} \; p_{12} \; p_{23} \; p_{34} \; p_{45} \; p_{56}]$$

$$\Rightarrow \Pi = [p_{11} \; p_{12} \; p_{13} \; p_{14} \; p_{25} \; p_{36} \; p_{47} \; p_{58}]^t$$

Third step:

$$\Pi := [p_{11} \; p_{12} \; p_{13} \; p_{14} \; p_{25} \; p_{36} \; p_{47} \; p_{58}] * diag\{[1 \; 1 \; 1 \; 1 \; p_{11} \; p_{12} \; p_{13} \; p_{14}]\}$$

$$\Rightarrow \Pi = [p_{11} \; p_{12} \; p_{13} \; p_{14} \; p_{15} \; p_{16} \; p_{17} \; p_{18}]^t$$

5.1.3   The independent time-step method.

It may be pointed out that [83] contains just a short implementation remark but it is indeed an important one. Besides, it is also true that quite often obvious facts are hard to be seen by oneself.

For some problems it is possible to divide the domain in such a way that different time steps Δt are suitable for each sub-region. This certainly depends on how the sub-regions are related among themselves, and on each problem. The authors have managed to apply this idea to a problem from fluid dynamics simulation. Such a method may be thought of as an "adaptive grid technique" applied to the time variable.

The two main ideas behind it are:

- to avoid any non-essential synchronization,

- whenever possible, to rest upon local information.

5.1.4   Windowed iterative methods.   To make simpler the explanation we will concentrate on Jacobi method, just like done in [84], but all that will be said can be applied to Gauss-SeidelGauss-Seidel, SORSOR, *et al.*

Consider the system

$$My = b$$

with y and b n–vectors, M an $n \times n$ matrix. The Jacobi methodJacobi is defined by

$$m_{\ell\ell} y_{\ell\ell}^k = - \sum_{\substack{j \neq \ell \\ j=1}}^{n} m_{\ell j} y_j^{k-1} + b_\ell, \ell = 1, ..., n$$

where the upper index is the iteration level label. This scheme is actually the point-Jacobi method. Now if M, y, b are partitioned in blocks

$$M = \begin{bmatrix} M_{11} & ... & M_{1,q} \\ ... & ... & ... \\ M_{q,1} & ... & M_{qq} \end{bmatrix}, y = \begin{bmatrix} y_1 \\ ... \\ y_q \end{bmatrix}, b = \begin{bmatrix} b_1 \\ ... \\ b_q \end{bmatrix}$$

and all M_{ss} blocks are non-singular, then we may re–write the above expression as

$$M_{\ell\ell} y_{\ell\ell}^k = - \sum_{\substack{j \neq \ell \\ j=1}}^{q} M_{\ell j} y_j^{k-1} + b_\ell, \quad \ell = 1, ..., q$$

It is seen that these block iteration methods are always faster than the point method.

Now, when these systems are obtained while numerically solving an evolution equation, we must wait the convergence at each time level before proceeding to the next one. The authors introduce a window which contains data from a fixed number of time-steps.

The algorithm takes off with all time-steps inside a given window initialized alike, *i.e.*, initial data are assigned as initial iteration guesses. Each processor takes care of a time-step and the window remains unchanged until convergence has been yielded for the first time-step. Whenever this occurs the window advances one time-step upward and for the newly included time step, the latest iteration value at the previous time step is employed to initialize the corresponding iteration process. And so on.

Let us write the evolution equation as

$$My(t_r) = Py(t_{r-1} + b(t_r), \quad r = 1, 2, ..., m.$$

We must observe that the term $Py(t_{r-1})$ is computed only once for the time-step at the bottom of the window, but for all others it must be updated with every new iterate value obtained. This puts an overhead on the method, surely.

It is important to observe that windowed iterative methods have an iteration matrix with the same spectral radius as the corresponding iterative method – for point as well as for block iteration – so that the convergence properties remain unchanged.

Its real advantage are:

- to accelerate the iteration process by using better initialization;

- to let no processor idle;

- in the whole, to advance more quickly on time.

5.1.5   The parallel time-stepping method.

The idea behind this method [85] is actually very close to that in the previous technique, the main difference remaining in the way the initialization is improved. We will stick to author's notation.

Assume the PDEPDE under study has been approximated by the sequence of linear systems

$$A_n u_n = f_n + B_n u_{n-1}, \quad n = 1, ..., N$$

with A_n, B_n $m \times m$ matrices, $u_n, f_n \in \mathbf{R}^m, u_0$ given. Of course u_n approximates $u(n\Delta t)$.

Suppose that we use an iterative method given by

$$u_n^k = Q_n(u_n^{k-1})$$

(which is the case for multigrid or SORSOR, for example). The number of iterations depends upon Q_n and $u_n^{(0)}$. If we have already chosen Q_n (the best choice, as far as we have figured out) then improvement is only possible by refining $u_n^{(0)}$. For this we may have a function $R_n(u_n^{(0)}, u_{n-1})$.

The following strategy is then followed: as soon as a new iterate is computed at a fixed time level, send it along to an idle processor, which will take it as the initial guess for the following time level.

And this pattern is then carried out also by other occasionally idle processors:

P_1 starts solving for u_1, with Q_1
As soon as an iterate is gotten, P_1 sends it to P_2
P_2 uses this iterate in R_2, just like as if it were indeed u_1
As soon as it gets an improvement, P_1 sends it to P_3
P_3 uses this improvement in R_3, just as if it were u_2.

We observe that R_n might be taken as a multigrid algorithm with a set of grids different from those for Q_n. This may be useful because different frequencies (for the error) are eliminated and propagate at different speeds. But a practical choice is $Q_n = R_n$ because it makes the algorithm more balanced. Its implementation becomes easier, too.

At least in the linear case, convergence may be rigorously shown, cf. [85].

5.1.6   BBN butterfly implementation: time-step overlap.   The most important result on this work [61] that is not really restricted to the BBN architecture and that has not become out-of-date is the *time-step overlap* strategy. Just like the ones presented above, it is a technique to get back to work processors that had become idle. Such an idle state will show up when the number of elements E is much greater than the number of processors P – which happens in most cases.

We proceed according to the scheme

> **serial loop on time-steps**
> > **do**
> > > **par-loop:** solve frozen element equations
> > **enddo**

At the end of each time-step serial *do* we will face a ramp-down in the number of active processors, which may even range from P to 1. The prescribed medicine is to make inactive processors to start the computing needed for the following time-step, and the best way to carry this out is as follows.

Take the assembling and the calculations for nodes and elements in a physical neighborhood to the same processor, so that it is possible to advance in a front-like way, the solution at the following time-step sweeping the domain, leaving back the parts of the region still to be computed. This strategy brings down the level of inter-processors communication required.

An almost linear speed-up is reported.

5.2.   Explicit schemes and high accuracy

The availability of parallel machines has somehow recovered finite-difference explicit methodsFinite difference for partial differential equations[13]. In particular, high precision algorithms have come back to attention, as it is shown in [86], or coupled to different strategies, as in [87, 88, 89, 90, 91, 92, 93, 94]. In [95] symbolic logic is taken as a tool to derive high-order compact implicit schemes which are solved with the multi-grid method. See also [96].

We illustrate some of the ideas that show up in this framework with a quite simple example, cf. [97], which deals with the linear, constant coefficient, one-dimensional diffusion equation. But be aware that even harder equations, like those with the additional difficulty of dealing with crossed derivatives, may be treated with this – or a similar – approach, cf. [94].

---

[13]H. Scholl, C_3AD – *Colloquia* on High Performance Scientific Computing, LNCC, Rio, 1993

A two-step explicit finite-difference schemeFinite difference (Disguised Euler Scheme – DES) is studied which features an accuracy of order $O(\Delta t)^2 + O(\Delta x)^4$, even though the simulation for the 4th order derivative is 2th order accurate. It has

$$\nu \frac{\Delta t}{\Delta x^2} \le \frac{2}{3}$$

as its CFL stability constraint, which imposes a small time step, but the more processors we dispose of, the fastest the loop on all the nodes may be run, due to the scheme simplicity and it may be advantageous with respect to implicit schemes. Indeed, take for example Crank-Nicolson (CN) method and ask the same accuracy from both schemes, say 10^{-4n}. The number of nodes required in each dimension for them is

$$N_{CN}^x = N_{CN}^t = O(10^{2n}) = N_{DES}^t$$

while

$$N_{DES}^x = O(10^n).$$

If we have a massively parallel environment, the solving times behave as:

$$T_{CN} = O(10^{4n}), \quad T_{DES} = O(10^{4n}).$$

Besides, it would be interesting to set up some comparison data for DES and DuFort-Frankel (DFF):

$$N_{DFF}^x = N_{DES}^x = O(10^{2n}) = N_{DES}^t,$$

$$N_{DFF}^t = o(N_{DES}^t).$$

Worst of all, we should emphasize the dangers of dealing with a scheme which is unconditionally stable but conditionally convergent (because conditionally consistent), as it occurs for DFF, or rather with a scheme under a not clear-cut convergence constraint such as the one required for DFF, namely $\Delta t/\Delta x \to 0$. Let $\{\Delta t_i, \Delta x_i\}$ be a sequence of grids and $U^{(i)}$ be the corresponding solutions. The stability condition assures the existence of a subsequence $U^{(i^j)}$ that converges (at least weakly). But it may as well happen that for this subsequence we can not assure that the above relation for the grid holds, since we are dealing with only a finite number of its terms. The subsequence (or some of its subsequences) may be approaching a solution of another equation, due to the scheme inconsistency.

To finish let us remark that high accuracy may be a property indeed sought in real life applications. Of course there exist practical problems for which one does not have access to data known to a high precision – these may be material parameters, initial or boundary conditions. But, on the other hand, for other problems or algorithms, a high level accuracy is quite a need, v.g. inverse problems, where solutions of the direct problem for many distinct set of data must be obtained before getting the answer, and this one will carry much less precision than the incoming data do.

As our very last remark, let us point out that a special care must be taken with high precision algorithms for the points close to the boundary, so as not to loose the sought accuracy. And as regards to programming in a SIMD environment (for which explicit schemes are well suited), one has either the choice of using "masks" or to incorporate into the coefficients the different treatment for the "inner", the "close-to-boundary" and the "boundary" points.

5.3. Piecewise parabolic method

Important applications of parallel algorithms are the huge numerical problems associated to the set of Grand Challenges, proposed by K.Wilson [98]. And one of the powerful tools to deal with some of these is the piecewise parabolic method, researched primarily by Paul Woodward and his collaborators. Particularly relevant is the report in [99] about a hand-tailored cluster of workstations assembled, during a pre-determined time span, for a particular set of calculations, which could not be carried otherwise. This method is very well suited for large computations, as those arising in plasma physics, turbulent flows, astrophysical modeling, see [100, 101, 102].

References

[1] J. M. Ortega and R. G. Voigt. Solution of PDE on vector and parallel computers. *SIAM Rev.*, 27(2):149–240, June 1985.

[2] J. L. Gustavson *et al.* Development of parallel methods for a 1024-processor hypercube. *SIAM Jour. Stat. Scient. Comp.*, 9(4):594–638, 1988.

[3] G. R. Joubert. Foreword. *Parallel Computing*, 23(1-2):vii, 1997.

[4] J. R. Rice. Parallel methods for PDE's. in *The Characteristics of Parallel Algorithms*, pages 209–231, 1987. Jamiesson, Gannon and Douglass (eds.).

[5] D. E. Keyes. DD: a bridge between nature and parallel computers. *ICASE Report*, 27(92/44):149–240, Sep.'92.

[6] C. A. de Moura. *Parallel Algorithms for Differential Equations*. LMGC, Univ. de Montpellier II, 1994. Rev. Edition: *Parallel Numerical Methods for Differential Equations – a Survey*. CIMPA Int. School on Adv. Algor. Tech. for Par. Comp. with Applic., Natal, Sep-Oct.1999, 38+ii pages. Rel. 08/99, IC-UFF, 11 Nov. 99, **http://www.caa.uff.br/reltec.html**.

[7] C. A. de Moura. Massive parallelism as the new track for scientific computing. *Proceedings Symposia Gaussiana, Conf. A, München, August 1993*, pages 591–598, 1995. Behara, M. *et al.* (eds.).

[8] H. A. Schwarz. *Gesammelte Mathematische Abhandlungen*, 2:133–143, 1890. prev. published in Vierteljahrsschrift der Naturforschenden Gesellschaft, Zürich, v. 15, 272–286, 1870.

[9] M. Konchady *et al.* Implementation and performance evaluation of a parallel ocean model. *Parallel Computing*, 24(2):181–203, 1998.

[10] M. Marrocu *et al.* Parallelization and performance of a meteorological limited area model. *Parallel Computing*, 24(5-6):911–922, 1998.

[11] D. Dabdub *et al.* Parallel computation in atmospheric chemical modelling. *Parallel Computing*, 22(1):111–130, 1996.

[12] S. Lanteri *et al.* Parallel solution of compressible flows using overlapping and non-overlapping mesh partitioning strategies. *Parallel Computing*, 22(7):943–968, 1996.

[13] P. LeTallec. DD in computational mechanics. *Comp.Mech.Adv.*, 1:121–220, 1994.

[14] P. LeTallec *et al.* Balancing DD for plates *in* DD methods in Sc. and Eng. Computing. *Contemporary Math.*, 180:515–524, 1994.

[15] P. LeTallec *et al.* A Neumann-Neumann DD algorithm for solving plate and shell problems. *SIAM J. Numerical Analysis*, 35:836–867, 1998.

[16] S. C. Brenner *et al.* Balancing DD for non-conforming plate elements. *Numerische Mathematik*, 82(1):25–54, 1999.

[17] D. C. Hodgson *et al.* A DD preconditioner for a parallel finite element solver on distributed unstructured grids. *Parallel Computing*, 23(8):1157–1181, 1997.

[18] T. F. Chan and D. C. Resasco. Analysis of DD preconditioners on irregular regions. *Advances in Computer Methods for PDE*, 1987. R. Vichnevetsky and R. Stepleman (eds.).

[19] P. Fisette *et al.* Contribution to parallel and vector computing in multi-body dynamics. *Parallel Computing*, 24(5-6):717–728, 1998.

[20] P. E. Bjørstad and O. Widlund. To overlap or not to overlap: a note on a DD method for elliptic problems. *SIAM J. Sci. Stat. Comput.*, 10(5):1053–1051, Sept. 1989.

[21] M. Sultan, M. Hassan, J. Calvet, and A. Bilal. A first-order parallel decomposition method for parameter estimation in large scale systems. *Large Sc. Syst.*, 9:51–62, 1985.

[22] H. J. Kushner. DD method for large Markov chain control problems and nonlinear elliptic like equation. *SIAM Jour. Scient. Comp.*, 18(5):1494–516, 1997.

[23] A. Baronio *et al.* A DD technique for spline image restoration on distributed memory systems. *Parallel Computing*, 22(1):1–18, 1996.

[24] E. P. Stephan *et al.* DD for indefinite hypersingular integral equation: The *h* and *p* version. *SIAM Jour. Scient. Comp.*, 19(4):1139–1153, 1998.

[25] C. A. de Moura, and M. Lentini. A parallel multi-shooting algorithm for TPBVP problems. *II Simp. Bras. Arq. Computadores: Proc. Paralelo*, 1988. also: 34º. S. Bras. Análise, IME-USP.

[26] P. S. Rao *et al.* Data communication in parallel block pre-conditioned methods for solving ODE's. *Par. Comp.*, 23(13):1889–1907, 1997.

[27] A. Quarteroni. DD and Parallel processing for the numerical solution of PDE. *Surv. Math. Ind.*, 1:75–118, 1991.

[28] T. P. Mathew *et al.* DD operator splitting for the solution of parabolic equations. *SIAM Jour. Scient. Comp.*, 19(3):912–932, 1998.

[29] F.-X. Roux. DD methods for static problems. *Rech. Aérosp.*, 1:38–48, 1990.

[30] P.-L. Lions. On the Schwarz alternating method. *Proc. 1 Int. Symp. D. D. Meth. for PDE*, pages 1–42, 1988. R.Glowinsky, G.H.Golub, G.A.Meurant and J.Périaux (eds.).

[31] P.-L. Lions. On the Schwarz alternating method. *DD Methods*, pages 47–70, 1989. T.Chan, R.Glowinsky, J.Périaux and O. Widlund (eds.).

[32] M. Djryja and O. Widlund. Towards a unified theory of DD algorithms for elliptic problems. *Ultracomputer*, Note #167, Dec.1989.

[33] O. Widlund. Some Schwarz methods for symmetric and nonsymmetric elliptic problems. *Tech. Report*, 581, Sep. 1991.

[34] M. Djryja and O. Widlund. DD algorithms with small overlap. *Tech. Report*, 606, May 1992.

[35] C.-C. J. Kuo and T. F. Chan. Two-color Fourier analysis of iterative algorithms for elliptic problems with red/black ordering. *SIAM J. Sc. Stat. Comput*, 11(4):767–793, 1990.

[36] T. F. Chan & D. Goovaerts. A note on the efficiency of incomplete factorizations. *SIAM J. Sci. Stat. Comput.*, 11(4):794–803, Jul'90.

[37] D. Amitai *et al.* Implicit-explicit parallel asynchronous solvers of parallel PDE's. *SIAM Jour. Scient. Comp.*, 19(4):1366–1404, 1998.

[38] W. Hackbush. *Multigrid Methods with Applications.* Springer-Verlag, N.Y., 1985.

[39] W. Hackbush and U. Trottenberg (eds.). *Multigrid Methods*, Proc. Conf. at Köln-Porz, Nov. '81. Springer-Verlag, Berlin, 1982. Lect. Notes Math. # 960.

[40] C.-A. Thole and U. Trottenberg. A short note on standard Multigrid algorithms for 3-d problems. *Appl. Math. Comp.*, 27:101–115, 1988.

[41] K. Solchenback, C.-A. Thole, and U. Trottenberg. Parallel Multigrid methods: Implementation on Suprenum like architetures and applications. *Rapport de Rech. 746, INRIA (Sophia-Antipolis)*, Nov. 1987.

[42] M. Alef. Concepts for efficient Multigrid implementation on Suprenum like architectures. *Paral. Comp.*, 17:1–16, 1991.

[43] T. F. Chan and Y. Saad. Multigrid algorithms on the hypercube multiprocessor. *IEEE Trans. on Comp*, C-35(11):969–977, 1986.

[44] D. Gannon and J. van Rosendale. On the structure of parallelism in a highly concurrent PDE solver. *J. Paral. Distr. Comp*, 3:106–135, 1986.

[45] P. O. Fredericksen and O. A. McBryan. Parallel superconvergent multigrids. *Lect. Notes Math. Appl. Series*, # 110, 1988. Multigrid: Theory, Applications and Supercomputing, S. McCormick (ed.).

[46] O. A. McBryan and *et al.* . Multigrid methods on parallel computers – a survey. *Impact of Comp. in Sc. and Eng.*, 3:1–75, 1991.

[47] O. A. McBryan. New architetures: performance highlights and new algorithms. *Par. Comp.*, 7:477–499, 1988.

[48] H. Ritzdorf and K. Stüben. Adaptive multigrids on distributed memory computers. *Preprint, GMD*, 1993. Postfach 1316, 53731 S. Augustin, Germany.

[49] P. Bastian. Load balancing for adaptive multigrid method. *SIAM Jour. Scient. Comp.*, 19(4):1303–1321, 1998.

[50] D. Braess *et al.* A cascade multigrid algorithm for the Stokes equation. *Numerische Mathematik*, 82(2):179–192, 1999.

[51] L. Ferm. The number of coarse-grid iterations every cycle for the 2-grid method. *SIAM Jour. Scient. Comp.*, 19(2):493–501, 1998.

[52] C. C. Douglas. Madpack: a family of abstract multigrid or multilevel solvers. *Comp. & Appl. Math.*, 14(1):3–20, 1995. Sp. Issue on High Performance Scientific Computing, C.A. de Moura (ed.)

[53] K. Brand, M. Lemke, and J. Linden. Multigrid bibliography. *Arbeitspapier der GMD*, 206, 1986.

[54] J. Linden and K. Stüben. Multigrid methods: an overview with emphasis on grid generation processes. *Arbeitspapier der GMD*, 207, 1986.

[55] W. Briggs. *A Multigrid tutorial*. SIAM, Philadelphia, 1987.

[56] S. Kim. DD iterative procedure for solving scalar waves in the frequency domain. *Numerische Mathematik*, 79(2):231–260, 1998.

[57] C. Carsteusen *et al.* Fast parallel solvers for symetric boundary element DD equation. *Numerische Mathematik*, 79(3):321–348, 1998.

[58] R. Hess *et al.* A computation of parallel MG and a FFT algorithm for the Helmholtz equation in numerical weather prediction. *Par. Comp.*, 22(11):1503–1512, 1997.

[59] B. Heise *et al.* Parallel solver for non linear elliptic problems based on DD ideas. *Par. Comp.*, 22(11):1527–1544, 1997.

[60] J. Y. Berthon *et al.* Which approach to parallelize S.C. codes, that is the question. *Par. Comp.*, 23(1-2):165–179, 1997.

[61] H. Allik and S. Moore. Finite element analysis on the BBN Butterfly multiprocessor. *Computer and Structures*, 27(1):13–21, 1987.

[62] C. Farhat and F.-X. Roux. A method of finite element tearing and interconnecting and its parallel solution algorithm. *Int. Jour. Num. Meth. Engin.*, 32:1205–1227, 1991.

[63] M. J. Daydé *et al.* Element by element preconditioners for large partially separable optimization problems. *SIAM Jour. Scient. Comp.*, 18(6):1767–1787, 1997.

[64] M. Amendola. *Cavitation in hydrodynamic lubrication: a parallel algorithm for a free-boundary problem* (in Portuguese). PhD thesis, IMECC-Unicamp, 1996.

[65] B. Radi *et al.* Adaptive parallelization techniques in global weather models. *Par. Comp.*, 24(8):1167–1176, 1998.

[66] I. T. Foster *et al.* Parallel algorithm for the spectral tranform method. *SIAM Jour. Scient. Comp.*, 18(3):806–837, 1997.

[67] A. Averbuch *et al.* Two dimensional parallel solver for the N-S equation using ADI on cells. *Par. Comp.*, 24(8):673–699, 1998.

[68] J. Douglas Jr., J. E. Santos, D. Sheen, and L. S. Bennethum. Frequency domain treatment of 1-d scalar waves model for semiconductors. *Math. Models and Meth. in Appl. Sciences*, 3(2):171–194, 1993.

[69] P. Gauzellino and J. E. Santos. Numerical methods for wave propagation in elastic and anelastic media. *Comput. Appl. Math*, 12(2):95–111, 1993.

[70] C. A. Ravazzoli and J. E. Santos. Consistency analysis for a model for wave propagation in anelastic media. *Latin Am. Appl. Research*, 25:141–151, 1995. Tech. Report # 173, Purdue Univ., D. Math., 1993.

[71] J. Douglas Jr., J. E. Santos, and D. Sheen. Approximation of scalar waves in the space-frequency domain. *Math. Models and Meth. in Appl. Sciences*, 4:509–531, 1994.

[72] J. E. Santos and P. M. Gauzellino. Parallel algorithms for wave propagation in fluid–saturated porous media. *Comput. Mech., New Trends and Applications*, pages 1–12, 1998. S.R. Idelsohn *et al.* (eds.).

[73] J. E. Santos and P. M. Gauzellino. Nonconforming iterative domain decomposition procedures for the simulation of waves in fluid–saturated porous solids. *Mecánica Computacional*, 18:657–664, 1997.

[74] J. Douglas Jr., F. Pereira, and J. E. Santos. Parallel numerical simulation of waves in dispersive media. *1st Conf. and Exp. Lat. Am. Geoph. Union, 4th Int. Congr. Braz. Geoph. Society*, 417–419, 1995.

[75] J. Douglas Jr., F. Pereira, and J. E. Santos. A parallelizable approach to the simulation of waves in dispersive media. *3rd Int. Conf. Math. & Num. Asp. Wave Propagation*, 673–682, 1995. G. Cohen (ed.).

[76] J. E. Santos. Space-frequency domain approximation of waves in dispersive media. *Mecánica Computacional*, 14:106–113, 1994.

[77] P. M. Gauzellino, and J. E. Santos. Parallel algorithms for the numerical simulation of waves. *Proc. "Italian-Latinamerican Conference on Appl. and Ind. Math.(ITLA'97)"*, 219–222, 1997. Rome, Italy.

[78] S. Petrova. Parallel implementation of fast elliptic solver. *Par. Comp.*, 23(8):1113–1128, 1997.

[79] D. J. Evans. Parallel algorithm design. *Meeting on Parallel Computing, Verona, '88*, Ch. I:1–24, 1989.

[80] C. A. de Moura. Parallel numerical schemes for evolutive differential equations. *IV ICIAM*, July 1999. Book of Abstracts, p.252.

[81] C. A. de Moura. Non-sequential numerical algorithms for time-marching models. 50th SBA, USP, S. Paulo, 699–703, Nov. 1999. *RT06/99, IC-UFF*, 1999. **http://www.caa.uff.br/reltec.html**.

[82] H. Stone. An efficient parallel algorithm for the solution of tridiagonal linear systems of equations. *J. ACM*, 20:27–38, 1973.

[83] R. Babb II, L. Storc, and P. Elgroth. Parallelization schemes for 2-d hydrodynamic codes using the independent time-step method. *Par. Comp.*, 8:85–89, 1988.

[84] J. H. Saltz and V. K. Naik. Towards developing robust algorithms for solving PDE's on MIMD machines. *Par. Comp.*, 6:19–44, 1988.

[85] D. E. Womble. A time-stepping algorithm for Parallel Computing. *SIAM J. Sci. Stat. Comput.*, 11(5):824–837, 1990.

[86] G. Cohen and P. Joly. Fourth order schemes for the heterogeneous acoustics equation. *Comp. Meth. in Appl. Mech. and Engin.*, 80:397–407, 1990.

[87] J. Tuomela. A note on high order schemes for the 1-d wave equation. *BIT*, 35:394–405, 1995.

[88] J. Tuomela. On the construction of arbitrary order schemes for many-dimensional wave equation. *BIT*, 36:158–165, 1996.

[89] J. B. Cole. A nearly exact second-order finite-difference time-domain wave propagation algorithm on a coarse grid. *Comp. in Phys.*, 8(6):730–734, 1994.

[90] J. B. Cole. High-accuracy solution of Maxwell's equations using nonstandard finite differences. *Comp. in Phys.*, 11(3):287–292, 1997.

[91] L. Tang *et al.* Uniformly accurate finite difference schemes for p-refinement. *SIAM Jour. Scient. Comp.*, 20(3):1115–1131, 1999.

[92] W. F. Spotz *et al.* Iterative and parallel performance of high-order compact systems. *SIAM Jour. Scient. Comp.*, 19(1):1–14, 1998.

[93] J. M. Rosinski *et al.* The accumulation of rounding error and post validation for global atmospheric models. *SIAM Jour. Scient. Comp.*, 18(2):552–564, 1997.

[94] E. Carbajal Peña. *A high precision algorithm for linear thermoelasticity and its parallel implementation* (in Portuguese). PhD thesis, IM/UFRJ, 1996.

[95] I. Altas *et al.* Multigrid solution of automatically generated high-order discretization for the biharmonic equation. *SIAM Jour. Scient. Comp.*, 19(5):1575–1585, 1998.

[96] S. Vandewalle, R. V. Driesschie, and R. Piessens. The parallel performance of standard parabolic marching schemes. *International Journal of Super Computing*, 3(1):1–29, 1991.

[97] C. A. de Moura. DES – an explicit really quadratic 2-level scheme for the diffusion equation. *J. Comp. Inf., Toronto*, 3(1):99–115, 1994. Avail. as Preprint 42/92, LNCC.

[98] K. Wilson. Grand challenges to computational sciences. *Future Generation Computer Systems*, 5:171–189, 1989.

[99] P. R. Woodward *et al.* Parallel computation of turbulent flow. *Comp. Appl. Math.*, 14(1):97–105, 1995.

[100] P. R. Woodward, and W. Dai. A high-order Godunov-type scheme for shock interactions in ideal MHD. *SIAM Jour. Scient. Comp.*, 18(4):957–981, 1997.

[101] P. R. Woodward, and W. Dai. An iterative Riemann solver for relativistic hydrodynamics. *SIAM Jour. Scient. Comp.*, 18(4):982–985, 1997.

[102] P. R. Woodward, and W. Dai. A high order iterative implicit-explicit hybrid scheme for MHD. *SIAM Jour. Scient. Comp.*, 19(6):1827–1846, 1998.

Index

Applied Optimization

1. D.-Z. Du and D.F. Hsu (eds.): *Combinatorial Network Theory.* 1996
   ISBN 0-7923-3777-8

2. M.J. Panik: *Linear Programming: Mathematics, Theory and Algorithms.* 1996
   ISBN 0-7923-3782-4

3. R.B. Kearfott and V. Kreinovich (eds.): *Applications of Interval Computations.* 1996 ISBN 0-7923-3847-2

4. N. Hritonenko and Y. Yatsenko: *Modeling and Optimization of the Lifetime of Technology.* 1996 ISBN 0-7923-4014-0

5. T. Terlaky (ed.): *Interior Point Methods of Mathematical Programming.* 1996
   ISBN 0-7923-4201-1

6. B. Jansen: *Interior Point Techniques in Optimization.* Complementarity, Sensitivity and Algorithms. 1997 ISBN 0-7923-4430-8

7. A. Migdalas, P.M. Pardalos and S. Storøy (eds.): *Parallel Computing in Optimization.* 1997 ISBN 0-7923-4583-5

8. F.A. Lootsma: *Fuzzy Logic for Planning and Decision Making.* 1997
   ISBN 0-7923-4681-5

9. J.A. dos Santos Gromicho: *Quasiconvex Optimization and Location Theory.* 1998
   ISBN 0-7923-4694-7

10. V. Kreinovich, A. Lakeyev, J. Rohn and P. Kahl: *Computational Complexity and Feasibility of Data Processing and Interval Computations.* 1998
    ISBN 0-7923-4865-6

11. J. Gil-Aluja: *The Interactive Management of Human Resources in Uncertainty.* 1998
    ISBN 0-7923-4886-9

12. C. Zopounidis and A.I. Dimitras: *Multicriteria Decision Aid Methods for the Prediction of Business Failure.* 1998 ISBN 0-7923-4900-8

13. F. Giannessi, S. Komlósi and T. Rapcsák (eds.): *New Trends in Mathematical Programming.* Homage to Steven Vajda. 1998 ISBN 0-7923-5036-7

14. Ya-xiang Yuan (ed.): *Advances in Nonlinear Programming.* Proceedings of the '96 International Conference on Nonlinear Programming. 1998 ISBN 0-7923-5053-7

15. W.W. Hager and P.M. Pardalos: *Optimal Control.* Theory, Algorithms, and Applications. 1998 ISBN 0-7923-5067-7

16. Gang Yu (ed.): *Industrial Applications of Combinatorial Optimization.* 1998
    ISBN 0-7923-5073-1

17. D. Braha and O. Maimon (eds.): *A Mathematical Theory of Design: Foundations, Algorithms and Applications.* 1998 ISBN 0-7923-5079-0

Applied Optimization

18. O. Maimon, E. Khmelnitsky and K. Kogan: *Optimal Flow Control in Manufacturing.* Production Planning and Scheduling. 1998　　　ISBN 0-7923-5106-1

19. C. Zopounidis and P.M. Pardalos (eds.): *Managing in Uncertainty: Theory and Practice.* 1998　　　ISBN 0-7923-5110-X

20. A.S. Belenky: *Operations Research in Transportation Systems:* Ideas and Schemes of Optimization Methods for Strategic Planning and Operations Management. 1998　　　ISBN 0-7923-5157-6

21. J. Gil-Aluja: *Investment in Uncertainty.* 1999　　　ISBN 0-7923-5296-3

22. M. Fukushima and L. Qi (eds.): *Reformulation: Nonsmooth, Piecewise Smooth, Semismooth and Smooting Methods.* 1999　　　ISBN 0-7923-5320-X

23. M. Patriksson: *Nonlinear Programming and Variational Inequality Problems.* A Unified Approach. 1999　　　ISBN 0-7923-5455-9

24. R. De Leone, A. Murli, P.M. Pardalos and G. Toraldo (eds.): *High Performance Algorithms and Software in Nonlinear Optimization.* 1999　　　ISBN 0-7923-5483-4

25. A. Schöbel: *Locating Lines and Hyperplanes.* Theory and Algorithms. 1999　　　ISBN 0-7923-5559-8

26. R.B. Statnikov: *Multicriteria Design.* Optimization and Identification. 1999　　　ISBN 0-7923-5560-1

27. V. Tsurkov and A. Mironov: *Minimax under Transportation Constrains.* 1999　　　ISBN 0-7923-5609-8

28. V.I. Ivanov: *Model Development and Optimization.* 1999　　　ISBN 0-7923-5610-1

29. F.A. Lootsma: *Multi-Criteria Decision Analysis via Ratio and Difference Judgement.* 1999　　　ISBN 0-7923-5669-1

30. A. Eberhard, R. Hill, D. Ralph and B.M. Glover (eds.): *Progress in Optimization.* Contributions from Australasia. 1999　　　ISBN 0-7923-5733-7

31. T. Hürlimann: *Mathematical Modeling and Optimization.* An Essay for the Design of Computer-Based Modeling Tools. 1999　　　ISBN 0-7923-5927-5

32. J. Gil-Aluja: *Elements for a Theory of Decision in Uncertainty.* 1999　　　ISBN 0-7923-5987-9

33. H. Frenk, K. Roos, T. Terlaky and S. Zhang (eds.): *High Performance Optimization.* 1999　　　ISBN 0-7923-6013-3

34. N. Hritonenko and Y. Yatsenko: *Mathematical Modeling in Economics, Ecology and the Environment.* 1999　　　ISBN 0-7923-6015-X

35. J. Virant: *Design Considerations of Time in Fuzzy Systems.* 2000　　　ISBN 0-7923-6100-8

Applied Optimization

36. G. Di Pillo and F. Giannessi (eds.): *Nonlinear Optimization and Related Topics*. 2000
ISBN 0-7923-6109-1

37. V. Tsurkov: *Hierarchical Optimization and Mathematical Physics*. 2000
ISBN 0-7923-6175-X

38. C. Zopounidis and M. Doumpos: *Intelligent Decision Aiding Systems Based on Multiple Criteria for Financial Engineering*. 2000   ISBN 0-7923-6273-X

39. X. Yang, A.I. Mees, M. Fisher and L.Jennings (eds.): *Progress in Optimization*. Contributions from Australasia. 2000   ISBN 0-7923-6286-1

40. D. Butnariu and A.N. Iusem: *Totally Convex Functions for Fixed Points Computation and Infinite Dimensional Optimization*. 2000   ISBN 0-7923-6287-X

41. J. Mockus: *A Set of Examples of Global and Discrete Optimization*. Applications of Bayesian Heuristic Approach. 2000   ISBN 0-7923-6359-0

42. H. Neunzert and A.H. Siddiqi: *Topics in Industrial Mathematics*. Case Studies and Related Mathematical Methods. 2000   ISBN 0-7923-6417-1

43. K. Kogan and E. Khmelnitsky: *Scheduling: Control-Based Theory and Polynomial-Time Algorithms*. 2000   ISBN 0-7923-6486-4

44. E. Triantaphyllou: *Multi-Criteria Decision Making Methods*. A Comparative Study. 2000   ISBN 0-7923-6607-7

45. S.H. Zanakis, G. Doukidis and C. Zopounidis (eds.): *Decision Making: Recent Developments and Worldwide Applications*. 2000   ISBN 0-7923-6621-2

46. G.E. Stavroulakis: *Inverse and Crack Identification Problems in Engineering Mechanics*. 2000   ISBN 0-7923-6690-5

47. A. Rubinov and B. Glover (eds.): *Optimization and Related Topics*. 2001
ISBN 0-7923-6732-4

48. M. Pursula and J. Niittymäki (eds.): *Mathematical Methods on Optimization in Transportation Systems*. 2000   ISBN 0-7923-6774-X

49. E. Cascetta: *Transportation Systems Engineering: Theory and Methods*. 2001
ISBN 0-7923-6792-8

50. M.C. Ferris, O.L. Mangasarian and J.-S. Pang (eds.): *Complementarity: Applications, Algorithms and Extensions*. 2001   ISBN 0-7923-6816-9

51. V. Tsurkov: *Large-scale Optimization – Problems and Methods*. 2001
ISBN 0-7923-6817-7

52. X. Yang, K.L. Teo and L. Caccetta (eds.): *Optimization Methods and Applications*. 2001   ISBN 0-7923-6866-5

53. S.M. Stefanov: *Separable Programming Theory and Methods*. 2001
ISBN 0-7923-6882-7

Applied Optimization

Applied Optimization

72. A. Hassan Siddiqi and M. Kočvara (eds.): *Trends in Industrial and Applied Mathematics*. 2002   ISBN 1-4020-0751-5

KLUWER ACADEMIC PUBLISHERS – DORDRECHT / BOSTON / LONDON